In 1914, the armies and navies that faced each other were alike down to the strengths of their companies and battalions and the designs of their battleships and cruisers. Differences were of degree rather than essence. During the interwar period, the armed forces grew increasingly asymmetrical, developing different approaches to the same problems. This study of major military innovations in the 1920s and 1930s explores differences in innovating exploitation by the six major military powers. The comparative essays investigate how and why innovation occurred or did not occur, and explain much of the strategic and operational performance of the Axis and Allies in World War II. The essays focus on several instances of how military services developed new technology and weapons and incorporated them into their doctrine, organization, and styles of operations.

MILITARY INNOVATION
IN THE
INTERWAR PERIOD

MILITARY INNOVATION IN THE INTERWAR PERIOD

Edited by

WILLIAMSON MURRAY
Horner Professor,
Marine Corps University

ALLAN R. MILLETT
Mason Professor,
The Ohio State University

CAMBRIDGE
UNIVERSITY PRESS

CAMBRIDGE
UNIVERSITY PRESS

32 Avenue of the Americas, New York NY 10013-2473, USA

Cambridge University Press is part of the University of Cambridge.

It furthers the University's mission by disseminating knowledge in the pursuit of education, learning and research at the highest international levels of excellence.

www.cambridge.org
Information on this title: www.cambridge.org/9780521637602

© Cambridge University Press 1996

First published 1996
First paperback edition 1998
21st printing 2009

A catalogue record for this publication is available from the British Library

ISBN 978-0-521-55241-7 Hardback
ISBN 978-0-521-63760-2 Paperback

To Professor Donald Kagan, Yale University, and the late Professor Harry L. Coles, Jr., who taught us the challenges and satisfactions of historical inquiry into the mysteries of mankind's military affairs

CONTENTS

ACKNOWLEDGMENTS

There are many individuals to whom we are indebted in our efforts to put together this volume. We are most indebted to Dr. Andrew Marshall of the Office of Net Assessment for his interested and enthusiastic support of this project from its inception through to its conclusion. This is the third major project of ours that he has supported with wisdom as well as financial resources, so we owe him a triple debt of gratitude. Our editor at Cambridge University Press, Frank Smith, has also been particularly helpful in supporting this project since it came to his attention; the readers for Cambridge also helped to make this a finer manuscript than it might have been otherwise, with their sharply focused and intelligent criticism. We would like to thank our production and copy editors and our proofreader, Janis Bolster, Winifred M. Davis, and Robert E. Graham, for their splendid work in tightening the manuscript and removing inconsistencies and errors. The Mershon Center provided consistent and helpful support, with Beth Russell, Christopher Ives, David Thompson, and Albert Palazzo all making considerable contributions to the smooth running of the project. Barry Watts also deserves special thanks for his advice in crafting a number of the essays in addition to his own participation in the writing of the final summary chapter.

Finally, we need to thank our wives, Martha E. Farley-Millett and Lesley Mary Smith, for their unswerving support as well as their sharp and clear criticism.

But in the end we must take responsibility for whatever weaknesses yet remain. We can only hope that this volume makes some small contribution to a complex subject: how do armed forces change?

MILITARY INNOVATION
IN THE
INTERWAR PERIOD

INTRODUCTION

WILLIAMSON MURRAY AND ALLAN R. MILLETT

When the West began its ascent to world supremacy in the sixteenth century, military institutions played a crucial role in its drive to power. Recent historical work suggests that the Western military framework has undergone cyclical periods of innovation beginning in the early fourteenth century and continuing to the present and that such periods have resulted in systemic and massive changes to the basic nature of warfare and the organizations that fight.[1] The military history of the twentieth century indicates that this pattern has continued unbroken except that the periods between major innovations have been decreasing even as the complexity of innovation has increased.

A number of factors have driven innovation in military affairs: the rapid pace of technological change, the vast sums spent on military research, and the increasing sophistication with which military organizations evaluate their performance and that of their weapons systems. The fusion of technology and potent management skills that mobilize mass organizations makes military change inevitable. If anything, the technologies influencing civilian life in the next century may have even greater impact on military force than has been true in this century.

The history of the years since 1939 has been one, for the most part, in which military organizations have had access to unparalleled levels of funding and resources. World War II was a great contest for world dominion, but the Cold War, which provided ideological fervor and motivation for vast defense budgets on both sides of competing alliance systems, followed almost immediately on the heels of the great conflict. Thus,

---

[1] Clifford J. Rogers, "The Military Revolutions of the Hundred Years' War," *The Journal of Military History,* vol. 57, No. 2, April 1993, p. 277.

motivation and financial support have been a basic factor in the military equation over the last half of this century.

But the period we are now entering appears to be one in which there is no clear threat to the United States, nor is one likely to appear for the foreseeable future. In such an environment, it is inevitable, particularly in democratic nations, that one will see continuing and substantial declines in defense spending. In contrast to U.S. defense spending during the Cold War, the decline will most likely not be cyclical in nature but instead represent a steady erosion of support. At the same time, scientific advances and technological innovations in society at large will confront military institutions with another period of great changes, but one in which there will be much lower levels of support.

The emerging strategic environment in which our military institutions will have to operate suggests a number of similarities to the period between the great world wars of the first half of this century. During this timeframe, military institutions had to come to grips with enormous technological and tactical innovation during a period of minimal funding and low resource support. Some succeeded, creating a huge impact on the opening moves in World War II. Others were less successful and some institutional innovation resulted in dismal military failure.

One must stress that in spite of low military budgets and considerable antipathy towards military institutions in the aftermath of the slaughter in the trenches, military institutions *were* able to innovate in the 1920s and 1930s with considerable success. And these innovations *were not on the margin:* The U.S. and Japanese navies changed the equation of war at sea with their creation of naval air power based on carriers that accompanied their fleets into battle. Similarly, the Germans developed an armored force, based on a combined-arms concept, that overthrew the entire balance of power in Europe by its breakthrough on the banks of the Meuse and the exploitation of that success to the English Channel. In air war, Air Marshal Sir Hugh Dowding set the technological specifications for the Hurricane and Spitfire, supported the initial research into the possible use of radio waves to detect aircraft, and then created an air defense network based on these innovations; his system innovation altered the *entire context* within which air war was to take place and enabled the RAF to triumph in the Battle of Britain. These innovations were thus of great moment; they represented fundamental, basic changes in the *context* within which war takes place. The factors that contributed to innovation in the interwar period are not always easy to discern, but they do provide a measure of how one might best think about innovation, particularly in the coming decades.

Consequently, the editors of this study have asked our authors to examine seven specific areas of innovation during the interwar period: armored warfare, amphibious warfare, strategic bombing, tactical bombing, submarine warfare, carrier aviation, and the development of radar. In each case, the authors were to compare and contrast the different experiences of three or more national military institutions. The disparity in the effectiveness of military organizations during this period demonstrates the wide spectrum constituting failure or success in innovation. At the end of this volume the two principal investigators have attempted to draw together their perspectives on innovation in this period and the possible lessons those interested in innovation might consider.

The authors of this volume, as practicing historians, are reluctant to draw "lessons learned" for the innovators of the future. History, reflecting the nature of the world from which it is drawn, is an inexact discipline. It does not and cannot offer clear answers. This study asks how one might study innovation and what patterns may characterize successful, guided changes. As Clausewitz has suggested:

> Whenever an activity deals primarily with the same things again and again – with the same ends and the same means, even though there may be minor variations and in infinite diversity of combinations – these things are susceptible of rational study. It is precisely that inquiry which is the most essential part of any *theory*, and which may quite appropriately claim that title. It is an analytic investigation leading to a close *acquaintance* with the subject; applied to experience – in our case to military history – it leads to thorough *familiarity* with it. The closer it comes to that goal, the more it proceeds from the objective form of a science to the subjective form of a skill, the more effective it will prove in areas where the nature of the case admits no arbiter but talent.[2]

The purpose of this study is to provide insights into the nature of the processes involved in major innovation and change in military organizations during the interwar period and to highlight those factors that encourage success as well as those that inhibit innovation. Among the crucial issues this study seeks to explore are the problems involved in doctrinal, technological, and weapons innovation in a period of severe budget constraint and revolutionary technological change. In particular the authors have attempted to address in a comparative sense the differences among their military organizations, to bring to the fore why some

---

[2] Carl von Clausewitz, *On War*, ed. and trans. by Michael Howard and Peter Paret (Princeton, NJ, 1976), p. 141.

succeeded and some failed. This is, of course, an historical study, but we are looking at history to suggest possible paths for the future.

To achieve this objective, the co-principal investigators asked the participants to structure their essays around three concepts: the strategic framework of the period, the organizational factors of the institutions under study, and the doctrinal framework of the services. We asked the essayists to pose a series of questions for each concept. The authors were to consider what the general strategic framework was within which military institutions had to function in the 1920s and 1930s, how military and political leaders viewed potential enemies, what the services viewed as their overall strategic mission, and how they conceived the next year. The essayists were also to investigate the domestic political environment of the services, especially the issues of funding, access to technological resources, and congruence between political objections and force capabilities. Finally, the authors were to comment on the impact of World War I on both the strategic environment and preparations for a future conflict.

When looking at organizational factors, the essays were to address the efficacy of the services' internal administration in accepting or discouraging innovations. They were to examine such aspects as differences in the pattern of administration between the services compared, national cultural preferences which influenced the services' preparations for war, and the role of professional military education in the capacity to adapt.

The issues addressed in the area of doctrinal framework were to include an assessment of the services' commitment to the problem of doctrine and of their rigor in drawing and learning the lessons of World War I. Other doctrinal concerns were the services' evaluation of their doctrine in light of their potential opponents, and the seriousness with which they incorporated doctrinal principles into training.

This study will neither attempt to establish any grand theory of innovation nor create a model for explaining innovation. Stephen Rosen has already demonstrated the difficulties of such a task. In his book *Winning the Next War*, he has analyzed the existing literature on innovation and failed to find any patterns, in either the military or the far more common industrial studies, that would support such claims. In fact, he has shown that it is far more common for innovation theorists to advance conflicting ideas than to agree on causal relationships.[3]

Rather, this work emphasizes the complexities and ambiguities inherent in innovation, which defined its framework during the interwar peri-

---

[3] Stephen Peter Rosen, *Winning the Next War: Innovation and the Modern Military* (Ithaca, 1991), pp. 1–53.

od. Thus, the essays will address the failures as well as the successes of the 1920s and 1930s. The study assumes that innovation is natural and the result of a dynamic environment in which organizations must accept change if they are to survive. While the period of 1918 to 1939 was technically one of peace, the future combatants engaged, especially as war approached, in intellectual and technological jockeying and sought advantages in material and doctrine. It is important to discover what innovative military organizations look like, what their characteristics are, and what actions can be taken to encourage innovation. This study, then, aims at providing a guide to what the past experiences of military organizations have been and how one might best think about innovation in the future. It does not, however, provide guarantees, only the surety that thinking about the past is indeed the only path to the future.

1

ARMORED WARFARE
The British, French, and German experiences[1]

WILLIAMSON MURRAY

Tanks had appeared in large numbers only at the end of the Great War. Their first success had come in the Battle of Cambrai in fall 1917. They had also played a major role in the Australian success at Amiens in early August 1918, a battle that Ludendorff characterized as "the darkest day for the German Army in the war." Whatever their promise, the performance of those ungainly vehicles in World War I was spotty. A slow, difficult-to-maneuver weapon of war, the armored fighting vehicle of 1917 and 1918 offered its crews minimum vision, maximum discomfort, and general mechanical unreliability.

It was a weapon designed for one simple task: crossing the killing zone between trench lines and breaking into enemy defenses.[2] Neither its developers nor operators had moved beyond that role when the war ended in November 1918. Admittedly, J.F.C. Fuller, at that point working in the War Office, had conceived an ambitious plan, Plan 1919, to use tanks in the next year to attack German headquarters – up to corps level – to paralyze the enemy's command and control. But peace came before the British Army could attempt such an ambitious conception. Consequently, as Eu-

---

[1] I am indebted to Barry Watts of the Northrop Corporation for his thorough, intelligent, and *imaginative* critique of earlier versions of this draft.

[2] For a recent discussion of the problems involved in the British Army's adaptation of the tank in World War I see: Stephen Peter Rosen's *Winning the Next War, Innovation and the Modern Military* (Ithaca, 1991). For the most thorough discussion of the role of the tank in the British war effort see Timothy Travers, *How the War Was Won: Command and Technology in the British Army on the Western Front, 1917–1918* (London, 1992). See also Shelford Bidwell and Dominick Graham, *Firepower: British Army Weapons and Theories of War, 1904–1945* (London, 1982); and B.H. Liddell Hart, *The Tanks, 1914–1939* (New York, 1959).

rope entered into two decades of peace, the tank, like the airplane, represented a weapon of potential and promise rather than performance on the battlefield.

The rush to judgment in the aftermath of World War II has fundamentally distorted our understanding of the process of innovation in armored warfare in the 1920s and 1930s. On one side, military reformers, in particular B.H. Liddell Hart and J.F.C. Fuller, used the disastrous defeats of the French Army in May 1940 and of the British Army during the succeeding two years to promote their own personal agendas.[3] On the other side, professional historians have emphasized the economic and strategic aspects of German rearmament to explain operational and tactical factors that occurred for reasons unrelated to and independent of the strategic framework.[4]

In sum, the traditional picture has explained the German success in the following fashion: the Germans, reacting to defeat in World War I, developed a *revolutionary* approach to war, one that emphasized maneuver and armored war as a means to escape the strategic and political consequences of their defeat in 1918. Their opponents, the stodgy, unimaginative officer corps of France and Britain, refused to learn the obvious lessons of the last war and went down to defeat in the great battles that occurred in May 1940 because the conceptions and approaches of the prior war had thoroughly muddled their thought processes.[5] However, the picture that has evolved over the past fifteen years has substantially altered the traditional view. The real explanation lying behind the catastrophe of May 1940 is more complex and opaque. And the development of armored forces in the interwar period, the demand for innovation that new weapons like the tank required, and the problems in developing historically grounded, yet relevant doctrine provide instructive lessons for the present day.

---

[3] For the best discussion and critique of Liddell Hart see Brian Bond, *Liddell Hart: A Study of his Military Thought* (London, 1972). See also John Mearsheimer, *Liddell Hart and the Weight of History* (Ithaca, 1988) which contains useful discussions. For Fuller see Anthony John Trythall, *"Boney' Fuller: The Intellectual General, 1878–1966* (London, 1977).

[4] The classic expositions of this line of argument are Burton Klein, *Germany's Economic Preparations for War* (Cambridge, MA, 1957) and Alan Milward, *The German Economy at War* (London, 1965). See also Larry Addington, *The Blitzkrieg Era and the German General Staff, 1865–1941* (New Brunswick, NJ, 1971).

[5] Even before World War II J.F.C. Fuller had provided a devastating, if unfair, critique of the leadership of the British Army and its capacity to innovate. See J.F.C. Fuller, *Generalship, Its Diseases and their Cure: A Study of the Personal Factor in Command* (London, 1936).

The processes of innovation that created armored forces in the interwar period are anything but clear. Idiosyncratic issues entered the picture to one degree or another because personalities, intellectual trends, societal influences, and the position of military organizations in society all affected innovation and adaptation to new technologies. Moreover, development of armored capabilities took place within a larger framework of doctrinal change, modernization, and technological innovation that affected all military capabilities. In that light a narrow focus on the peculiarities involved in developing armored or panzer divisions would entirely miss the larger problems involved in innovation and distort the actual factors that are of importance in explaining success or failure.

This case study aims to address the issue of innovation by examining the experiences of the British, French, and German armies in developing conceptions of armored warfare in the 1920s and 1930s. It is a story that suggests the inherent difficulties involved in any successful innovation. In a period of substantially reduced military budgets and great distrust of military institutions – at least in the democracies – military institutions had to develop forces that took into account a whole host of changes. Moreover, few tactical and operational lessons from the last war were clear to anyone, including historians.[6] But all reacted to the terrible experiences of World War I through which they had recently passed. How military organizations would perform on the battlefield of the next war very much reflected how they adapted the murky and unclear lessons of the preceding conflict to the process of innovation.

THE STRATEGIC AND POLITICAL FRAMEWORK

Military innovation does not occur in a political vacuum. The military organizations under study here existed within different political and strategic environments despite the fact that they were neighbors in Western Eu-

---

[6] Perhaps the greatest disservice that Fuller and Liddell Hart rendered their nation was the idea that the single, obvious, explanation for the hideous slaughter of the trenches had lain in the stupidity of the generals. The complexities involved in understanding the World War I battlefield are best suggested by the fact that only in the 1980s with the publication of Timothy Lupfer's *The Dynamics of Doctrine: The Changes in German Tactical Doctrine During the First World War* (Leavenworth, KS, 1981) and Timothy Travers' *The Killing Ground: The British Army, the Western Front, and the Emergence of Modern Warfare, 1900–1918* (London, 1987) and *How the War Was Won* that we have finally begun to understand the World War I battlefield. We still do not have an equivalent work for the French, Italian, or Russian armies. If historians who possess the documents and unlimited time have taken seventy years to unravel the changing face of the battlefield, one should not be surprised that the generals had some difficulty during the war.

rope. Moreover, their experiences were distinctly different in the decades under consideration. The 1920s were a relatively peaceful period, while the 1930s were a time of increasing tension. But even then, throughout the interwar period the German Army felt that it possessed a clear mandate to prepare for a war on the continent, while the British Army until March of 1939 never received firm direction from the national government to prepare for such a contingency.[7]

The British political and strategic environment

From 1920 until early 1939 the British Army existed in an antimilitary milieu, one in which *all* the democratic parties rejected the experience of World War I. Liddell Hart gave voice to the national feelings with his argument for "limited liability." Such a strategic approach, he argued, was the traditional British way of war – an approach which had allowed Britain to escape the heavy casualties associated with continental war. By using naval supremacy and a small army for attacks on the periphery (as well as on colonies), the British had influenced the course of continental conflicts, while at the same time capturing a great world empire.[8]

Liddell Hart's arguments had little basis in historical fact, but they appeared as an attractive alternative to the prospect of another blood bath on the Western Front. Ironically, his arguments only served to undermine whatever rationale might have existed to create armored forces in the British Army of the 1930s. The attitudes among the *literati* reinforced a national mood of isolationism; novels and war reminiscences such as Frederick Manning's *The Middle Parts of Fortune*, Guy Chapman's *Passionate Prodigality*, Robert Graves' *Goodbye to All That*, and the novels of Siegfried Sassoon – among innumerable others – reinforced antiwar and antimilitary attitudes that more and more characterized the conventional wisdom. By the mid-1930s much of the educated population in Britain fervently believed that nothing was worth the price of war.[9]

Not surprisingly this sentiment created a hostile political environment for the army, particularly since it received much of the blame for the war's casualty bill. Despite an increasingly threatening international environ-

---

[7] This was equally true throughout the history of the Weimar Republic as in the period of the Third Reich. The only difference lay perhaps in the level of immediacy with which the officer corps lived with the threat of war.

[8] See B.H. Liddell Hart, *The British Way in Warfare* (London, 1932).

[9] The surest indication of the national mood in the early 1930s was the infamous Oxford resolution of 1933, a resolution that declared that Britain's best and brightest would not fight for either king or country at a time that Hitler was already in power in Germany.

ment in the 1930s, governmental and national willingness to expend financial and other resources on defense remained minimal.[10] Even more distressing for the army was the fact that its sister services received priority for funding, personnel, equipment, and training. Army leaders, at least from the mid-1930s, recognized that Germany was their most probable enemy and that such a conflict would require the commitment of its troops to the continent. But they made little headway against political and popular perceptions.

In spring 1937 Neville Chamberlain became prime minister, and his government wholeheartedly embraced a strategy of "limited liability." A series of defense reviews emphasized that Britain would not commit an army to the continent under any circumstances. As the prime minister told his colleagues in spring 1937, he

> did not believe that we could, or ought, or in the event, would be allowed by the country to enter a Continental war with the intention of fighting on the same line as in the last war. We ought to make up our minds to do something different. Our contribution by land should be on a limited scale. It was wrong to assume that the next war would be fought largely by ourselves alone against Germany. If we had to fight we should have allies who must . . . maintain large armies. He did not accept that we must also send a large army.[11]

As a direct result, work to prepare the army for a continental role halted. In March 1938, Lord Gort, chief of the imperial general staff (CIGS), told the Committee of Imperial Defense that "in the circumstances it would be murder to send our soldiers overseas to fight against a first-class power."[12]

Essentially, the army represented to British politicians no more than a colonial police force, aimed at controlling the colonies. The government underlined Gort's testimony by assigning the following priorities to the army: 1) to protect the British Isles; 2) to guard the trade routes; 3) to garrison the empire and 4) to cooperate in the defense of Britain's allies – but only after it had met its other commitments.[13]

Chamberlain imposed a limit on defense spending in 1937 that cut

---

[10] For a closer examination of these issues see Williamson Murray, *The Change in the European Balance of Power, The Path to Ruin* (Princeton, NJ, 1984), chapter 2.

[11] PRO CAB 23/88, Cab 20 (37), Meeting of the Cabinet, 5.5.37., p. 180.

[12] PRO CAB 2/7, CID, Minutes of the 313th Meeting held on March 17, 1938.

[13] PRO CAB 24/275, CP 72 (38), 19.3.38., CID, "The Organization of the Army for its Role in War."

£70 million from the army budget.[14] As a consequence, design work for production of a modern tank ceased entirely.[15] The government described the army as "general purpose" force, a nebulous formula that made it difficult to request supplies or equipment for any theater of operations. The army had no tank in production, its artillery was antiquated, its anti-tank gun obsolete, and its vehicular support inadequate. These deficiencies showed up in the Czech crisis in September 1938, when Britain possessed only two ill-trained, ill-equipped, and ill-prepared infantry divisions for deployment on the continent. Nevertheless, even at the end of September, with a German attack on Czechoslovakia looming, Chamberlain refused to commit British forces to the continent and the defense of France.[16]

Moreover, the extraordinary events of Munich failed to precipitate any significant changes in defense policy, especially with regards to the army.[17] Chamberlain refused to raise the army's funding level, even though the War Office warned the government that the army was so badly prepared that if a war came, "our soldiers would be put to unjustifiable and avoidable risks if adequate arrangements are not authorized for their proper equipment."[18]

Chamberlain finally relented in February 1939 and admitted to the Cabinet that "he came to this conclusion with some reluctance, [but] he saw no alternative." The Cabinet authorized the army to prepare two divisions for deployment to the continent within twenty-one days, and another two for deployment within sixty days.[19] Then in April 1939 in reaction to Hitler's occupation of the remainder of Czechoslovakia, the

---

[14] PRO CAB 53/43, COS 811, 19.12.38., COS Sub-Committee, "The State of Preparedness of the Army in Relation to the Present Situation," Memorandum by the CIGS, p. 42.

[15] PRO CAB 2/7, CID, Minutes of the 380th Meeting held on January 27, 1938, p. 5.

[16] See the report of the conversation between the French commander-in-chief, General Maurice Gamelin and his British counterparts: *Documents Diplomatiques Français*, 2nd Ser., vol. XI, Doc 376, 26.9.38., "Compte Rendu des Conversations techniques de General Gamelin au Cabinet Office 26 September 1938."

[17] Some historians have argued that the time Chamberlain gained by Munich allowed the British to repair their deficiencies. Such an argument would require that a significant increase in rearmament programs after Munich take place but such did not occur until March 1939. See Murray, *The Change in the European Balance of Power*, pp. 269–274, 294–297.

[18] PRO CAB 27 /648, D (38), 3, Calvert: Committee on Defense Programs and Their Acceleration, "The Role of the Army in light of the Czechoslovakian Crisis," Memorandum by the Secretary of State for War, October, 1938.

[19] PRO CAB 23/97, CAB 8 (39), Meeting of the Cabinet, 22.2.39., p. 306.

government totally reversed the position that it had maintained since 1920. Chamberlain announced introduction of conscription and an increase of the regular army to sixteen divisions, while the Territorial Army would have sixteen additional divisions in reserve.[20]

This strategic and political environment had considerable impact on the British Army's ability and willingness to address the questions raised by war on the continent. First, political direction indicated that the army would *never, under any circumstances*, find employment on the continent again. This directing principle changed only in the months immediately prior to the war. Then, when the regular army had at last received the charge to prepare itself for a continental role, for political reasons the government demanded that it accept a huge influx of conscripts and build a great *mass* army. Moreover, military reformers, led by Fuller and Liddell Hart, complicated the situation by heaping scorn on the officer corps and generals, which hardly made the latter receptive to new ideas.

As early as 1934 some in the War Office recognized that the army would have to fight the Germans on the continent in the near future. Nevertheless, the army leadership confronted insoluble problems owing to political uncertainties under which its political leaders operated and the restrictions that those governments placed on the army. As one defense planner put the issue in 1935:

> Here is, of course, the salient difference between us and Germany . . .
> [T]hey know what army they will use and, broadly, how they will use
> it and can thus prepare . . . in peace for such an event. In contrast, we
> here do not even know what size of army we are to contemplate for pur-
> poses of supply preparations between now and April 1939.[21]

The French political and strategic environment

In many respects the French experience mirrors that of the British. Nevertheless, France's geographic position as the only major power with a common frontier with Germany gave a peculiar and tragic twist to its fate in 1940. World War I had left a hideous scar across the collective memory of the nation.

---

[20] Murray, *The Change in the European Balance of Power*, pp. 296–297. For a full discussion of the role of conscription in British defense debates throughout the period see Peter Dennis, *Decision by Default, Peacetime Conscription and British Defense* (London, 1972).

[21] PRO CAB 63/14, Letter from Sir A. Robinson to Sir Thomas Inskip, Minister for the Coordination of Defense, 19.10.36.

With a male, military-age population of 13,500,000 [France] mobilized an army of 8,410,000 [for World War I], including North African and colonial forces. Of these, 1,122,400 were killed or died during the war, 3,594,889 were wounded, and 260,000 were missing. Hence 1,382,400 French soldiers, or 16.4 percent of those mobilized, were dead or missing in the war. Of those killed or missing, 36,600 were officers, and of the 100,600 infantry officers mobilized, 29,260 (or 29 percent) died.[22]

Even if the French dreaded another war, they never lost their sense that military force and power were the coins of international relations. As Winston Churchill has suggested: "Worn down, doubly decimated, but undisputed masters of the hour [in 1919], the French nation peered into the future in thankful wonder and haunting dread. Where then was that SECURITY without which all that had been gained seemed valueless, and life itself, even amid the rejoicing of victory, was almost unendurable? The mortal need was security."[23]

Consequently, interwar France remained split between a desperate desire to avoid war, and a realistic sense – as compared to Britain – that Germany, especially Adolf Hitler's Third Reich, would at some time in the future unleash another conflict.[24] For the most part, the French Army existed within a political environment in which the government gave considerable support to its military institutions. Disagreements existed in the 1920s over the length of time that conscripts should serve, but the appearance of a more violent German threat in the 1930s settled the dispute. Throughout the interwar period civil-military relations were surprisingly good, especially considering French traditions.[25] Financial data from the 1930s does suggest that the French began their rearmament late. In fact, from 1935 through 1938, the Italians spent more on their armed forces in actual dollars than did the French.[26] The problem, however, was not

[22] Robert Doughty, *The Seeds of Disaster, The Development of French Army Doctrine, 1919-1939* (Hamden, CT, 1985), p. 72.

[23] Winston Churchill, *The Gathering Storm* (Boston, 1948), p. 6.

[24] For a general discussion of how French strategy and policy worked during this period see Anthony Adamthwaite, *France and the Coming of the Second World War, 1936-1939* (London, 1977). For a specific discussion of how the French assessed the strategic environment see Stephen Ross, "French Net Assessment in the 1930s," in *Calculations, Net Assessment and the Coming of World War II*, edited by Williamson Murray and Allan R. Millett (New York, 1992).

[25] Robert Doughty, "The French Armed Forces, 1918-1940," in *Military Effectiveness*, vol. 2, *The Interwar Years*, edited by Allan R. Millett and Williamson Murray (London, 1988), pp. 40-41.

[26] MacGregor Knox, *Mussolini Unleashed, 1939-1941* (Cambridge, 1982), pp. 294-295.

so much a lack of funding, but rather how the French prepared; here, as we shall see, they made serious mistakes.

Nevertheless there were difficulties in the strategic and political environment within which the French Army operated. The conflict over the terms of military service and traditional hostility between politicians and generals created areas where the leadership, military as well as civilian, refused to argue about certain subjects on their military merits. Two cases illustrate this point. The initial conceptions for the Maginot Line drew heavily on the battlefield experiences of 1918; many, including Marshal Pétain, argued for creation of a deep fortified zone, a defense in depth.[27] But such an approach would have placed much of France's industrial capacity within the defensive fortified zone; such an impairment of war industries was unacceptable, especially given French beliefs that the next war would be long. The result was the thin, single line of powerful fortifications.[28]

The second case where political factors interfered with military analysis had to do with the response to Charles de Gaulle's book, *Towards the Army of the Future,* by both generals and politicians.[29] The arguments over de Gaulle's work had little to do with the military utility of his proposed professional armored corps. Unfortunately, de Gaulle had managed to open up the ancient conflict between the military and the left. For politicians like Léon Blum, de Gaulle's ideas sounded all too much like an effort to wrest control of the army from the Republic.[30] On the other hand, senior army leaders preferred not to resurrect the quarrel with the left over a radical military proposal, especially one for which they possessed little sympathy.

More importantly, the delicacy of civil–military relations may have made the military unwilling to challenge the Popular Front's refusal to increase defense spending in the crucial years of 1936 and 1937. This action had a most unfortunate impact on the French Air Force and resulted in the failure of French industry to field a fighter aircraft equivalent to the Bf 109 in sufficient numbers in 1940.[31] For the army, budgets in those

---

[27] The following discussion is drawn from Doughty, *Seeds of Disaster,* chapter 3.

[28] It is worth noting that the decision had little impact on the events of 1940 since the Maginot Line achieved its goal and the Germans went around it. In fact, the Germans never broke through the main line of fortifications in the fighting of May–June 1940.

[29] Charles de Gaulle, *Vers l'armée de métier* (Paris, 1934). Martin Alexander, *The Republic in Danger, General Maurice Gamelin and the Politics of French Defence, 1933–1940* (Cambridge, 1992) is particularly good on the political ramifications of de Gaulle's proposals for an all-professional tank force. (See particularly ibid., pp. 123–125, 145–146.)

[30] See the excellent discussion of this point by Philip Bankwitz, *Maxime Weygand and Civil Military Relations in Modern France* (Cambridge, MA, 1967), pp. 152–155.

[31] See Williamson Murray, *Luftwaffe* (Baltimore, MD, 1985), p. 40.

years impinged on development and production of a modern tank for the ground forces.[32] Whether the army would have utilized additional funding for development and production of more modern tanks is, however, open to question.

On the whole the strategic environment should have favored development of tanks and armored forces. Unfortunately, French military leaders *never* confronted political leaders about the serious imbalance between requirements and funding levels.[33] Moreover, French commitments to the defense of Germany's eastern neighbors suggested the need for a mobile, well-prepared force that would be capable of moving against the Rhineland. But French doctrine throughout the interwar period remained a conception of tightly controlled combat that left little room for such a force. For their part, left-wing politicians treated modern war as a matter of the workers downing tools to defend the nation (the *levée en masse*, which already had had a pernicious impact on French military history); they thereby removed whatever existed of maneuvering space that might have allowed reformers to bring sufficient political pressure to force the army to move in new directions.[34]

The German political and strategic environment

Not surprisingly, the German Army existed within a political and strategic environment quite different from that of the British and French. But, however impressive German innovation in mechanized warfare, the German military mislearned the larger strategic and political lessons that World War I should have suggested to a discerning observer of the Wilhelmine Reich's defeat.[35] To a great extent, the German military reflected the faults of the society at large – a political world that in the case not

---

[32] For the French Army's efforts to develop tanks and armored forces throughout this period see Doughty, *Seeds of Disaster*, chapters 7 and 8.

[33] Neither of the two apologias for Maurice Gamelin's leadership address this issue in detail, nor do the authors compare and contrast French military expenditures with those of their rivals. The above-mentioned fact that Mussolini's Italy with a far smaller economic base managed to spend more than did France suggests that something was seriously out of alignment, especially considering the relative silence of the French high command. See Alexander, *The Republic in Danger* and Robert J. Young, *In Command of France, French Foreign Policy and Military Planning, 1933–1940* (Cambridge, MA, 1978).

[34] See Alexander Werth, *The Twilight of France, 1933–1940* (London, 1942), p. 44.

[35] For the best single explanation of why this was so see Holger Herwig, "Clio Deceived, Patriotic Self-Censorship in Germany after the Great War," *International Security* (Fall 1987). See also Williamson Murray, *German Military Effectiveness* (Baltimore, MD, 1992), pp. 3–17.

only of Nazi Germany but also of the Weimar Republic as well misread the strategic and political lessons of the last conflict.

The literature of post-World War I Germany suggests how differently Germans viewed their wartime experience from the French and British. Only two antiwar novelists of note emerged in Germany (one Jewish), compared to a flood of writers in France and Britain. On the other hand, the British and the French produced no equivalent to the *ferocious* novels of Ernst Jünger, one of the great stylists of the German language in the twentieth century (he was a winner of the Goethe Prize *and* the *pour le mérit*), and Jünger thought that World War I had represented a thoroughly ennobling experience through which every generation should pass.

As a result, the German Army never had to sell itself to the public, the majority of whom respected and admired things military. For the 1920s and the early 1930s the problem was not parliamentary or governmental limitations on spending, but rather the imposition of strict military controls by the Treaty of Versailles. That treaty limited the army to 100,000 men and forbade possession of tanks, aircraft, and other weaponry. While throughout the 1920s there was general and consistent cheating on the Treaty's provisions by the German military, aided and abetted by the republic's politicians, there were limits as to what it could get away with.[36] In the end, these limitations affected the development of the army and its ability to innovate, particularly in regard to innovating in technological fields.[37]

Political leaders allowed the German Army much latitude throughout the life of the republic. The *Reichswehr* ran its own independent strategic policy, a case in point being the secret agreements to test tanks and aircraft in the Soviet Union. Moreover, the election of the army's World

---

[36] A British diplomat in Berlin at the end of 1938 related the following story to illustrate to his government what the Nazis meant by the word "truth:" "A German officer in conversation with a former military attaché was at pains to prove that English officers were gentlemen and French officers were not. To illustrate the point he related that the Allied Control Commission had visited his garrison after the war and had demanded that he pull down a certain wall on the ground that according to information received arms were concealed there. 'I gave my word of honor as a German officer that no arms were concealed behind that wall. The Englishman believed me and was prepared to go away. The Frenchman did not with the result that the wall was pulled down and that my arms were taken away.'" Beside a revealing glimpse at the German mentality in the 1930s the story underlines the wide-scale recourse to cheating on the terms of Versailles. PRO FO 371/21701, C 15228/528/18, Kirkpatrik (Berlin): "Herr Hitler's Word," 7.12.38.

[37] According to Guderian, the first tank he ever saw was when he was sent to Sweden in 1928. Guderian, *Panzer Leader* (New York, 1980), p. 12.

War I leader, Paul von Hindenburg, as president of the republic provided the army with a position of considerable power within the political system.

Hitler's appointment as Reich's Chancellor in January 1933 soon robbed the army of its political independence. But in return the Nazi regime provided broad public support and unlimited funding, which allowed for a massive rearmament program. By 1936, the implications of that program were becoming clear to the army leadership: either war in the near future or national bankruptcy.[38] But then Hitler had always made clear to his military leadership the nature of his goals for the new German state.[39] Those strategic goals, however, did not guide operational and tactical developments. Hitler played little role in development of the panzer forces except to make a few favorable remarks to Guderian in 1934.[40] In fact, army leaders pushed a conservative program that emphasized creation of traditional formations and placed development of armored forces within the framework of regular army doctrine and conventional military forces. Nevertheless, such an approach had generally beneficial consequences in development of German armor and its doctrine in the late 1930s.

A comparison of the German, French, and British strategic environments reveals a wide disparity among the three nations, but those differences do not explain the variances in innovation. Britain and Germany had enormously differing reactions to the experiences of World War I; not surprisingly, both nations came to different perceptions about the utility of force in advancing national goals. But the strident antiwar and anti-army sentiments in Britain do not by themselves explain the inability to innovate, especially since the British had pioneered armored warfare in World War I.

In Germany, no matter how favorable the climate, innovation could not take place without other factors coming into play. However, Hitler's extraordinary goals – nothing less than the subjugation of the entire continent – did not lead to a demand for some specialized new form of warfare.[41] It was the influence of other factors beyond the strategic and po-

---

[38] Wilhelm Deist, *The Wehrmacht and German Rearmament* (London, 1981), pp. 47–51.

[39] In early February 1933 within a week of becoming Chancellor, Hitler had spelled many of those goals out to the Reich's military leadership. "Aufzeichnung Liebmann," *Vierteljahrshefte für Zeitgeschichte*, 2, no. 4 (October 1954).

[40] See Friedrich Hossbach, *Zwischen Wehrmacht und Hitler, 1934–1938* (Hannover, 1949), p. 39.

[41] It is worth noting that Mussolini's goals, considering Italy's resources, were as megalomaniacal as those of the Führer, but they hardly resulted in the creation of innovative and imaginative new military conceptions at any level of warfare. See Knox, *Mussolini Unleashed*.

litical environment that allowed the Germans to innovate in such devastating fashion.

Like the British, the French failed to innovate in the field of armored warfare. While their strategic situation less obviously called for aggressive, offensive forces than did that of Germany, France's strategic framework did suggest the need for a force capable of aiding its Eastern European allies. But the French failure did not lie in developing an exclusively defensive doctrine; in fact, French doctrine was offensively minded, and it paid little attention to the problems of defensive warfare. Unfortunately, doctrinal innovation, influenced strongly by misreadings of the last war, particularly the lessons of 1918, led to the creation of a stylized, tightly controlled conception of tactics that proved unsuited to the modern battlefield, particularly one inhabited by Germans. Robert Doughty's severe critique of French interwar doctrinal developments is more than supported by his evidence:

> The inordinate emphasis on firepower prevented the French from understanding how improvements in tactical mobility coupled with tactical methods originating from the German infiltration tactics of World War I could overturn many accepted and combat-tested methods. The notion of a carefully controlled and tightly centralized battle belonged to another era, and the sense of chaos, lack of control, and futility which emerged from the French participation in the 1940 campaign demonstrated the inability of the French to force their method of fighting onto the Germans.[42]

Given the experiences of these three countries, one can suggest that the political and strategic environment can exercise considerable influence over some military innovation, but it is hardly a sufficient or complete explanation for success or failure in that process.

WAR, INDIVIDUALS, ORGANIZATIONS, AND DOCTRINE

The strategic and political environment can indeed create a climate conducive to innovation. The elements in such change, however, occur within organizations themselves. It is the interplay between past experiences, individual leaders and innovators, and the cultural climate within military organizations that determines how successfully innovation proceeds.

Organizations and technology were simpler in the first half of this cen-

---

[42] Doughty, *Seeds of Disaster*, p. 179.

tury; yet their simplicity can be deceptive. Few officers in the period possessed the technical background to judge the technological questions which they confronted.[43] Even more important, the modern battlefield had undergone a drastic alteration between 1914 and 1918 and few lessons of that experience were yet clear. And so innovation took place in response to the tactical revolution that had taken place on the battlefields of the last war as a crucial determinant but in an atmosphere where the experts were still developing the possible lessons of that conflict.

The British

For the British, the battlefield victories of 1918 provided no clear path.[44] In retrospect, the stunning victory of 8 August 1918 had combined the use of tanks and infantry to blast a break in the German lines; clearly that triumph presaged development of armored exploitation warfare.[45] Unfortunately for British innovation, the remaining British victories of 1918 rested on an infantry-artillery paradigm similar to the Somme and Passchendaele.[46] As the historian Timothy Travers notes:

> It can be seen, therefore, that the BEF [British Expeditionary Force] really conducted two kinds of warfare in the second half of 1918: first, mechanical warfare in July and August; and secondly, traditional or semi-traditional open warfare, from the end of August to the Armistice. Yet the fact that the large scale mechanical attacks did not take place in the last months, while the war was won by means that were familiar to most officers, did strongly influence the way that mechanization and mechanical warfare were debated in the 1920s and 1930s. It is important, however, to note that this debate did not actually start *after* the war, but in fact commenced in late 1917 and early 1918.[47]

The difference in the degree of success between 1917 and 1918 lay in the fact that an exhausted German Army, having suffered nearly 1 million

---

[43] This was true even for airmen. One of the crucial questions that should have been addressed in the interwar period was that of developing a long-range escort fighter to protect bombers on deep penetration missions. Yet right through 1942, senior Allied airmen argued that such aircraft were technologically impossible. Not until the P-51 appeared in late 1943 were such arguments revealed for what they were: the opinions of men who without the necessary technical and engineering background decreed what they felt to be the case. See Williamson Murray, *Luftwaffe*, p. 214.

[44] The discussion in this essay on what was happening in the British Army in the year of its triumph rests largely on the ground-breaking study by Travers, *How the War Was Won*.

[45] Ludendorff referred to the success of 8 August as the "blackest day of the German Army."

[46] Ibid., p. 179.   [47] Travers, *How the War Was Won*, p. 179.

casualties between mid-March and mid-July, broke apart in the late 1918 battles.

The British never established a coherent doctrine in 1918. Field Marshal Douglas Haig's General Headquarters failed to exercise appreciable control over its army and corps commanders; as a result, each of the separate branches as well as individual units fought as independent entities. This was not entirely a bad thing for it allowed considerable latitude to innovators such as Rawlinson, Monash, Maxse, and those within the new tank corps. But for most, the 1918 battles resembled a more sophisticated replay of 1916 and 1917 except that the German Army was completely breaking down. Ironically British casualties were considerably higher during this period of victory than they had been at Passchendaele: 271,000 for the blood bath in Flanders, but 314,200 for the equivalent period in 1918.[48]

The fact that the British Army failed to establish a committee to examine the war's lessons until 1932 magnified the lack of clarity in its understanding of World War I's lessons. The CIGS at that time, Lord Milne, gave the committee's members broad powers and requested that they "study the lessons of the late war, as shown in the various accounts, and to report whether these lessons are being correctly and adequately applied in our manuals and in our training generally."[49] In general, the committee's report was highly critical of the army's performance throughout the war and made a number of recommendations. Its proposals should have provided the basis on which substantial reforms of both doctrine and professional military education might have taken place.

Unfortunately, the report arrived early in the tenure of the next CIGS, Field Marshal Archibald Montgomery-Massingberd, a man who had no intention of allowing uncomfortable assessments to circulate through the officer corps. Consequently, only the most senior army commanders received copies of the original report, while the War Office prepared a condensed version with *significant* omissions and changes that distorted its message into a more favorable and unrealistic appraisal of the army's performance in the war.[50]

---

[48] The disparity in Canadian losses was even greater: 29,725 for late summer and fall 1917 and 49,152 for the equivalent period in 1918. The great difference in the collective memory of the survivors was the fact that Haig's offensive in Flanders achieved no discernable gains, while the attacks in 1918 made advances never before seen by any other army – including the German – on the Western Front during the course of the war. Ibid., p. 181.

[49] Quoted in Harold R. Winton, *To Change an Army, General Sir John Burnett-Stuart and British Armored Doctrine, 1927–1938* (Lawrence, KS, 1988), p. 127.

[50] Ibid., p. 131.

The watered-down version was what the officer corps received. In fact, this incident fell into a general pattern in which the army's leadership cooked the books. To compound the disregard for the study of recent military experience, the volumes in the official British history not only arrived late (the last ones appeared after World War II), but deliberately skewed the historical record to present the army's performance in the most favorable light.[51] Finally one must note that in 1920 Liddell Hart, then a captain on active service, had been charged to rewrite the basic infantry manual. The War Office disliked one of his proposed chapters, deleted the offending section, and simply replaced it with a chapter from the 1911 manual.[52]

One of the major contributors to such unwillingness to countenance critical analysis was the army's leadership. In the immediate postwar period, General Sir Henry Wilson was the CIGS; his record suggests that he possessed considerable interest in the potential for innovation and adaptation to the challenges of the last war. But Wilson, besides being one of the brightest individuals in the British Army, was also a violent and unstable Ulsterman.[53] His assassination by the IRA led to Lord Cavan's appointment as his replacement. Not only had Cavan's experience in 1918 been in Italy, where conditions had been quite different than those on the Western Front, but he lacked imagination and interest in ideas.[54]

Cavan's successor, Lord Milne, was the most forward-thinking and progressive of interwar CIGSs. The criticism that both Fuller and Liddell Hart heaped on him both then and later was grossly unfair. Milne had the courage to allow considerable experimentation with armored and mechanized warfare. In fact, it was almost entirely during his tenure that the army executed the experiments that laid out the road to the future. In 1927 Milne went so far as to suggest that increasing firepower as well the notable changes in the technology of the gasoline-powered engine would eventually dominate the battlefield and that one would see armored divisions at some future date. But he also warned that financial constraints were imposing considerable limitations on what the army could develop for the foreseeable future.[55]

However, Milne's successors proved less far-sighted. Montgomery-

---

[51] Travers, *The Killing Ground*, pp. 203–249.

[52] B.H. Liddell Hart, *Memoirs*, vol. 1 (London, 1966), p. 49.

[53] An IRA gunman ended Wilson's army career and life in 1921.

[54] Cavan did not believe that serving officers should publish books and ordered Fuller not to publish *The Foundations of the Science of War* in 1923. J.F.C. Fuller, *Memoirs of an Unconventional Soldier* (London, 1936), p. 420.

[55] Winton, *To Change an Army*, p. 81.

Massingberd was an out-and-out opponent of serious innovation; his successor Field Marshal Cyrill Deverell was little better. Lord Gort, the final prewar CIGS, came to the position with a mandate for reform, but having achieved office, stolidly set himself against substantive reforms or innovations. As he told Liddell Hart in 1938: "We mustn't upset the people in the clubs by moving too fast – we must get over the shock of the Army Council changes" (i.e., Deverell's firing and replacement by Gort).[56]

Yet, if the army were unfortunate in those selected to guide its course in the 1920s and 1930s, its structure and cultural values contributed to a pervasive disinterest in innovation. To a great extent its postwar administrative structure, culture, and class attitudes remained those of the prewar army.[57] As a senior officer in the interwar period suggested about the War Office:

> I have, as an individual, no personal grudge against the War Office. I served two four-year periods as a member of it, and knew it to be full of capable men and good fellows both on the military and on the civil side. But I always had the feeling which was shared by many others of my generation, that there was something wrong with it. It was in itself too top-heavy; it was constantly immersed in detail which could so easily have been dealt with by commanders, in their different degrees, had reasonable powers and financial responsibility been allowed them; it was out of touch with the real army, especially with the troops overseas; . . . and above all, it had no military head.[58]

Granted, organizational as well as political constraints impinged on the ability to innovate. The Cardwell system was the most glaring inhibitor to innovation by its demand that units at home and forces in the empire match up on a one-to-one basis.[59] But there was another side to the coin. As Michael Howard has suggested, "the evidence is strong that the army was still as firmly geared to the pace and perspective of regimental soldiering as it had been before 1914; that too many of its members looked on soldiering as an agreeable and honorable occupation rather than a serious profession demanding no less intellectual dedication than that of the doctor, the lawyer, or the engineer."[60]

---

[56] Liddell Hart, *Memoirs*, vol. 2, p. 88.

[57] For a short, crisp summation of the history of the British Army between the wars see Brian Bond, *British Military Policy Between the Two World Wars* (Oxford, 1980), especially chapter 2, "The Character and Ethos of the British Army Between the Wars."

[58] Ibid., p. 39.

[59] See Winton, *To Change an Army*, pp. 174, 184, 189–90, and 229–31.

[60] Michael Howard, "The Liddell Hart Memoirs," *Journal of the Royal United Services Institute* (February, 1966), p. 61.

Before the war Fuller's fellow officers had regarded him as unusual because of his penchant for reading books. Despite the appearance of some students of history such as Wavell and Slim, the army of the 1920s and 1930s was little more involved in serious study than its prewar predecessor. The evidence suggests a general disparagement of study in the army's cultural system.[61] "Judging by the space the [memoirs of the interwar period] allotted to [activities in peacetime] – under such titles as *Big Game, Boers, and Boches* (the order is interesting) – the outstanding peacetime memories of many officers were wrapped up with sports, games, and killing all manner of beasts."[62] In 1939 Percy Hobart, trying to prepare what eventually became the 7th Armored Division in the Western Desert, remarked in a letter to his wife:

> I had the cavalry CO's in and laid my cards on the table. They are such nice chaps, socially. That's what makes it so difficult. But they're so conservative of their spurs and swords and regimental tradition, etc., and so certain that the good old Umpteenth will be all right . . . , so easily satisfied with an excuse if things aren't right, so prone to blame the machine or machinery. And unless someone upsets all their polo, etc., for which they have paid heavily – it's so hard to get anything more into them or any more work out of them. Three days a week they come in six miles to Gezirah Club for polo. At 5 pm it's getting dark: they are sweaty and tired. Not fit for much and most of them full up of socials in Cairo.[63]

Such attitudes existed throughout the army, largely as a result of a military culture that rested on the insularities of the regimental system. That system did provide the *esprit de corps* and *élan* so essential to the battlefield; however, even at its best, it engendered a "muddy boots" approach to soldiering, one that regarded intellectual effort with contempt and retarded an understanding of operations beyond battalion level.[64] Moreover, the British professional military education system was incapable of lifting officers' understanding out of the concerns of regimental soldiering. Some commanders even regarded junior officers who sought

---

[61] Of all the things that are crucial to innovation, the cultural values of military organizations may be the most important. But the value system is precisely the most difficult to change. In the early 1960s the U.S. military gave little weight to the physical condition and appearance of its senior officers. To varying degrees each service then embarked on major campaigns to change those cultural attitudes. It was not an easy matter. I am indebted to Lt. Gen. Paul Van Riper, USMC, for this point.

[62] Bond, *British Military Policy Between the Two World Wars*, p. 64.

[63] K. Macksey, *Armored Crusader* (London, 1967), p. 159.

[64] See the letter by Brigadier George Lindsay to Knox, 29 Oct 1932. Lindsay Papers, WO 32/2820. WO163/38, "Report of Committee on Regimental Officers," May 1932.

positions at the staff college as deserters, disloyal to the regiment, and overly ambitious.[65] Finally, education at Camberly (the staff college) rarely stretched student minds or encouraged them to examine the operational and tactical lessons of the last war or of technological developments.[66]

Despite such inhibiting factors, it is indeed astonishing how much the British Army accomplished in the interwar period. In fact, the British experiments with armored warfare between 1926 and 1934 contributed to a considerable extent to the creation of the German panzer forces after Hitler came to power.[67] How to explain this ironic result? Clearly Milne backed the armored experiments with a substantial portion of the army's scarce funding throughout his tenure as CIGS.[68] Unfortunately for the British Army, he received little support from the innovators both within the army and on the outside. In retrospect, the critics paid no attention to the financial constraints, the political realities, or the strategic commitments that Milne confronted throughout his tenure as CIGS. In 1926 Milne offered the command of the initial experimental force to Fuller. But Fuller had the extraordinary temerity to turn the assignment down, because he, as a lieutenant colonel, failed to receive all of the demands that he made on the CIGS – a choice his biographer aptly characterized as "probably the worst decision of his life."[69] Fuller probably also refused the assignment because he disapproved of the decision to operate the experimental force in conjunction with the more traditional branches, the artillery and the infantry.

Both Fuller and his compatriot Liddell Hart launched increasingly vitriolic attacks on the British Army's leadership in the 1930s. The consequence was that they exercised a decreasing influence over the thought processes of the army itself. They rejected not only the help of those whom they regarded as troglodytes in the army's upper levels, but also the overtures of a number of forward-thinking officers in the traditional branches. To a great extent, the result of their strident advocacy served to exacerbate the split between innovators (by far the smaller group) and the

---

[65] Bond, *British Military Power Between the Two World Wars*, p. 64.   [66] Ibid., p.37.

[67] General Oswald Lutz, the chief of armored development in Germany until 1938, told Sir John Dill during the latter's visit to Germany in 1935 "with considerable pride that the German tank corps had been modelled on the British." PRO CAB 16/112, DRC 31, 9.10.35, p. 271.

[68] Milne went so far in a 1927 speech as to suggest that someday in the future the British Army would be an all-armored force. Murray, *The Change in the European Balance of Power*, p. 85.

[69] Trythall, *Boney' Fuller: Soldier, Strategist and Writer, 1878–1966*, pp. 136–141.

great mass of professional soldiers and assured that their ideas played a decreasing role in the preparation of British ground forces for war. Even more disturbing, the wounds they inflicted prevented the innovators from exercising any influence even after the German victories of 1939–1940 should have suggested that the army reconsider its tactical approach to war.[70]

During the period when Milne was CIGS, experiments with armored forces played a crucial role in turning theory into reality. By 1923 the British Vickers medium tank had achieved speeds of twenty miles per hour.[71] For the army maneuvers of 1927 and 1928, Milne allowed a concentration of armored and motorized forces to emphasize experimentation and the potential for innovation. Underlining the fact that considerable support existed in the army for major innovations, the division commander Sir John Burnett-Stuart, who was to control the mechanized force, wrote to the War Office in summer 1926:

> [W]hat help are you going to give me in organizing, launching and guiding this experiment? It is no use handing it over to an ordinary Divisional Commander like myself. You must connect with it as many enthusiastic experts and visionaries as you can; it doesn't matter how wild their views are if only they have a touch of divine fire. I will supply the commonsense of advanced middle age.[72]

The 1926 maneuvers did indicate considerable problems with a mechanized force. But they also underlined the potential of such an approach. Early in the maneuver, the light scouting group, under a sophisticated tank pioneer, Lt. Col. Frederick Pile, carried out a stunning twenty-five-mile march that brought the exercise to a halt. Burnett-Stuart, instead of displaying displeasure, commented:

---

[70] The fate of Hobart is instructive. In 1938 he went out to Egypt to establish what eventually became the 7th Armored Division. But in the process he fell afoul of the army hierarchy and was fired from his division command just as the war was breaking out. Summer 1940 found him serving as a corporal in the home guard, when Churchill took an interest in his fate and demanded that the War Office bring him back onto active duty despite the strenuous objections of the CIGS, Sir John Dill, who argued that during "various occasions" in Hobart's career, he had been "impatient, quick-tempered, hotheaded, intolerant, and inclined to see things as he wished them to be instead of as they were." Churchill prevailed and Hobart returned to active duty. But Hobart's position remained limited to commanding and training two outstanding armored formations, the 11th Armored and the 79th Armored, the latter responsible for the innovative armored vehicles that played such an important role in the British success on D-Day. Martin Gilbert, *Winston S. Churchill*, vol. 6, *Their Finest Hour* (Boston, 1983), pp. 861–863.

[71] B.H. Liddell Hart, *The Tanks*, vol. 1, p. 218.

[72] Quoted in Winton, *To Change an Army*, p. 76.

I know that a lot of you will not like the tactics which you saw employed
by the Light Group in these manouevers. You will think them risky. But
I assure you that in armored war these things will be tried, they will probably come off, there will always be people who will chance their arm this
way, and you have got to be prepared to meet them when they do.[73]

The 1927 and 1928 maneuvers suggested the wider potential of armored warfare.[74] They underlined the tactical advantages of maneuver
and firepower that armored formations would possess. They also suggested the considerable logistical and communication support that such a
form of warfare required. Particularly in 1927, the experimental armored
force achieved stunning successes against traditional infantry and cavalry
formations. Unfortunately, the general in charge of Southern Command
in 1928, where the maneuvers took place, was Montgomery-Massingberd.
In particular, he emphasized the negative impact that the new formations
were having on the traditional branches by their successes in exercises;
therefore, he argued that it was wrong to create a force equipped with
new armament. Rather, he suggested, mechanization and motorization
should take place throughout the whole army. His arguments combined
with the shortages of resources to disband the experimental armored force
until 1931.[75]

Nevertheless, in 1931 further experiments took up where the innovators had left off in 1928. In that year, a resurrected experimental armored
force practiced tactics for deep penetration raids.[76] Even more important
was the recognition of the importance of radios to maneuvering armored
formations.[77] The 1934 maneuvers were, however, the capstone for the
development of the army's armored conceptions. Throughout the maneuvers, the flexibility, mobility, and firepower of armored forces consis-

---

[73] Ibid., pp. 80–81.
[74] In reports as early as 1926 the Germans were learning a great deal about the potential
of the tank from experiments that the British Army was carrying out. See Reichswehrministerium, Berlin, 10.11.26, "Darstellung neuzeitlicher Kampfwagen," National
Archives, T-79/62/000789.
[75] The Germans immediately noted the impact that distributing tanks among regular infantrymen moving on foot had on the exercises. A Reichswehr report at the end of the
year pointed out that such an approach had hindered the armored troops without speed-
ing up the advance of the infantry. Reichswehrministerium, "England: Die Manöver mit
motorisierten Truppen, September 1929," National Archives, T-79/30/000983.
[76] Liddell Hart, The Tanks, vol. 1, pp. 292–294.
[77] Winton, To Change an Army, p. 118. This would not be clear to the French Army even
in 1940.

tently proved superior to those of more conventional forces. Burnett-Stuart, now commanding Southern Command, designed a complex and difficult set of tactical and operational problems as a final test and those problems failed to display armored forces to best advantage.[78]

Yet within the tightly structured framework of the exercise's problem, armor displayed great flexibility. But many senior officers came away from the exercises with the impression that the tank arm had not lived up to the expectations of its advocates. The impression undoubtedly affected thinking in the army until the German victories of May 1940. In his memoirs Liddell Hart takes Burnett-Stuart severely to task. But what Liddell Hart's severe criticisms miss is that the purpose of maneuvers is to train troops, *not* to put on shows for visiting firemen. Burnett-Stuart was doing what a good general *must* do in training: test his troops under unfavorable circumstances rather than favorable circumstances. In this case the result was unfortunate for its political impact on much of the senior leadership, but that is a risk that good commanders must run in doing their job properly. And one must note that what played such a crucial role in the German Army's rapid expansion in the 1930s was its extraordinary emphasis on harsh, demanding training in all exercises.

The tale of armored development for the remainder of the 1930s was one of missed opportunities.[79] The CIGSs after Milne displayed scant interest in pursuing armor's potential, but rather tied its development entirely to the infantry and cavalry establishments. In an army possessing a coherent doctrine, such a path might not have been entirely bad. But the British Army remained tied up within its regiments and separate constituencies of infantry, cavalry, and artillery. Of the three, the cavalry dis-

[78] Harold Winton makes a compelling argument that Burnett-Stuart was not attempting to discredit the armored forces as Liddell Hart and others have suggested, but rather was exercising his inclination as a great trainer. Unfortunately, he missed the political importance of these maneuvers for persuading senior officers of the direction in which the British Army needed to go. Winton, *To Change an Army*, pp. 180–83.

[79] Of all concerned in the late 1930s only Percy Hobart kept his attention on the issue of military innovation and the importance to Britain's military contributions on the next war. Fuller retreated into fascism; Liddell Hart's efforts focused on preventing the British government from committing the army to the continent, and the army's leadership withdrew into a self-imposed shell. In October 1937 Hobart commented in a memorandum (one that the CIGS severely took him to task for): "The greatest threat from our point of view is a German drive to secure advanced air bases. German armored formations, leading such a drive, can be countered by defense where it is ready, but only by armored counterattack otherwise. Our necessarily small contribution in the first phase must be of maximum value. In relation to the forces of our allies the greatest value we can provide is offensive armored formations to assist to counter the German drive." Quoted in Bond, *British Military Policy Between the Two World Wars*, p. 190.

played the most resistance to new ideas and technology.[80] Representing its views, Alfred Duff Cooper, Secretary of State for War, announced in the House of Commons in 1935 that asking the cavalry to give up horses for trucks "was like asking a great musical performer to throw away his violin and devote himself in the future to the gramophone."[81]

Ironically, the army's leadership decided in the late 1930s that rather than increase armored forces by expanding the Royal Tank Regiment, it would motorize the cavalry.[82] The reasons lying behind the decision were peculiarly British; as the War Office argued in 1937:

> Various proposals were considered including one for an army of a more highly mechanized nature than that decided for the regular army. . . . The Chiefs of Staff have stated that in their opinion the present would be a most unfortunate moment to disturb an organization which has valuable traditions and has survived the lean years through which it has passed since the war.[83]

Perhaps, given governmental unwillingness to begin serious rearmament of the army even in the late 1930s, more enlightened leadership would not have had British forces in much better shape to deal with the battlefields of 1940 and 1941. At the least, however, an army that had prepared intellectually in the 1930s would have had more realistic expectations about the face of war in the early 1940s. But a chiefs-of-staff report suggested in 1938 that a German advance through Belgium and Holland "despite mechanized forces, would be considerably slower than in the last war."[84]

---

[80] Field Marshal Haig's justly infamous quote ("I feel sure that as time goes on you will find just as much use for the horse – the well-bred horse – as you have ever done in the past.") reflected the views of a retired officer firmly rooted in the past. But the commander of the cavalry, General Sir Alexander Godley, noted in the 1920s: "On the other hand, if I were asked, 'Will you go to war with a mobile force composed [of] armored cars, tanks, and such-like?' I think I should refuse to go! I should say that I would not go without a force of cavalry, I should want, and should insist on having, an ample portion of mounted troops." Winton, *To Change an Army*, p. 29.

[81] Liddell Hart, Memoirs, vol. 1, p. 227.

[82] The cavalry regiments in this process of motorization in the late 1930s maintained their regimental badges and traditions and were only to be "trained" by the Royal Tank Corps. That latter force was soon reorganized into battalions (and then regiments) of the Royal Tank Regiment, further adding to the tribalization of the British Army.

[83] PRO CAB 24/269, C.P. 115 (37), 23.4.37., "The Organization, Armament, and Equipment of the Army," Memorandum by the Secretary of State for War, p. 138.

[84] PRO FO 371/22915, C 1503/30/17, COS 833, CID, "The Strategic Position of France in a European War," late 1938. One cynical postwar writer has commented that "fortunately. . . .the army, if not equipped with tanks or guns or anti-aircraft that enabled them to fight, had more mechanical transport than the French or Germans and were equipped to run away." Lt. Col. C.E. Carrington, "Army Air Cooperation, 1939–1943," *Journal of the Royal United Services Institute* (December 1970), p. 38.

Where the British Army paid the heaviest price for "the locust years" lay in its inability to adapt to the battlefields in the later war years. This unfortunate phenomenon was very much the result of Field Marshal Alan Brooke's leadership throughout World War II.[85] In particular, he insured that *none* of the innovators in armored warfare reached senior positions – division level and above – of armored forces committed to battle. Consequently, the British Army lost the hard-earned lessons of the 1920s and 1930s for almost the entire course of World War II.

In retrospect, Milne received little help from the innovators in substantially advancing armored warfare in the experiments that occurred during his tenure as CIGS. Instead the innovators exacerbated the gap between themselves and the rest of the army. Had they been willing to work with Milne, they might have preserved some of the advances that occurred during his years. But one must also recognize that the British government chose very badly in picking Milne's successors. These men comfortably assumed that what was good enough for the old army was good enough for them and quite simply failed to address any of the substantial problems that war in the twentieth century has raised. Once on the wrong track, the British Army would never fully recover from the cultural and intellectual mistakes made in the period between 1934 and 1940.

The French

On no other nation had the direct burden of World War I lain more heavily than on France,[86] and the experiences of that war twisted and distorted French doctrine and their army's capacity to learn. The French Army had entered World War I with the most offensive-minded doctrine of any of the combatants.[87] But the failed offensives of 1914 and 1915 underlined how unrealistic were the tactical assumptions on which French doctrine rested.

---

[85] Unfortunately, British historians have generally been unwilling to admit the scandalously bad performance of British forces in World War II, particularly in its later years. They have overestimated the competence of Alan Brooke at every level of war, while at the same time denigrating the leadership that Churchill provided at the strategic level. The problems that British military forces confronted in the war were the result of bad generalship, not Churchill's mistaken directions. For a discussion of British military effectiveness in World War II see Allan R. Millett and Williamson Murray, *Military Effectiveness*, vol. 3 (London, 1988), chapter 3.

[86] In terms of its indirect political burdens (with the creation of the Bolshevik regime), one might argue that they lay most heavily on the peoples of the Soviet Union as the post-1989 period is underlining even for academics.

[87] For the most thorough discussion of military doctrine before the outbreak of World War I see Michael Howard, "Men Against Fire: The Doctrine of the Offensive in 1914," in *The Makers of Modern Strategy*, ed. by Peter Paret (Princeton, 1985), pp. 510–526.

In 1916, using the innovative thoughts of a junior officer, Captain An-dré Laffargue, the French made substantial progress in adapting to the re-alities of the World War I battlefield. On the Somme and the fall battles at Verdun the French displayed substantially more flexibility and devolu-tion of authority to those on the sharp end of battle. But General Nivelle's disastrous offensive of spring 1917, when French attacks ran into Ger-man defenses based on revolutionary changes in their defensive doctrine, ended whatever inclination existed within the French Army for aggressive offensive tactical conceptions.[88] From spring 1917, the French proved un-willing to consider tactical or operational concepts that exposed more than minimum numbers of their own troops to the enemy's firepower.[89]

On the administrative side during the interwar period, the French Army possessed a nightmare of misassigned and misdirected organiza-tions. Its structure contributed to a paralysis of indecision that its last commander-in-chief before the defeat, General Maurice Gamelin, only ex-acerbated.[90] Theoretically, the general staff commanded the army, but in

---

[88] For the development of tactical doctrine in the German Army see Timothy Lupfer, *The Dynamics of Doctrine, The Changes in German Tactical Doctrine during the First World War* (Fort Leavenworth, KS, 1981) and James Hogue, "Puissance de Feu: The Struggle for New Artillery Doctrine in the French Army, 1914–1916," Ohio State University Mas-ter's Thesis, 1988.

[89] In fairness to the French there was nothing in the German offensive tactics of 1918 that suggested that the Germans had found a form of warfare that would reduce the casual-ties of attacking forces. As we shall point out below the German offensive tactics of 1918 were extraordinarily costly to the attackers; where their results differed from previous tactical experiences on the Western Front lay in the fact that the Germans had consider-able territorial gains to show for the heavy losses that they had suffered.

[90] Over the past twenty years two major works have examined French strategy and Gen-eral Maurice Gamelin's role in the army's higher leadership in particular: Young, *In Com-mand of France, French Foreign Policy and Military Planning, 1933–1940* and Alexan-der, *The Republic in Danger, General Maurice Gamelin and the Politics of French Defence, 1933–1940*. Both authors go far in extending our understanding of the strate-gic and political context within which French politicians and military operated. But nei-ther work discusses in *any* detail the operational and particularly the tactical framework within which the French Army prepared itself for war. While this author will not deni-grate the importance of an understanding of grand strategy, in the final result military organizations are paid to kill the enemy in as effective a fashion as possible. The absence of any discussion of the operational and tactical dimensions of preparing for the next war (there is not a single reference in the indices of either book to doctrine, operations, or tactics) leaves their analyses fundamentally flawed. In fact the greatest failing of these two works lies in the lack of understanding on the part of the authors of what effective military institutions are. The clearest example of Alexander's misunderstanding of oper-ational matters is his comment on Gamelin's decision to take the French Army's entire strategic reserve and place it on the far left of the Allied line for a misplaced rush to Hol-land: "The plan involved taking a gamble on temporarily depleting the French strategic

fact it possessed no control over finances, personnel, or administration. The secretary general ran the administration, but had no command responsibilities. Consequently, no coherent or consistent organization existed to address substantive issues – a state of affairs fully in accord with the instincts of all too many politicians and generals. As Robert Doughty has suggested,

> In the final analysis, the French High Command lacked a clear chain of authority and responsibility that could provide the army a firm sense of direction for developing its doctrine and designing its weapons. While over-centralization may stifle initiative, the fragmented organization of the French High Command also stifled creative solutions to doctrinal problems.[91]

Divisions of class were less important to the French than to the British, but a substantial rift within the army existed between the general staff, selected to run the army from the top, and those engaged in regimental soldiering. In effect the French created a military organization that was the reverse of the British Army's heavy emphasis on regimental soldiering. In 1915 Ferdinand Foch had picked up on Captain André Laffargue's ideas for a radical transformation of the army's tactical doctrine. Foch then sent the young captain to Joffre's headquarters as a means to influence the development of a more realistic doctrine in the staff. But Gamelin, at that time Joffre's *chef de cabinet*, simply shunted Laffargue into the map room because, as he suggested, he had no intention of allowing a junior captain to influence such important questions.[92]

Throughout the interwar period, the French did take their preparations for a ground war seriously. However, their doctrine fell victim to the organizational vacuum; administrative chaos allowed a small group to hijack doctrine and control doctrinal issues as their own particular turf. Moreover, there was no large-scale examination of the lessons of the last war by a significant portion of the officer corps.

The *École Superieure de Guerre* (the French War College) played the dominant role in forming and articulating the doctrine of the "methodi-

---

reserves." (Alexander, *In Defence of France*, p. 398). Robert Doughty's killing depiction of Gamelin as "the quintessential staff officer and bureaucrat" underlines the difference in understanding and emphasis in his approach from those of Young and Alexander. Doughty, *Seeds of Disaster*, p. 185.

[91] Doughty, *Seeds of Disaster*, p. 128.

[92] For Laffargue's contribution to the development of German doctrine see Lupfer, *The Dynamics of Doctrine*, pp. 38–39. I am indebted to James Hogue for the notes of his meeting with General André Laffargue during the summer of 1987.

cal battle." The intellectual direction of that institution reflected the idio-syncratic views of General Eugène Debeney and his leadership of the First Army in summer 1918.[93] Consequently, while the French studied the war seriously throughout the interwar period, they did not broaden their study beyond a narrow focus on a few carefully selected battles that had oc-curred in 1918. Admittedly, French experiences in the war reinforced the proclivities of a school system that emphasized tight control over maneu-ver. The changes in French offensive doctrine that had occurred in late 1916 *had not* resulted in stunning advances – as was the case with the Germans in 1918 – but rather *had resulted* in catastrophic defeat in 1917, a defeat that had almost broken the army and nearly resulted in the de-feat of France.

One must, therefore, see development of French doctrine during the in-terwar period within the framework of the army's World War I experi-ences. As a result, doctrine emphasized tightly controlled operations, in which artillery would dominate both battlefield and movement forward. French tactical manuals stressed that "firepower had given a remarkable strength of resistance to improvised fortifications." The French Army would only go over to the offensive "in favorable conditions after the as-sembling of powerful material means, artillery, tanks, munitions, etc."[94] The doctrine had a number of consequences: on the one hand it made the French incapable of understanding the principles underlying German doc-trine; on the other, it gave them an overconfidence in the ability of their firepower to contain enemy attacks.[95]

Such a doctrine did not mean that the French were disinterested in tanks. They put considerable effort into developing the tank as a weapon; in 1940 they possessed more tanks than the Germans, and some of their tanks were superior in speed, armor, and firepower to those of the *Wehrmacht*.[96] But since French armored development occurred within a doctrinal straitjacket that aimed at maximum control over operations, the French failed to see the tank's potential. Above all, they viewed armored vehicles as weapons that only supported the "methodical battle."[97]

---

[93] For Debeney's influence over French doctrine as well as his influence over the army see Doughty, *Seeds of Disaster*, pp. 6, 77, 112, 118–120, 185. Debeney does not appear in Young's book, *In Command of France*, while he appears hardly at all in Alexander's. But Doughty's *Seeds of Disaster* underlines how extraordinarily important a role he played in the casting of a fundamentally flawed French doctrine.

[94] Quoted in Paul-Marie de la Gorce, *The French Army* (New York, 1963), p. 271.

[95] Doughty, *Seeds of Disaster*, pp. 86, 92.

[96] See R.H.S. Stolfi, "Equipment for Victory in France in 1940," *History*, February 1970.

[97] In 1937 a German officer commenting on infantry-tank cooperation during a French ex-ercise noted that a tank force pushing forward in one of the tightly controlled "bounds"

In the critical years before the war, the French made slow progress toward establishing an armored force. There existed, especially among lower ranks, some interest in experimenting with new forms of warfare and a partial recognition of the possibilities that armored warfare offered. A memorandum by the general staff suggested that armored divisions could have a mission analogous to that of the cavalry, but would have the additional advantage of being constituted as a shock force capable of rupturing an enemy's front. Armor, with its strategic and tactical mobility, would also have the possibility of swift and far-reaching action.[98] But the doctrinal framework as well as the high command's indecisions, kept development of such a force within narrow parameters.[99] Discussions in the Council of War, a governing body of the army, examined limited, technical questions. Gamelin invariably argued that each proposal needed further study, rather than precipitate action. A 1937 meeting of the council established a general inspector of tanks to study exercises with armored forces the following summer.[100]

But a meeting of the same body in December 1938 underlines how little progress had occurred over the course of a year. Gamelin admitted that mobilization for the Czech crisis had upset plans to study the establishment of an armored division. His colleagues agreed to authorize establishment of two armored divisions in 1939, but those divisions would contain only four weak tank battalions. What exactly this agreement meant is difficult to see, since Gamelin and the other participants postponed deciding on the table of organization for such units to a later date.[101]

---

had found itself in an action that involved "seven minutes of attack and seventy minutes of waiting for the arrival of the infantry." Quoted in Doughty, *Seeds of Disaster*, p. 159.

[98] DDF, 2nd Series, I, Doc. 3 82, 18.1.36, "Note de l'État-major de l'Armée pour le Haut-Comité militaire."

[99] Robert Young admits quite accurately: "The 1936 review commission of General Georges, for example, undeniably paid tribute to the importance of tanks in modern warfare. But it did so within the orthodox context of a tank force that was vulnerable to antitank weaponry and limited by considerations of terrain, that was consequently dependent on infantry and artillery support, and that could operate behind the enemy's lines only *after* his defenses had been `sufficiently disrupted.'" Young, *In Command of France*, p. 180.

[100] Assemblée Nationale, *Rapport au nom de la commission chargée d'enquêter sur les événements survenus en France de 1933 à 1945*, vol. 2, (Paris, 1947), "Séance du Conseil de la Guerre tenue le 15 décembre 1937 sous la présidence de M. Daladier, Ministre de la Guerre," p. 183. Martin Alexander argues that Gamelin actively supported the creation of armored divisions in the late 1930s, and indeed he did in the bureaucratic sense of putting such a proposal on meeting agendas. But nowhere does Alexander provide evidence that Gamelin made the creation of such armored divisions a priority.

[101] Ibid., vol. 2, "Séance du Conseil superieur de la Guerre, tenue le 2 décembre 1938 sous la présidence de général Gamelin."

Beyond doctrinal problems, one must emphasize a general failure of the army's senior leaders. As the minutes of the above December 1938 Council of War meeting underline, Gamelin remained wedded to the doctrine of the "methodical battle." He provided neither leadership nor direction to the army.[102] A staff officer best summed up his qualities by commenting: "Well let me tell you, the general has no guts."[103] Like Montgomery-Massingberd, Gamelin refused to stomach dissenting opinions; in 1935 he established the high command as the sole arbiter for doctrine. From that point on, all articles, lectures, and books by serving officers had to receive approval by the high command before publication. As the French general André Beaufre noted in his memoirs: "Everyone got the message, and a profound silence reigned until the awakening of 1940."[104]

After the catastrophe, the French historian Marc Bloch accurately summed up the army's failures.

> Our leaders, or those who acted for them, were incapable of thinking in terms of a new war. . . . The ruling idea of the Germans in the conduct of war was speed. We, on the other hand, did our thinking in terms of yesterday or the day before. Worse still: faced by the undisputed evidence of Germany's new tactics, we ignored, or wholly failed to understand, the quickened rhythm of the times. . . . Our own rate of progress was too slow and our minds too unelastic for us ever to admit the possibility that the enemy might move with the speed which he actually achieved.[105]

The Germans

Historians have increasingly focused on the German Army during the last two years of World War I.[106] In those two years the Germans invented

---

[102] Admittedly he was an immensely successful political general, who also possessed an alluring command of the syntax of the French language. His skills as an essayist eventually got him elected to the Académie Française and have intrigued and misled a whole generation of historians. But the language of war in May 1940 underlined his complete incapacity to deal with the world on the hard edge of battle. See Robert Doughty, *The Breaking Point, The Defeat of France, May 1940* (Hamden, CT, 1940) for the general failure of the French high command as well as Gamelin's enormous and unfortunate (from the perspective of the French high command) contribution to the catastrophe.

[103] André Beaufre, *1940, The Fall of France* (New York, 1968), p. 43.

[104] Ibid., p. 47. •

[105] Marc Bloch, *Strange Defeat* (New York 1968), pp. 36–37, 45.

[106] The work responsible for forcing military historians to focus on the revolutionary

modern war. Nevertheless, one needs a cautionary note; the Germans achieved their operational and tactical victories in spring 1918 at enormous cost. Between mid-March and July they lost approximately 1 million men.[107] While they achieved unheard-of advances, in effect they destroyed their army, not only in a physical sense but also in terms of its morale. By late summer they had over half a million deserters and had lost the cream of their attack divisions.[108] In effect, Ludendorff sealed the Reich's strategic fate by the very success of his tactical innovations. And his comment to Crown Prince Rupprecht of Bavaria that "we will punch a hole into [their line]. For the rest, we shall see" underlines the extent of the strategic bankruptcy of his offensives.[109]

However, the new German offensive tactics did not solve the problem of firepower and maneuver beyond the range of supporting artillery fire. On 28 March the "Michael" offensive came to a dramatic halt before the BEF's machine guns, which blew away attacking German infantry who had outrun their artillery.[110] Admittedly the Germans had worked out how to break into and through enemy defenses, and then how to exploit that opening in the immediate battle area. But without motorized support, breakthroughs in 1918 had scant opportunity to achieve strategic or even operational freedom; in the end they involved the same high rate of attrition that had marked the great battles of 1916 and 1917.[111]

The crucial decisions for the Germans that resulted in their battlefield triumphs early in World War II came in the immediate aftermath of World

development of tactics in the last years of the war was Lupfer's *Dynamics of Doctrine*. For a further consideration of the development of German infantry doctrine see Bruce Gudmunsson, *Stormtrooper Tactics* (New York, 1989).

[107] Travers, *How the War Was Won*, p. 179.

[108] Wilhelm Deist, "Der militärische Zusammenbruch des Kaiserreichs. Zur Realität der 'Dolchstosslegende,'" in *Das Unrechts-Regime*, ed. by Ursula Buttner (Hamburg, 1986).

[109] Crown Prince Rupprecht, *Mein Kriegstagebuch*, vol. 2, p. 372.

[110] Travers, *How the War Was Won*, p. 179.

[111] One of the ironies of the two world wars is the perception that the battles in World War I involved far higher losses. It is now questionable whether they did, at least on a day-to-day or month-to-month basis. The essential difference is the perception of gain. The Allied advance through Normandy and then across France in the period from June through September 1944 involved horrendous casualties for the units that participated in those battles – equal at least on a unit-by-unit basis to the worst killing battles of 1916 or 1917. The difference, of course, lay in the gains of those 1944 battles, which indicated that the losses suffered were bringing the war closer to conclusion. Also, the advance spread the casualties over a wide swath of territory as opposed to confining the losses on both sides to a narrow fringe of territory, such as happened at Verdun, the Somme, and Passchendaele.

War I. There was nothing inevitable about those decisions; as with much of history they involved chance and the interplay of personalities. The selection of General Hans von Seeckt, first as chief of the general staff and then as commander-in-chief of the army was the most important factor. Seeckt's contribution resulted from the fact that he made the decisions about the selection process for the new officer corps, the focus of that officer corps, and the value system and culture of the Weimar Republic's military arm.[112]

Among the issues confronting Seeckt in 1919 was the demand by the victorious powers that Germany reduce its 350,000-man postwar army, with approximately 15,000 officers, to 100,000 men and only 4,000 officers.[113] There were a number of constituencies within the existing German Army, all of whom had their claim to control the new army. For example, the army's first commander, General Walther Reinhardt, favored giving front line officers, the *Frontsoldaten*, pride of place in the new *Reichswehr*.[114] It is also well to remember that the traditionalists in the army, the nobility, had maintained a powerful lock on decisions throughout the war.[115] And those traditionalists would have remained in control of the Imperial Army's administration had the Germans won.

But Germany had lost and therefore confronted the victorious Allies' demand that it reduce the number of officers on active duty. However unpalatable the reduction, Germany's unsettled situation, the discrediting of traditional leadership, and his own perceptions as a general staff officer led Seeckt to bring the whole body of the general staff into the new army and to place that organization and its officers in virtually all the major command and staff positions. Thus, Seeckt created a very different officer corps from what had existed before World War I, one whose cultural ethos emphasized intellectual as well as tactical and operational excellence.

Along with positioning the general staff as the heart and brain of the new army, Seeckt initiated a comprehensive program to examine the

---

[112] The most recent biography of Seeckt in English, which possesses some considerable strengths as well as weaknesses is James S. Corum, *The Roots of Blitzkrieg, Hans von Seeckt and German Military Reform* (Lawrence, KS, 1992).

[113] Rudolf Absolom, ed., *Die Wehrmacht im Dritten Reich* (Boppard am Rhein, 1969), pp. 25–26.

[114] Corum, *The Roots of Blitzkrieg*, p. 33.

[115] It is well to remember that Ludendorff and the general staff had wanted to promote the war's great expert on artillery tactics, Lt. Col. Georg Bruckmüller, to Colonel in 1918. The Prussian War Ministry flatly refused. Bradley Meyer, "Operational Art and the German Command System in World War I," PhD Dissertation, The Ohio State University, 1988, p. 296.

lessons of World War I. He established no fewer than fifty-seven committees, formed of general staff officers and experts in particular areas, to
examine the broad and specific questions that the war had raised.[116]
Seeckt's charge to those examining the war was that they produce

> short, concise studies on the newly gained experiences of the war and
> consider the following points: a) What new situations arose in the war
> that had not been considered before the war? b) How effective were our
> prewar views in dealing with the above situations? c) What new guide-
> lines have been developed from the use of new weaponry in the war?
> d) Which new problems put forward by the war have not yet found a
> solution?[117]

For the most part, general staff officers led these committees and, in the
end, over 400 officers became involved in this work.[118] Because the majority of these officers had first-hand experience with the tactical and doctrinal developments of 1917 and 1918, their reports rested on solid, realistic assessments of what had actually occurred, not on what generals
might have believed to have happened. The result was the extraordinary
Army Regulation 487 ("Leadership and Battle with Combined Arms"),
Part I of which appeared in 1921 and Part II in 1923.[119]

Post-World-War-I German doctrine consequently emphasized conceptions that were starkly different from those of the British and French. The
first was a belief in maneuver.[120] The second emphasized an offensive
mind set; the third demanded that commanders decentralize operations
to the lowest level possible. The fourth required officers and NCOs to use

[116] As Seeckt noted: "It is absolutely necessary to put the experience of the war in a broad
light and collect this experience while the impressions won on the battlefield are still
fresh and a major proportion of the experienced officers are still in leading positions."
Quoted in Corum, *The Roots of Blitzkrieg*, p. 37.

[117] Ibid., p. 37.

[118] The contrast between the development of German doctrine and British postwar efforts
to develop their tactical conceptions could not be more graphic. Whereas the Germans
involved over 400 officers (over 10 percent of their officer corps) in the effort, the British
turned the task of rewriting their infantry tactical manual over to B.H. Liddell Hart,
who was then a twenty-four-year-old infantry lieutenant of limited military experience
(Ibid., p. 39).

[119] Ibid., p. 39.

[120] As Seeckt emphasized to his Reichswehr commanders: "Operation ist Bewegung." Still
it is well to remember that Seeckt was not always as far-seeing as his recent biographer
has suggested. In 1928 he wrote: "Many prophets already see the whole army equipped
with armored vehicles and the complete replacement of the horseman by the motorized
soldier. We are not yet that far." Bundesarchiv/Militärarchiv, W 10 - 1/9, Oberstleutnant Matzky, "Kritische Untersuchung der Lehren von Douhet, Hart, Fuller, und
Seeckt," Wehrmachtakademie, Nr. 90/35 g.K., Berlin, November 1935, p. 44.

their judgment on the battlefield; the fifth stressed that leadership at all levels must always display initiative. Moreover, *all* officers had to be thoroughly familiar with army doctrine and that doctrine was to form a coherent framework within which the whole army operated.

Through 1933, the *Reichswehr*, due to its small size and elite status, trained its officers to an extraordinarily high level. With massive rearmament in 1933, that sharp edge dulled somewhat and affected the level of initiative that the army expected officers to display. But through the first years of World War II, the Germans inculcated their officer corps in all the branches to display initiative; above all, officers were to exploit the fleeting opportunities that the battlefield offered, and to lead from the front.[121] It was within this framework that the development of the Germany panzer forces took place. This development reflected the general principles of initiative, exploitation, and maneuver that lay at the heart of the *army's basic* doctrine – not some revolutionary development pushed by a few radicals.[122]

The tactical and conceptual progress that the *Reichswehr* made in the 1920s was not necessarily daring when compared, for example, to the pioneering efforts of the British in armored warfare during the same period. However, the Germans were more realistic and imaginative in tying what actually happened in exercises during the 1920s to the general understanding and tactical conceptions of their entire army. A report on the army's first experiments with motorized troops in the Harz mountains in 1922 underlines this capacity to feed the experience gained in exercises to the army as a whole. As Seeckt noted in his introduction:

> I fully approve of the Harz exercise's conception and leadership, but there is still much that is not clear about the specific tactical use of mo-

---

[121] Towards the end of the seven-month period in which the German troops in the west had been preparing for the offensive against the French, *Generaloberst* Fedor von Bock, commander of Army Group B, warned his senior officers in April 1940: "In many exercises recently, particularly at the battalion and regimental levels, an inclination to caution and circumspection has appeared. Therein lies the danger that on one side German leadership will pass up opportunities to seize favorable situations . . . while on the other hand the enemy will be allowed time to recognize an intention. Once a commander has decided to attack, so must everything that he orders be established that the eyes, heart, and sense of the troops are directed to the front." Heeresgruppe B, IaNr.2211/40, "Bemerkungen zu den Truppenübungen in Fruhjahr, 1940," NARS T-312/752/8396741.

[122] That is the central explanation for why the generals that led the panzer divisions in World War II rarely came from an armor background. Most were infantry, artillery, or cavalry officers, but they possessed the same coherent approach to war that the armored officers possessed.

tor vehicles. I therefore order that the following report be made available by all staffs and independent commands as a topic for lectures and study. Troop commanders must see to it that experience in this area is widened by practical exercises.[123]

Because of Versailles' ban against tanks in the *Reichswehr*, there were considerable hurdles in the way of progress, even towards such simple things as gaining insight into how vehicles actually behaved and functioned on the battlefield.[124] However, German doctrinal conceptions, by emphasizing exploitation, speed, leadership from the front, and combined arms, provided a solid framework for thinking through not only how the *Reichswehr*, if it possessed tanks, might employ them against an enemy, but how a potential opponent might utilize armor against German forces.[125]

After all, during World War I the Germans had accumulated greater experience in coping with tanks than had the Allies. From 1929 to 1933 they were also able to accumulate some firsthand experience with tanks at the Kazan tank school in Russia.[126] Nor were the Germans reluctant to learn from their former adversaries. A *Reichswehr* report examining the British maneuvers of 1926 noted that armored vehicles with their increased speed now possessed the capacity to strike out independently and that motorized infantry and artillery might accompany the tanks. The report further added:

In addition, with existing models, one can now clarify what will happen with tanks behind the enemy's main line of resistance after a successful breakthrough. Tanks can be used: for attacks on the enemy's rear positions, against advancing reserves, as well as against command posts and artillery emplacements. For such tasks, present-day tanks are far more capable than older models. We, therefore, recommend that, in ex-

---

[123] Reichswehr Ministerium, Chef der Heeresleitung, Betr: "Harzübung, 8.1.22," NARS T-79/65/000622.

[124] Guderian himself claimed after World War II that he had never seen the inside of a tank when he was tasked to teach tank tactics in 1928; the general staff rectified this weakness by packing him off to Sweden for four weeks' service with a Swedish tank unit. Guderian, *Panzer Leader*, p. 12.

[125] German tank experience during World War I was much more limited than the Allies', although the Germans created some nine-tank companies in 1918 (Corum, *The Roots of Blitzkrieg*, p. 122). In the judgment of Ernst Vockheim, one of the German tank officers during World War I, the Germans' "most successful" tank battle occurred in October 1918 north of Cambrai when ten German tanks managed to plug a hole in the line created by a British tank-led advance (ibid., p. 123).

[126] Corum, *The Roots of Blitzkrieg*, pp. 191–195.

ercises, armored fighting vehicles [or mock representations of tanks since the Germans had no armor at this time] be allowed to break through repeatedly in order to portray this method of fighting and thus to collect added experience.[127]

This report is interesting for a number of reasons. First, it extrapolated from British experience to suggest how the *Reichswehr* might use tanks to develop the Germans' own conceptions of penetration and exploitation. Second, it directed that future maneuvers take into account possibilities that the British maneuvers had revealed. Finally, it recommended that the *Reichswehr* must acquire a wide base of experience in its future exercises to evaluate the possibilities that mobile, armored warfare offered. Significantly, the Germans were seeking not only to better themselves, but working on their insight of what others might try to do to them.

One should not take these observations as suggesting that the Germans embraced the tank's potential with starry-eyed enthusiasm. In truth, there was considerable skepticism about the potential of panzer units up to the 1939 Polish campaign. As Rundstedt, who led Army Group A's drive (and the panzer forces) through the Ardennes in May 1940, commented to Guderian at the end of an armor exercise in the late 1930s: "All nonsense, my dear Guderian, all nonsense."[128] Nevertheless, Rundstedt's skepticism did not prevent him – and officers like him – from recognizing that tanks might extend the infantry's capacity to exploit tactical situations on the battlefield.

On the other hand, German armor advocates remained fully within a combined-arms framework in which infantry and artillery would extend the tank's potential. This larger framework was crucial to the Germans' deadly effectiveness on World War II battlefields, while their British counterparts consistently proved incapable of integrating armor within a larger framework – a judgment that includes British armor officers as well as

---

[127] Reichswehrministerium, Berlin, 10.11.26, "Darstellung neuzeitlicher Kampfwagen," NARS T-79/62/000789. The general accuracy and perception of this report on the British maneuvers suggest a number of interesting possibilities. Since the Germans did not have a military attaché in London until the early 1930s and German intelligence was to prove so abysmal throughout World War II, this author is inclined to believe that the Germans were reading the reports in the British press with great care – reports that went well beyond what British Army evaluators of these maneuvers were suggesting. One of the reporters of these maneuvers was B.H. Liddell Hart and it is probable that his articles, which were forward-thinking in their interpretation of what had happened on the Salisbury plain, formed much of the basis for the German Army's analysis of British experiments with armor.

[128] M. Plettenburg, *Guderian: Hintergründe des deutschen Schicksals, 1918–1945* (Düsseldorf, 1950), p. 14.

those of the artillery and infantry.[129] The Germans watched British experiments with armor well into the 1930s; as late as 1934 General Ludwig Beck, chief of the general staff, introduced an extensive report of that year's maneuvers that circulated throughout the army.[130] Guderian himself, not one to give others credit, admitted in his memoirs that the Germans had translated the current British field manual on the employment of armored fighting vehicles and used it as the basic primer for developing armored warfare.[131]

But the most important single factor in German innovation was the fact that they possessed a coherent doctrine based on a thorough and honest reading of the evidence. Moreover, they had considerable knowledge of what others had tried as they began their massive rearmament program in 1933. As the army's doctrinal manual underlined, "when closely tied to the infantry, the tanks are deprived of their inherent speed" – at a time when the Germans possessed not a single tank on German territory.[132]

The crucial players in development of the panzer force were neither Hitler nor a group of maverick junior officers like Heinz Guderian.[133] Rather, Beck and the army's commander, General Werner von Fritsch, provided the support and resources for the new form of warfare. After World War II, Guderian depicted Beck as thoroughly hostile to armored

[129] See in particular Williamson Murray, "British Military Effectiveness in World War II," in *Military Effectiveness*, vol. 3, *The Second World War*, ed. by Allan R. Millett and Williamson Murray (London, 1988), pp. 107–121.

[130] His introduction noted the following: "The maneuver caused a great sensation in England and resulted in many challenges. Once again results showed that the leadership of large mechanized formations, even in an army as motorized as the English, which has the most extensive experience in the world, nevertheless, ran into considerable difficulties and that reality can be disappointing. On the other hand, one gains the impression that English tank troops were brilliantly led and performed in outstanding fashion. British military leadership seems completely aware of present difficulties. Therefore, it will be especially instructive to watch the direction which the next development of the English Army will take in the area of motorization." Der Chef des Truppenamts, Dez. 1934: "England; Manöver des Panzerverbandes, 18. bis 21.9.34," NA T-79/16/000790.

[131] Heinz Guderian, *Panzer Leader* (London, 1952), pp. 20–22. The manual to which Guderian refers is probably Broad's pamphlet "Mechanized and Armored Formations," which a certain Capt. Stuart sold to the Germans. Kenneth Macksey, *Tank Warfare* (London, 1971), pp. 85–86.

[132] Chef der Heeresleitung, "Truppenführung," 1933, 1:133.

[133] This is not meant to suggest that Guderian had no role, but rather to underline his considerable overestimation and exaggeration of his importance in his memoirs *Panzer Leader*. In fact, the reader should be warned that of all the memoirs by senior German generals (and there are many) Guderian's number among the most dishonest *and* misleading.

warfare, but the former's testimony reflected a Nazi ideologue's outrage at the latter's participation in the 20 July 1944 plot to kill Hitler. As early as 1935, Beck conducted a general staff tour on how to utilize a panzer corps – no panzer divisions yet existed. In 1936 the general staff, at his direction, examined potential operations for a hypothetical panzer army.[134]

The performance of experimental armored units in the Wehrmacht's 1935 maneuvers impressed Beck and Fritsch sufficiently to authorize establishment of three panzer divisions.[135] The rapidity with which the Germans moved to test armored formations stands in stark contrast to the French in the late 1930s, when the latter took over four years to authorize such formations.[136] By the end of 1935 Beck was recommending panzer divisions for attacks against long-range objectives and suggesting their use as an independent force "in association with other motorized weapons."[137]

The development of panzer divisions took place within the broad framework of German doctrine. Both Lutz and Guderian – Lutz's deputy during the mid-1930s – emphasized that panzer units must include an integral force of motorized infantry, artillery, engineers, and signal troops as well as armor.[138] Doctrine stressed the use of such a combined arms force in close and explosive cooperation; not surprisingly, discussion of armored employment focused on ruthless, mobile, and rapid exploitation of breakthroughs.[139] Significantly, such thinking was already inherent in the doctrine of the other combat branches. So the new panzer divisions merely extended basic principles on which all German doctrine rested, an evolutionary rather than a revolutionary change.

Even so, as suggested above, the new tank troops ran into considerable skepticism among senior officers. General Otto von Stulpnägel remarked

---

[134] Erich von Manstein, *Aus einem Soldatenleben* (Bonn, 1958), p. 241.

[135] Robert O'Neill, "Doctrine and Training in the German Army," in *The Theory and Practice of War*, ed. by Michael Howard (New York, 1966), p. 157.

[136] Doughty, *Seeds of Disaster*, pp. 161–177.

[137] Wilhelm Deist, *The Wehrmacht and German Rearmament* (London, 1981), pp. 42–43.

[138] Guderian, *Panzer Leader*, Appendix 24.

[139] Among the many documents discussing the development of the armored forces see: Kommando der Panzertruppen, Ia op. Nr 4300, Berlin, 10.11.36, "Bemerkungen des kommandierenden Generals der Panzertruppen in Jahre 1936 und Hinweise für die Ausbildung, 1936–37," NA T-79/30/005913; Kommando der Panzertruppen, Ia Nr. 3770/37, Berlin, 15.11.37, "Besictigungsbemerkungen des kommandierenden Generals des kommandos der Panzertruppen in Jahre 1937," NA T-79/30/000937; and Generalkommando VII Armeekorps, Ia Nr. 59/38 g., "Verwendung von Panzerenheiten im Kampf der verbundenen Waffen," T-79/69/00317.

to Guderian that such ideas were utopian and that large armored units were militarily impossible. In 1936, one commentator on military affairs suggested that the tank was only a weapon of the present and because of antitank weapons would suffer the fate of the horse.[140] Yet opposition to armored warfare among senior officers differed significantly from what was occurring among British and French officers. German doubters questioned whether mechanized formations could make the deep penetrations that advocates like Guderian claimed, not whether tanks could substantially aid infantry and armor in pushing their way through enemy defensive positions and making possible the infantry exploitation that had occurred in the attacks in spring 1918.

In the late 1930s German mechanized forces continued the steady innovation that had marked German doctrine since 1920. It was not until Poland that the officer corps as a whole began to grasp the potential of armored exploitation on the operational level of war. Significantly, the Luftwaffe could only provide close air support in the breakthrough battle during both the Polish and French campaigns.[141] During exploitation in Poland close air support was often as dangerous to the German troops as it was to the enemy.[142] In April 1940, the Luftwaffe and the 1st Panzer Division carried out the first tests for radio controlled close air support during mobile operations, but the French campaign came too soon for panzers and Stukas to use the tests in the spring 1940 battles. Consequently, while the Luftwaffe provided crucial close air support at Sedan in the breakthrough, it provided little direct support throughout the remainder of the campaign.[143] Not until the Barbarossa campaign did the Germans work out the process of rendering close air support in a mobile environment. The point here is that the extension of close air support capabilities was typical of evolutionary innovation within the German military which allowed the Germans to fit new concepts and capabilities into a coherent, realistic framework.

The process of military innovation in armored war in the German

---

[140] M. Ludwig, "Gedanken über den Angriff im Bewegungskrieg," *Militärwissenschaftliche Rundschau* (1936), p. 154.

[141] See Williamson Murray, *German Military Effectiveness* (Baltimore, 1992), chapter 2 for a fuller discussion of Luftwaffe close air support.

[142] See particularly the after action report of the 10th Panzer Division. 10. Panzer Div., Abt. Ia Nr 26/39, "Erfahrungsbericht," 3.10.39, NARS, T-314/614/632. See also Williamson Murray, "The Luftwaffe Experience," in Case Studies in the Development of Close Air Support, ed. by Benjamin Franklin Cooling (Washington, DC, 1990).

[143] As Guderian attests in his memoirs the Luftwaffe was fully capable of bombing a German corps commander's headquarters in the French campaign. Guderian, *Panzer Leader*, p. 91.

Army does not appear, on the historical evidence, to have been linear or precisely predictable.[144] The leaders of the post-World War I German Army did not deliberately set out to create a new way of fighting, but rather aimed to build upon the operational and particularly the tactical lessons of 1914–1918 in a coherent and effective fashion. By casting their net widely over the experiences of the last war and examining that experience realistically (at least at the tactical and operational levels), they moved, in fits and starts, toward a new conception of fighting. They managed to do so, moreover, despite the constraints of the Treaty of Versailles and, until after Adolf Hitler's rise to power, meager defense budgets.

Indeed, it seems likely that, on the eve of the 1940 campaign, few of even those German officers involved in development of armored warfare during the interwar period had a firm belief that their efforts would transform land warfare. Vagueness and uncertainty in planning the 1940 campaign as to what the army high command should do if the three panzer corps got across the Meuse River before the eighth or ninth day of the offensive in the west confirms the inability of most German generals to predict in detail how a *blitzkrieg* campaign could play out against French and British opposition.

In his memoirs, even Guderian characterized his panzer corps' success in getting its rifle regiments across the Meuse on the fourth day as "almost a miracle."[145] Robert Doughty's meticulous reconstruction of the crossings supports Guderian's characterization. In fact, at numerous junctures, the operations of the three panzer corps that crossed the Meuse during the battles of 13–14 May 1940 succeeded by razor-thin margins and, in some cases, by sheer luck, while, in other cases, the initial attempts to cross completely failed.[146] Nevertheless, the cumulative result of these

---

[144] To deny that future events are precisely predictable is not, however, to deny causality. On this crucial point, see Heinz-Otto Peitgen, Hartmut Jürgens, and Dietmar Saupe, *Chaos and Fractals: New Frontiers of Science* (New York, 1992), pp. 9–14. Nor should the denial of precise or detailed predictability be taken to deny that some individuals can develop surprisingly sound *intuitions* about the future.

[145] Guderian, *Panzer Leader*, p. 84. For Balck's personal recollections of the crossing of the Meuse by 1st Motorized Infantry Regiment, 1st Panzer Division, XIXth Panzer Corps, on May 1940, see Balck, "Translation of a Taped Conversation with General Hermann Balck, 13 April 1979," pp. 2–8. Balck was the commander of 1st Panzer's infantry regiment at the time. What is clear from the historical accounts of the crossings by the three panzer corps is that Hoth's crossing only succeeded through Rommel's extraordinary leadership, Reinhardt's panzer corps never made a successful crossing until the panzer corps on its flanks had succeeded, and only three of the six crossings initially attempted by Guderian's panzer corps got across the Meuse.

[146] Doughty, *The Breaking Point*, pp. 83, 135, 139, 164–165, 321, 323–324, and 329–332.

small differences, was that, by the morning of 16 May, "the French Army teetered on the edge of collapse, and in subsequent days the Germans won one of the most decisive operational victories in military history."[147]

Continuing German success in World War II with armored warfare in North Africa and the Soviet Union reflected the capabilities of officers with a broad doctrinal framework; German doctrine allowed them to innovate successfully in armored warfare. In contrast the French never had time to incorporate innovations on the battlefield because their collapse came so swiftly; the British, who possessed the time due to geography, never displayed the talent.[148] Their defeats in North Africa reflected a lack of a doctrinal framework in both training and execution. Their officers were not capable of adapting to a larger framework of war. They remained prisoners of the compartmentalized conceptions of warfare that characterized their branches. They remained gunners and infantrymen to the end, while the CIGS, Field Marshal Alan Brooke, insured that no one with experience in armored warfare ascended the ladder to command of an armored division in combat.

THINKING ABOUT INNOVATION

There are, of course, simplistic solutions to the problems of innovation in military institutions. For example, after the Franco-Prussian War many military professionals believed that creation of efficient staffs would represent the difference between success or failure in the next conflict. One current political scientist has suggested that political pressure in support of military reformers is the crucial ingredient in innovation.[149] Unfortunately, there are no simplistic answers to the questions posed by innovation in an uncertain world. Innovation, like most complex human endeavors, occurs in military institutions in an opaque and unclear landscape. Historians, with the benefit of hindsight, can see the events and developments of the 1920s and 1930s with some clarity. But the importance of armored forces to the next war, however, was not so clear to the decision makers and military leaders at the time.

In such an atmosphere, armies had to innovate during periods when few resources were available and when the lessons of World War I re-

---

[147] Ibid., p. 331.
[148] The postarmistice army of Vichy France displayed no serious capacity to understand why it had been so badly beaten: see Robert Paxton, *Parades and Politics at Vichy* (New York, 1966).
[149] See Barry Posen, *The Sources of Military Doctrine: France, Britain, and Germany Between the World Wars* (Ithaca, 1989).

mained uncertain. Besides the difficulties involved in learning the lessons of past battles,[150] military institutions confronted a number of other problems. Above all, war presented challenges that peacetime training could not replicate. Military institutions consequently existed in an environment in which there was little evidence as to what paths could best further innovation.

To add to these complexities, military organizations had to establish doctrinal frameworks to deal with the issues that war raises. Unfortunately, they often have taken such doctrinal conceptions into war and, instead of innovating in response to the realities which they actually confronted, they molded conditions to fit peacetime perceptions and assumptions. Recent scholarship on topics as diverse as the British Army in World War I and the U.S. Army in Vietnam suggests that this pattern is pervasive in twentieth century military history.[151] In World War II, however, the Germans displayed considerable capability to adapt and innovate within their doctrinal framework in response to actual battlefield circumstances, but this example remains unique, with the possible exception of the Israelis in 1973.

Ironically, even direct experience in war is not a guarantee that doctrine will see essential points and move military organizations in the correct direction. The French, despite serious study and analysis after World War I, entirely misconstrued the paths toward which war was moving.

The armies of the interwar period also had to come to grips with the technological changes that the tank embodied as well as the larger problems of societal motorization. In 1914 armies moved on their feet with their supplies drawn by horse. By 1939 the British Army had entirely motorized its forces. The German Army bridged the gap by creating a relatively small force of mechanized units (less than 20 percent of their force structure in 1940) that fit within their doctrinal framework and which allowed them to gain operational freedom on the battlefield. Such freedom of maneuver had not existed in World War I, except in the opening weeks of the conflict. But the bulk of the German Army in 1939 and for the remainder of the war (nearly 80 percent for both French and Russian campaigns) remained a foot and horse-drawn army – tactically and opera-

---

[150] We might note here the number of so-called "experts" who in reaction to the Yom Kippur War decreed the death of the tank, not to mention all of those who concluded in the wake of the Iran-Iraq war that the Iraqi military represented some reincarnation of the Waffen SS.

[151] For the British Army in World War I see Travers, *The Killing Ground*; and for the U.S. Army in Vietnam Andrew Krepenevich, *The Army in Vietnam* (Baltimore, MD, 1988).

tionally modern, but obsolete in how it moved and was supplied. This dichotomy was a major factor in why the Germans lost the 1941 campaign against the Soviet Union; the logistical infrastructure simply could not support the drives deep into Soviet territory. Moreover, backward German attitudes towards logistics, maintenance, and production resulted in myriad runs of differing varieties of tanks and support vehicles, none of which the army was in a position to support. The result was a logistician's nightmare and another factor in the defeat of German forces on both Eastern and Western Fronts in 1943 and 1944.

Nevertheless, in an operational and tactical sense, the Germans innovated with armor better than anyone else. A number of factors beyond their doctrinal framework account for this success. Like the French, they studied the last war and interwar developments in considerable depth. Unlike the French, however, they tolerated a high degree of debate within the officer corps about war, tactics, and operations. The French on the other hand, especially under Gamelin, shut down such debate.

Moreover, the Germans proved surprisingly willing to tolerate outspoken officers, Guderian being a prime example. "Hammering Heinz" at one time or another antagonized virtually every senior officer in the army with little discernable impact on his career, at least until he ran afoul of his Führer in December 1941. Guderian appears even more obnoxious than Fuller and Hobart, yet the British Army misused both officers. Admittedly, General Erich von Manstein was eventually removed from his position as chief of staff to Army Group A in 1940 for arguing too much with the high command, but in the end he received command of an infantry corps and within two years was an army commander and field marshal.

The Germans were particularly adept in connecting the intellectual drive within their army to the operational world. The general staff provided an exceptional feedback loop about what was occurring at various levels in the army; it insured that officers in the field had a healthy respect for the doctrine within which they were to operate. Even more important, a high level of trust existed between the various levels of command on operational and tactical matters.

German professional military education pursued practices that differed radically with what passed for professional military education in British and French armies (and certainly with most armies in the world today). The system did not do a good job at training German officers to think about the political and strategic dimensions of war, not to mention its intelligence and logistical dimensions. But even the competition for admittance to the *Kriegsakademie* (staff college) forced junior officers to come to grips with their profession; the four days of exams to gain entrance to advanced schooling suggests how serious the army was about professional military education throughout an officer's career. Finally, having used intellectual attainment along with tactical proficiency to determine fitness at the junior officer level, the Germans then measured the few students who had made it to the *Kriegsakademie* against an extraordinarily high standard of excellence required to wear the crimson stripe of a general staff officer. The general staff officer was then protected for the rest of his career against the malice and jealousy of his fellow officers by the fact that his promotion and assignment were controlled by the general staff.

The resulting system did not create a factionalized army, perhaps because the general staff officer also had to perform in the regular army as a troop commander. Rather the system of intense competition for schooling raised the army's whole tone.[152] The results in terms of enthusiastic training, careful preparation, endless practice, and a thorough grounding in the basics produced an officer corps capable of innovating in peacetime more effectively and realistically than its opponents. Those qualities also made this German officer corps more capable of learning on the actual battlefield.

Blitzkrieg operations were not a fully developed concept before September 1939; the German Army put the pieces together in the battles from 1939 to 1941. In 1938 no panzer corps appeared in plans to invade Czechoslovakia. Panzer corps operated in the Polish campaign, but were subordinated to infantry armies. The Tenth Army under Generaloberst Walther von Reichenau contained significant panzer forces in addition to its infantry units. It allowed panzer units to operate in a fashion close to what Liddell Hart and Fuller had advocated. Nevertheless, panzer forces still operated only as corps in the French campaign, and the armies and army groups above them often retarded their operations. Yet, General Franz Halder, chief of the general staff, who consistently doubted the effectiveness of armored formations, took Manstein's proposal for a thrust of four divisions through the Ardennes and expanded it to place all the German armored formations on this crucial segment of the front. Not until the Soviet campaign did the Germans get their armor into panzer

---

[152] This emphasis on professional military education had an enormous impact on the culture of the whole German Army. Erwin Rommel is a particularly good example. Although he was *not* a general staff officer, Rommel was an avid student of military history and in addition wrote a very serious book about his experience as an infantry officer in World War I. See Sir David Fraser's *Knight's Cross; A Life of Field Marshal Erwin Rommel* (New York, 1993).

groups, which still did not have the status of armies – and would not achieve this status until the end of 1941.

The above suggests that the process of innovation is an immensely complex one. Individuals can, and often do, exercise great influence over the process. Seeckt's decisions in 1919 and 1920 played an essential role in creating an organization that could innovate within the realistic parameters of technology and tactical doctrine. A wide variety of factors that contribute to uncertainty influenced the military organizations of the interwar period. Recognition of the problems involved in change and innovation is one thing, but it is equally difficult to secure the necessary changes in organizational environments – enshrouded by the fog of uncertainty – that will allow innovation to take place. But the example of armored warfare in the 1920s and 1930s suggests the extent of the payoff for successful innovation and the penalties for failure.

2

ASSAULT FROM THE SEA
The development of amphibious warfare between the wars

THE AMERICAN, BRITISH, AND JAPANESE EXPERIENCES

ALLAN R. MILLETT

On the morning of 12 November 1921, Charles Evans Hughes, Secretary of State of the United States, surveyed the thousand expectant spectators with a coolness matched only by the autumn winds. Hughes came to the podium of Continental Hall in Washington, D.C., to begin the International Conference on Limitation of Armament. Much to the surprise of the delegations and audience that crowded the hall, Hughes proposed a comprehensive plan to limit naval forces and to create a regime of non-aggression in Asia, the most potent remaining arena of international rivalry. By the time the delegates had completed their work in February 1922, they had cobbled together three major, interrelated treaties that shaped the naval policies of Britain, the United States, and Japan through most of the interwar period.[1]

One of the three major treaties – the Treaty for the Limitation of Armament – carried important implications for amphibious operations, a form of naval warfare not addressed in the negotiations. Although the conferees discussed the development of battleships, naval aviation, and submarines, they failed to discuss the ability of their navies to conduct assaults from the sea. Ironically, the Japanese Army was the world's foremost amphibious force in 1921. Within twenty years Britain, the United States, and Japan all identified a need for amphibious capability, but Japan alone possessed the doctrine, tactical concepts, and forces for such operations in 1939.

The demand for such forces could be inferred from Article XIX of the "Five Power Treaty," which limited the size and modernity of the battle-

---

[1] Harold and Margaret Sprout, *Toward a New Order of Sea Power: American Naval Policy and the World Scene, 1918–1922* (Princeton, NJ, 1943), pp. 149–160.

fleets of all three great navies. The signatories of the Five-Power Treaty (which also applied to France and Italy) agreed to establish no new bases in Asia and the western Pacific outside of carefully defined "home lands" and to hold the military capability (coastal defenses and fleet repair and maintenance facilities) of existing bases within the treaty area to 1921 levels. This provision limited any further development of the American naval bases in the Philippines and Guam, the British base at Hong Kong, and the Japanese bases on the island of Formosa and in the central Pacific islands held under a League of Nations mandate. If any of the three navies fought one another, they would have to occupy and develop extemporized bases or capture the bases that survived the Five-Power Treaty.[2]

The British, American, and Japanese had a rich heritage of amphibious operations before World War I but had made no special effort to develop amphibious forces under naval control or as joint (army-navy) forces. By World War I each of these three naval forces had demonstrated on several occasions a similar basic concept for successful landing: land where there is no opposition from ground forces. The British had done so during the Egyptian campaign (1882) and the Punitive Expedition to China (1900); the Americans had invested Santiago, Cuba, and Manila, the Philippines (1898) after unopposed landings; and the Japanese had twice besieged Port Arthur (1894 and 1904–1905) after transporting armies across the Yellow Sea. The principal danger in such an operation was an attack upon the invasion force by the opposing navy. In the Japanese and American experiences, mishandled and inferior Russian and Spanish squadrons had perished in attempts to disrupt naval operations, and the entire history of the nineteenth century British empire demonstrated the naval impotence of Britain's conquests.[3]

The post-World War I treaty regime lasted well into the 1930s because of the burden of wartime governmental debt and the collapse of revenues during the Great Depression. Even after Japan repudiated the Washington Conference treaties in 1936, hope and bankruptcy combined to keep the Royal Navy and United States Navy under the treaty limits in terms of warship construction. Neither could afford a major base improvement program, especially far from their national shores. Japan, however, start-

---

[2] Article XIX, Treaty for the Limitation of Armament, February 6, 1922, reprinted in ibid., pp. 302–311. For an introduction to the naval history of World War I and the interwar period, see Clark G. Reynolds, *Command of the Sea: The History and Strategy of Maritime Empires* (New York, 1974), pp. 445–500.

[3] For the history of the Royal Navy at the height of the Victorian era, see Arthur J. Marder, *The Anatomy of British Sea Power: A History of the British Naval Policy in the Pre-Dreadnought Era, 1880–1905* (New York, 1940).

ed to improve and fortify its naval facilities in the Mandates in 1937 with the most ambitious effort focused on Yap in the western Carolines, "the Gibraltar" of the central Pacific. The Japanese tested Western resolve in 1931 by establishing a puppet regime in Manchuria and in 1932 by launching an air and land punitive expedition against the Chinese of Shanghai for abusing Japanese property and lives. In both "incidents" Western nations, collectively or singly, responded with sanctions insufficient to persuade the Japanese that imperialism did not pay. By 1937, when Japan started its full-scale war to dominate China, the three principal signatories of the Five-Power Treaty had embarked upon major naval rearmament programs. Yet the problem of defending and creating advanced naval bases for extended naval campaigns still remained a low priority for all three navies. This inattention had roots in World War I and in the tumult of the Russian, Greek, and Turkish civil wars that followed.[4]

World War I provided ample examples of the feasibility of amphibious operations, as well as evidence that *some kinds* of amphibious operations could result in prohibitive losses. The most ambitious operation of the war, the Allied expedition to the Dardanelles in 1915, was also its most celebrated failure. The controversies that clustered about the Dardanelles campaign made rational investigation difficult, for at issue were the political career and credibility of Winston Churchill, the expedition's sponsor as First Lord of the Admiralty; command relations between the Royal Navy and the British Army; the distrust between British commanders and the Australian–New Zealand expeditionary corps; the timid conduct of fleet operations before the landing, which might have made an amphibious assault unnecessary; and disagreements with the French forces assigned to the campaign. Tactical pundits claimed that the slaughter of the British 29th Division on "V" and "W" beaches proved that amphibi-

---

[4] For the interrelationship of British, American, and Japanese naval policy in the interwar years, see Harlow A. Hyde, *Scraps of Paper: The Disarmament Treaties between the World Wars* (Lincoln, NE, 1988); Arthur J. Marder, *Old Friend, New Enemies: The Royal Navy and the Imperial Japanese Navy* (Oxford, 1981); Gerald E. Wheeler, *Prelude to Pearl Harbor: The United States Navy and the Far East* (Columbia, MO, 1963); Stephen E. Pelz, *Race to Pearl Harbor: The Failure of the Second London Conference and the Onset of World War II* (Cambridge, MA, 1974); Mark R. Peattie, "The Nan'yo: Japan in the South Pacific, 1885–1945," in Ramon H. Myers and Mark R. Peattie, eds., *The Japanese Colonial Empire, 1895–1945* (Princeton, NJ, 1984), pp. 172–210; David C. Evans and Mark R. Peattie, *Kaigun: Doctrine and Technology in the Japanese Navy, 1887–1941* (in press, 1995), chapter 14; Michael A. Barnhart, *Japan Prepares for Total War: The Search for Economic Security, 1919–1941* (Ithaca, NY, 1987); Jonathan G. Utley, *Going to War with Japan, 1937–1941* (Knoxville, TN, 1985); Thaddeus Tuleja, *Statesmen and Admirals—Quest for a Far Eastern Naval Policy* (New York, 1963).

ous assaults could not prevail against prepared defenses armed with machine guns and artillery. Virtually every phase of the campaign received searching investigation after the war. The weight of the evidence seemed to give the Dardanelles all the elements of a fated Attic disaster, not a laboratory for identifying and correcting the operational flaws that might have swung the campaign to the Allies.[5]

World War I, however, also provided evidence that amphibious landings were not necessarily a forlorn hope. The Germans conducted successful assaults against the Russians on the Baltic islands of Courland and Oesel in the Gulf of Riga. Despite the Dardanelles failure, the Admiralty continued to give serious attention to amphibious operations. In 1918 it mounted several raids against German submarine bases in Belgium, the most dramatic being a night assault on Zeebrugge in April. The same year the Royal Navy examined the feasibility of amphibious landings along the Adriatic with the enthusiastic support of Admiral William S. Sims, commander of U.S. naval forces in European waters. After the war, seaborne expeditionary forces went to Murmansk, Vladivostok, the newly independent Baltic states, and cities along the Black Sea to try to confront the Red armies during the Russian civil war. Expeditionary forces also arrived by ship to provide safe havens (and some diplomatic leverage) during the war between Greece and Turkey. In a sense these expeditions repeated the nineteenth century pattern of avoiding defended ports or beaches, or ensuring through prior diplomacy that the disembarkation site would be in friendly hands. The British armed forces, in fact, viewed amphibious operations as a type of "combined operations," a campaign in which the navy, the army, and the newly independent Royal Air Force each might have an important role. This definition tended to burden any consideration of amphibious operations with all the interservice conflicts that beset the British armed forces.[6]

Of all the belligerents of World War I the Japanese showed they had not forgotten the amphibious concepts they had used against the Chinese in 1894 and against the Russians in 1904–1905. In September 1914 they sent an expeditionary force by sea to the Shantung peninsula to invest the Ger-

---

[5] Much of the best writing on Gallipoli predates World War II: John Masefield, *Gallipoli* (London, 1916); C.E.W. Bean, *The Story of Anzac*, 2 vols. (Sydney, 1921, 1924); Capt. William D. Puleston USN, *The Dardanelles Expedition* (Annapolis, MD, 1927). The author visited Gallipoli in July, 1993.

[6] Admiral Sir Herbert Richmond, *Amphibious Warfare in British History* (Exeter, 1941); Admiral the Lord Keyes, *Amphibious Warfare and Combined Operations* (Lee Knowles Lectures, 1943) (Cambridge, 1943); James L. Stokesbury, "British Concepts of Amphibious Warfare, 1867–1916," PhD dissertation, Duke University, 1968.

man leasehold of Tsingtao. Greatly outnumbered, the German garrison could do little, and after a brief siege it surrendered. A similar fate befell German colonial forces on the Pacific islands of Saipan, Palau, and Truk as well as on some atolls. In the Pacific islands the landing forces came from the ranks of the Imperial Japanese Navy rather than from the army. Four years after its whirlwind campaign of 1914 Japan showed it had lost none of its expertise in moving land forces by sea when it sent an army of 70,000 to Vladivostok as the leading edge of Allied intervention in Siberia.[7]

THE STRATEGIC CONTEXT OF
AMPHIBIOUS OPERATIONS

Britain, Japan, and the United States faced different strategic dilemmas in the 1920s and 1930s, all of which contributed to their varied development of amphibious forces. The British Army and Royal Navy had a minor interest in "combined operations," and the Royal Air Force had virtually none. For much of the interwar period the British armed forces struggled to justify their estimates under the "Ten Year Rule," a judgment by civilian leaders that the armed forces would not face a major war for at least ten years. When the possibility of war increased in the 1930s, the British had no choice but to view Germany as the principal enemy and Europe as the most important potential theater of operations. As the British Chiefs of Staff Committee and service staffs examined force requirements, structure, and service roles and missions, they envisioned a European war primarily of motorized armies and of air forces with wide capabilities for air defense, strategic bombardment, and ground force support.

Any modernization of the Royal Navy was distinctly secondary, let alone an investment in amphibious forces. The navy did not view amphibious capability as important as strengthening the battle line, carrier aviation, the destroyer force, or even the submarine force. A naval war would be fought from fixed bases under Allied control, sited either in the United Kingdom or along the route from Britain to India. Although the Japanese invasion of China in 1937 sent clear signals that a Far Eastern crisis had come, the Chiefs of Staff and their civilian ministers could not justify a higher priority for amphibious forces. They recognized that such forces might be essential to any naval campaign east of Suez, but even the

---

[7] For general histories of the Japanese armed forces, see Stephen Howarth, *Fighting Ships of the Rising Sun: The Drama of the Imperial Japanese Navy, 1895–1945* (New York, 1983) and Meiron and Susie Harries, *Soldiers of the Sun: The Rise and Fall of the Imperial Japanese Army* (New York, 1991).

defense of Hong Kong and Singapore did not look promising if Japan decided to extend its interests south from China. The fate of the British empire in East Asia – especially Malaya, Burma, and India – might rest in the hands of two uncertain allies, the United States and the Soviet Union. Nevertheless, the British had great trouble believing that the Royal Navy and Royal Air Force, backed by the armies of the Commonwealth, could not stop the Japanese armed forces along the Malay Barrier.[8]

The Japanese political and military elites believed that their nation's destiny, the very survival of the Japanese people, would require a major war for the domination of Asia. Such a contest should subjugate China and eliminate the Russians, the Americans, and the Europeans as regional economic and military powers. Some Japanese diplomats and civilian leaders believed that war was unnecessary for Japan's continued economic growth and regional hegemony, but their voices failed in the 1930s under extreme domestic pressure. More optimistic Japanese military leaders believed that a growing world crisis centered in Europe would allow Japan to reap further territorial gains.

The real barriers to expansionism were not the European colonial powers but the United States and the Soviet Union. Both played an unwelcome role in supporting an independent China, but Japan could not fight every potential foe at once, although the most emotional and naive officers thought otherwise. Essentially, Japanese planners had to join one of two geopolitical schools and a wrong guess not only endangered careers, but lives. One school might have been called the "continentalists," mostly army officers, who believed that Japan's future depended on its domination of mainland Asia, especially China, Manchuria, the Russian Maritime Province, and even Siberia. Any influence south of China (French Indochina, Siam, Burma) would be welcome, but not essential. The other school, the "imperialists," was most influenced by the British experience and envisioned an empire (the "co-prosperity sphere") that might include mainland enclaves in China, but which drew its strength from "the

---

[8] In addition to the sources already cited, see Paul Kennedy, "British 'Net Assessment' and the Coming of the Second World War," in Williamson Murray and Allan R. Millett, eds., *Calculations: Net Assessment and the Coming of World War II* (New York, 1992), pp. 19–59; Brian Bond and Williamson Murray, "British Armed Forces, 1918–1939," in Allan R. Millett and Williamson Murray, eds., *Military Effectiveness* (London and Boston, 1988), vol. 2, pp. 98–130; H. P. Willmott, *Empires in the Balance: Japanese and Allied Pacific Strategies to April 1942* (Annapolis, MD, 1982), pp. 39–66, 95–129; Peter Lowe, "Great Britain's Assessment of Japan Before the Outbreak of the Pacific War," in Ernest R. May, ed., *Knowing One's Enemies: Intelligence Assessment before the Two World Wars* (Princeton, NJ, 1984), pp. 456–475.

South Seas resource area." The prime targets were Malaya and the Dutch East Indies with French Indochina and the Philippines added principally for strategic security. These objectives implied a commitment to eliminate American bases west of Hawaii and to conquer the South Pacific islands under Australian mandate. Thereby the Japanese would establish a strategic outpost line that could confound a U.S. Navy counter-offensive through an "interception-attrition" strategy.

Japanese strategic planning, however, hinged not only on geopolitical considerations but also on timing. The Japanese decision to conduct an "imperial-navalist" campaign in 1941 stemmed from assessments impossible before that year. The planners realized that the Soviet Union could do little to oppose Japan while it fought for survival against Germany, and that the United States had decided to rebuild its armed forces – and deploy some of them – in anticipation of eventual involvement in the European war. (In 1941, for example, the U.S. Marine Corps deployed a brigade from California to Iceland so that it could keep a marine division on the East Coast intact for contingency missions in the Atlantic and Caribbean.) The coalition of Japanese army and navy leaders who eventually drove their government to war tested the Russians in the Nomonhan Incident of 1939 and found their own forces wanting. A campaign that began with surprise naval operations, a characteristic of earlier Japanese wars, promised better results. The Japanese planned for the "southward advance" in full confidence that they could move and land sufficient expeditionary forces.[9]

For the United States in the interwar period, Japan represented the most likely enemy and one that only an extended naval campaign across the vast spaces of the central Pacific Ocean could defeat. While political, business, educational, and religious leaders discounted the likelihood of war with Japan, senior officers of the American armed forces, especially those of the U.S. Navy, believed even before World War I that war with Japan had become inevitable. Part of this assumption rested, no doubt,

[9] Fujiwara Akira, "The Role of the Japanese Army," and Asada Sadao, "The Japanese Navy and the United States," in Dorothy Borg and Shumpei Okamota, eds., *Pearl Harbor as History: Japanese American Relations, 1931–1941* (New York, 1973), pp. 189–195 and 225–259; Louis Morton, "The Japanese Decision for War," *U.S. Naval Institute Proceedings*, 80, December 1954, pp. 1324–1334; Carl Boyd, "Japanese Military Effectiveness: The Interwar Period," in Millett and Murray, eds., *Military Effectiveness*, vol. II, pp. 131–168; Alvin D. Coox, "Japanese Net Assessment in the Era before Pearl Harbor," in Murray and Millett, eds., *Calculations*, pp. 258–298; Michael A. Barnhart, "Japanese Intelligence before the Second World War," in May, ed., *Knowing One's Enemies*, pp. 424–455; and Rear Adm. Yoichi Hirama, "Japanese Naval Preparations for World War II," *Naval War College Review*, 44, Spring 1991, pp. 63–81.

on racism, but most of it stemmed from a close examination of Asian politics and the obligation to defend the Philippines and the mid-Pacific islands of Guam, Wake, and Midway, all especially vulnerable after the Japanese gained control of Germany's Pacific colonies in World War I.

No responsible American military planner believed that a European power represented a danger until the late 1930s, as only Britain had the military capability to threaten the United States for many years after World War I. Japan, in contrast, had both a fleet and a motive to challenge the United States, champion of Chinese nationalism (within limits) and governor of the Philippines. With the navy taking the lead, the military planning committees attached to the Joint Army-Navy Board focused on one contingency plan after 1919, Plan ORANGE, the strategic conception for a conflict with Japan. Although it went through predictable alterations and controversies – for example, could the Philippines be held? – Plan ORANGE shaped the navy's force structure in the interwar period and also influenced organization and deployment of the army, especially those ground and air units assigned to China, the Philippines, and Hawaii.[10]

The senior officers of the U.S. Navy and Army differed from their contemporaries in Britain and Japan with their focus upon a single major war plan with a single potential foe. Not until 1938 did the military planning bureaucracy of the army and navy consider the possibility of a German threat to the Western Hemisphere; earlier contingency plans for operations against Canada, Mexico, and Cuba had faded into irrelevance in the 1930s. Planning a war with Japan gave both services ample roles and missions to justify budget requests to the White House and Congress, even if their requirements were never met in full. The army, for example, argued that the Japanese threat to the Philippines, Hawaii, Alaska, and the Canal Zone – not to mention the west coast of the United States – justified continued investment in long-range bombers, heavy coast artillery, tactical

---

[10] Edward S. Miller, *War Plan ORANGE; The U.S. Strategy to Defeat Japan, 1897–1945* (Annapolis, MD, 1991); David Kahn, "United States Views of Germany and Japan by 1941," in May, ed., *Knowing One's Enemies*, pp. 476–501; Fred Greene, "The Military View of American National Policy, 1904–1940," *American Historical Review*, 65, January 1961, pp. 354–377; Calvin Christman, "Franklin D. Roosevelt and the Craft of Strategic Assessment," in Murray and Millett, eds., *Calculations*, pp. 216–257; Russell F. Weigley, "The Role of the United States Navy," in Borg and Okamoto, eds., *Pearl Harbor as History*, pp. 165–188, 197–223; Philip T. Rosen, "The Treaty Navy, 1919–1937," and John Major, "The Navy Plans for War, 1937–1941," in Kenneth J. Hagan, ed., *In Peace and War: Interpretations of American Naval History, 1775–1984* (Westport, CT, 1984), pp. 221–262; and Ronald Spector, "The Military Effectiveness of the U.S. Armed Forces, 1919–1939," in Millett and Murray, eds., *Military Effectiveness*, vol. II, pp. 70–97.

aviation, and mobile defense forces of combined arms. Horse cavalry regiments were about the only element that the army could not easily relate to War Plan ORANGE. Those units the army deployed to protect the Mexican border from raiders, still a threat in the 1930s. Experimental armored forces also appeared to have limited relevance to a war with Japan, but they were so small (one brigade in the mid-1930s) that they were not a budgetary burden.

For the navy a possible war with Japan required development of forces capable of operating across 7,000 miles of ocean. The navy assumed that only a major fleet engagement in the western Pacific and an eventual sea-air siege of the Japanese home islands could determine the outcome of such a war. The campaign would require battleships, heavy and light cruisers, destroyers, and submarines, all capable of operating for weeks at sea. The surface fleet would need protection from island-based air attack, which meant development of fleet carriers and embarked air groups; the major debate about naval aviation was not its general utility, but its offensive capability away from surface combatants. The navy also required long-range amphibian patrol aircraft for reconnaissance duties, a function shared with submarines, and needed a mobile service force capable of fueling, arming, repairing, and reprovisioning the fleet from extemporized bases and anchorages.

After the Japanese seized control of the German central Pacific islands in 1914, navy planners assumed that the fleet would have to wrest bases from a Japanese defense force. Although various revisions of ORANGE differed in the priority given to the use of advanced bases, planners focused on several potential sites in the western Marshalls (with emphasis on Eniwetok) and Truk island group, whose special advantages for anchorages and airfields had not escaped the Japanese. Even if the fleet reached the Philippines without a major engagement, it would still have to establish a major base on the southern coast of Mindanao before reconquering Luzon, because a direct thrust into Manila Bay or Lingayen Gulf would give the Japanese unacceptable operational advantages for air and naval action. In addition to this consideration, army planners added another objective in the 1930s, the seizure of islands (especially the Marianas) suitable for long-range bombers.[11]

In sum, Britain, Japan, and the United States all had some strategic ra-

---

[11] Miller, *War Plan ORANGE*, pp. 100–249; Chief of Naval Operations (Adm. R.E. Coontz) to Major Gen. Cmdt. Marine Corps (Major Gen. G. Barnett), "Function of Marine Corps in Wartime," memo, 28 January 1920, File 221 (1920), Secretary of the Navy/CNO Correspondence, 1920, General Records of the Navy Department, RG 80, NA.

tionale for creating amphibious forces, or at least studying amphibious operations in the conceptual sense, but only the United States had a distinct requirement to create advanced naval bases and, if necessary, to seize such bases from a determined defender. All three nations, therefore, required amphibious capability, but the United States alone identified the need to conduct an *opposed landing*. Of course, a force that could capture a defended base could also land at an unopposed site because it had already mastered the intricacies of ship-to-shore movement under fire. The different paths of the three great naval powers in developing amphibious forces, however, rested not just in the geostrategic context in which they functioned but reflected as well a complex interaction of organizational politics, funding, operational doctrine, and technological challenges.

THE ORGANIZATION OF AMPHIBIOUS FORCES

Britain

Given its history of joint expeditions, naval campaigns, imperial defense, and disengagement from continental alliances, the British had every reason to maintain some sort of amphibious assault force. Although they recognized the likely wartime demand for such a capability, they did not have such a force. The shock and frustration of failure in the Dardanelles campaign provided some justification for the neglect, but this military trauma was not the only, nor the most significant, reason for the neglect of amphibious warfare. Britain's interwar military and naval staff committee system, along with its staff college adjuncts, did a good deal of thinking about joint operations. But the British failed to create a modest amphibious capability in the sense that the Royal Navy and army did not conduct exercises with amphibious shipping and landing forces on a regular basis, even to test some of the theories produced by the planning staffs. Instead the British military establishment made it difficult for operational development of any sort to receive a fair hearing, and the parliamentary form of government itself simply reinforced innate military conservatism. Budgetary restraints provided a convenient excuse for maintaining the status quo, but the British armed forces might have done more within the prison of fiscal restraint, as the American experience suggests. The Royal Navy and army simply assumed that amphibious operations presented too many problems for them to reach a solution in peacetime.[12]

---

[12] This analysis depends primarily upon the following sources: Kenneth J. Clifford, *Amphibious Warfare Development in Britain and America from 1920–1940* (Laurens, NY,

British interest in amphibious warfare after World War I began with the general consideration of interservice cooperation – complicated by the birth of the independent Royal Air Force and its capture of naval aviation – and the specific issue of naval base defense. In 1920 the Board of Admiralty, the navy's highest policy-making committee, ordered the naval staff to prepare a mobile base plan, with special consideration of air operations, because fixed bases might not be adequate in another war. This Admiralty interest reflected similar work by a special all-services syndicate of 169 officers, drawn from the three service staff colleges and convened for a week at the army's staff college in October 1919. The syndicate examined the defense of Singapore and Hong Kong and reported to their respective ministries that the existing guidance of joint operations, The Manual of Combined Naval and Military Operations (1913), needed thorough review and revision in light of the recent world war. The Admiralty, working through the Royal Naval Staff College, convinced the army and the Royal Air Force to participate in a joint committee that would review the staff colleges' recommended changes and publish a revised manual on joint operations. The Inter-Department Committee on Combined Operations met in June 1920 and ended its deliberations with a staunch defense of the status quo, service independence, and rejection of the importance of preparing for opposed landings.

Dissatisfied with the first committee effort, the Admiralty called for a second review in 1921, which produced a new publication, The Manual of Combined Naval Military and Air Force Operations-Provisional, 1922. The new manual reflected sound study at the Royal Naval Staff College about the inherent problems of amphibious operations but did not demonstrate any broad institutional commitment, only guidance for the staff colleges in planning their biannual academic exercises on joint operations. A 1925 revision of the manual incorporated the "lessons learned" from these problems but had no force outside the school system.

The operating forces contributed little to the limited operational debate in the staff colleges, and the two actual field exercises in the 1920s demonstrated little more than the initiative of local commanders. In 1924, Rear Admiral Sir Herbert Richmond, naval commander in East Indian waters, staged a landing in the Bay of Bengal with a scenario drawn from the plans to defend Singapore. The landing force, a reinforced battalion, operated under the supervision of the students of the staff college at Quet-

1983); Stephen Roskill, *Naval Policy Between the Wars*, vol. 1, *The Period of Anglo-American Antagonism, 1919–1929* (New York, 1968) and vol. 2, *The Period of Reluctant Rearmament, 1930–1939* (New York, 1976); Bernard Fergusson, *The Watery Maze: The Story of Combined Operations* (New York, 1961).

ta, and Richmond, a champion of amphibious operations, took this experience back to England when he became director of the senior officers' course at the Royal Naval Staff College. A 1928 landing in the Moray Firth, Scotland, originated with a Scots battalion commander who wanted to give his troops some novel training, in this case a trip to sea on three destroyers, an unopposed landing in ships' boats, and a dramatic assault on the Black Watch's own barracks at Fort George. The next exercise did not occur until 1934, but it benefitted from the interest of Admiral Lord Cork and Orrery, commander-in-chief, Home Fleet, who used forty-one of his own ships to land an army brigade group at the mouth of the Humber River against an opposing army brigade. This landing, however, provided little useful training, as its principal purpose was to serve as a communications test.[13]

Amphibious warfare development in Britain lacked any organizational commitment and institutional memory, both of which the Royal Marines might have provided. The Royal Marines, however, did not seize the limited opportunity to become the champions of amphibious operations. From a wartime high of 40,000 officers and other ranks, the peacetime Royal Marines shrank to 9,000 men. The predictable turmoil of force contraction grew more complicated after a 1922 decision to merge the Royal Marine Artillery and Royal Marine Light Infantry, a distinction that dated from 1855. This decision meant a merger of messes (always traumatic in British forces) and a reorganization of the marine barracks and training establishment.

Moreover, in 1923 the Admiralty called for a thorough study of the wartime functions of the Royal Marines and its training for wartime missions. An Admiralty committee, chaired by Admiral Sir Charles Madden and including ranking marine officers, studied the issue for six months and issued its report in August 1924. The Madden Committee reasserted the need for marines aboard warships for landing party and gun crew service – traditional duties – but urged that marines serving ashore (in "divisions" identified with each marine barracks) train to seize and defend temporary bases and to conduct amphibious raids upon enemy naval bases and coastal positions. To provide such a peacetime "striking force," the Royal Marines should be increased to approximately 16,000 officers and men. Without such an increase in strength, only about one-third of the marines would be available for amphibious warfare training.[14]

The Madden Committee report brought no substantial change to the

---

[13] Clifford, *Amphibious Warfare Development*, pp. 30–41.
[14] Roskill, *Naval Policy*, vol. 1, pp. 539–540; Fergusson, *Watery Maze*, p. 36; Clifford, *Amphibious Warfare Development*, pp. 48–49.

Royal Marines despite the fact that the Admiralty and senior marine officers agreed with its conclusions. The Admiralty decided that it could not ask for the money to increase the Royal Marines, except to provide detachments for warships. It faced no resistance from senior officers of the Royal Marines, who accepted their lot, happy or not. Overwhelmed with the requirement to provide elite detachments for ships and ad hoc battalions for colonial service, the leaders of the Royal Marines discouraged dissent among their own officers, some of whom urged a real commitment to the amphibious assault mission despite the limitations of manpower and funding. Moreover, so the admirals and marine generals argued, the Royal Marines had another mission: the preparation of their component of the Mobile Naval Base Defense Organization. This organization, conceived in 1920 but undeveloped until the 1930s, offered the Royal Marines a more attractive wartime function than the amphibious assault and built upon skills already present in the by then defunct Royal Marine Artillery because the operational core of the Royal Marines Group, Mobile Naval Base Defense Organization, was an anti-aircraft artillery brigade and a coast defense artillery brigade. A single Royal Marines infantry battalion was the only mobile defense component of the group in a total force of over 7,000 officers and men. To the degree that the Royal Marines interested themselves in amphibious operations into the late 1930s, they organized annual exercises to test landing craft, vehicles, and equipment needed to disembark a heavy artillery force over an undefended beach or at a friendly port. Although some of these experiments had applicability to offensive amphibious operations, their scope and scale provided little useful experience for an amphibious assault.[15]

In the meantime, the Admiralty remained restive about its lack of knowledge and experience in amphibious operations, a disquiet fed by reports from the staff college exercises and reports on parallel developments in the United States and Japan. The Japanese, for example, demonstrated at Shanghai in 1932 that they needed far fewer transports and landing craft than the British had estimated were necessary to put ashore a combined arms force of 12,000. (The Japanese used fourteen ships to do a job that British planners believed would require forty vessels.) In 1936 the Royal Naval Staff College, again preparing revisions for The Manual of Combined Operations, provided a persuasive report, organized by Captain Bertram Watson, RN, that allowed the Admiralty to persuade the War Office and Air Ministry to take action on studying amphibious operations.

The Watson memorandum could not have come at a more propitious

---

[15] Clifford, *Amphibious Warfare Development*, pp. 5–19.

time, for the Deputy Chiefs of Staff Committee working on the Manual had concluded that joint operations needed a permanent institutional advocate in the armed forces. In 1938, as recommended by the Watson memorandum and after much bureaucratic memo-passing, the Chiefs of Staff established the Inter-Service Training and Development Sub-Committee, which it made directly responsible to the Deputy Chiefs of Staff, and established the Inter-Services Training and Development Centre at Eastney Barracks, Portsmouth, a Royal Marines base. Captain L.E.H. Maund, RN, a Richmond protege, with an interservice staff of three (army, RAF, Royal Marines) and some clerks, opened the center in July 1938 with £30,000 to conduct research and development on amphibious operations. The Sino-Japanese war was already almost a year old, and the Munich crisis came only two months later. In the meantime, the Deputy Chiefs of Staff issued the 1938 edition of the Manual of Combined Operations, which still provided more wisdom on the problems of joint operations than decisions on their solution.[16]

In the short year before the outbreak of war, the Inter-Services Training and Development Center struggled to coordinate the theoretical studies of amphibious operations (still the province of the staff colleges and the Admiralty staff) and the development of prototype amphibious ships and landing craft. It also studied the relationship of airborne operations to landings and the special engineering requirements of defending and overcoming shore defenses. Perhaps its most important contribution was its educational effort within the highest circles of British defense planning, the Chiefs of Staff, and the Committee of Imperial Defense. Maund, who had seen Japanese amphibious operations in Shanghai in 1937, proved to be a forceful and persuasive advocate of amphibious development. He won an important convert, Major General Hastings Ismay, secretary to both the Deputy Chiefs of Staff and Committee of Imperial Defense, who reflected after one report that "as our strong suit was command of the sea, we should be making poor use of our hand if we relegated combined operations to the background of our war plans. Indeed it seemed that we should be laying ourselves open to justifiable criticism if we made no effort in time of peace to train on modern lines for a possible combined operation in time of war."[17]

---

[16] Donald F. Bittner, "Britannia's Sheathed Sword: The Royal Marines and Amphibious Warfare in the Interwar Years – A Passive Response," *Journal of Military History*, 55, July 1991, pp. 345–364; Maj. Gen. J.L. Moulton, vol. 1, *The Royal Marines* (Greenwich, London, 1982), pp. 86–90.

[17] Clifford, *Amphibious Warfare Development*, pp. 46–84.

As Ismay observed, the British armed forces had allowed themselves to fall behind Japan and the United States in amphibious development, and the situation could not be reversed overnight. By the time war came, Brigadier Bernard Law Montgomery was the only senior army officer who had commanded an amphibious landing force, his own 9th Infantry Brigade, which had made an opera-bouffe assault in July 1938, at Slapston Sands, Devonshire. The weight of British military opinion bore down the arguments of amphibious warfare enthusiasts. Only the navy had senior spokesmen in favor of the concept such as Admirals Sir Roger Keyes, Richmond, and Lord Cork and Orrery.

More importantly, the British political elite provided no external connection within Parliament for amphibious warfare, and the discipline of the party system and civilian control made it difficult for military missionaries to sidestep the bureaucracy. Winston Churchill remained in isolation, and no one took his place. In the world of military journalism and punditry, Captain Basil Liddell Hart flirted with the concept of the amphibious assault as part of his advocacy of the "indirect approach," but he failed to embrace the cause, probably because of his inherently inconsistent advocacy of British isolationism and mechanized warfare. With the exception of a few isolated and specialized shipbuilders, no important industrial interests embraced amphibious warfare; the Board of Trade, which exercised governmental power over the merchant marine, endorsed military views on transport requirements without much interest. The British armed forces could have emphasized amphibious operations but chose to direct little attention toward this type of complex military capability until 1940.[18]

Japan

Although the Japanese military establishment drew inspiration during the Meiji Restoration from the German Army and the Royal Navy, it surpassed its European models in attention to amphibious warfare after World War I. In terms of usable military capability Japan entered World War II as well prepared as the United States, both in terms of operational forces and published doctrine. Of course, the Japanese used their interventions in China as a testing ground for amphibious warfare, and they made major adjustments to their doctrine and forces as a result of those experiences, in particular after the Shanghai Incident of 1932 and the ini-

[18] Ibid., pp. 72–84; Rear Adm. L.E.H. Maund, *Assault From the Sea* (London, 1949), pp. 1–23.

tial landing operations of 1937 in support of their invasion of northern China.

As the threat of war with Russia and the press of operations in China consumed more of the attention of the Japanese Army, the leadership in refining amphibious warfare doctrine and planning shifted to the navy, which still focused upon American possessions in the Pacific. The navy's development of amphibious landing forces, therefore, concentrated on small, highly mobile, and lightly armed naval infantry regiments adequate for the capture of the weakly defended islands of Guam and Wake. Any major operations, i.e., those involving a division, would require the participation of the Imperial Japanese Army, which controlled not only the landing forces, but most of the landing craft and transports of the amphibious force itself. Japan, therefore, had two amphibious forces – one each for the army and the navy – capable in many ways of operating together, but also independent of one another. As these forces proved in 1941 and 1942, Japan could conduct amphibious landings with dazzling success, but these forces also had inherent weaknesses eventually exploited by the Allies after 1942.[19]

Like their Western counterparts, the general staffs of the Imperial Japanese Army and Imperial Japanese Navy – both of which operated independently from the Ministries of War and Navy – conducted a major reassessment of their contingency plans, embodied in the *Yōhei kōryō* ("Outline of the Employment of Forces") and its companion *Teikoku Kokubō Hōshin* ("Imperial Defense Policy"). The 1918 version of *Yōhei kōryō* stressed preparations for an opposed landing on Luzon as part of a general war plan for a conflict with the United States. Another review completed in 1923 retained the Luzon operation and added another landing to capture the American base at Guam. Having also studied the Gallipoli operation, general staff planners of the Japanese Army concluded that the army could no longer rely principally on the navy to organize amphibious operations. Even before they worried about the menace of airplanes and submarines to amphibious forces, army officers believed they would require more rapid landings than in the past and that they should have to come ashore prepared to fight. The general scenario of past operations would not meet the test of modern battle: light naval infantry securing an undefended landing site, soldiers towed to shore in unpowered barges by navy boats, the slow disembarkation of artillery and horse-drawn transport, the measured organization of army field forces ashore before an overland campaign against the enemy. Army planners reached

---

[19] Clifford, *Amphibious Warfare Development*, p. 76.

the peak of their concern in 1926 when they estimated that the conquest of the Philippines would require three divisions.[20]

The army, the dominant military service in Japan, already had an institutional base upon which to build an amphibious warfare force. Powerless to halt the army's amphibious warfare crusade, the navy provided substantial support when the army began to develop its force in the 1920s. The Army Transportation Department, located at Ujina in Hiroshima prefecture, provided all land and sea transportation services much as the Quartermaster Department did for the U.S. Army during the same period. This department [Rikugun un'yubu] contained a special bureau for sea transportation with its own fleet of transports and had well-developed ties to the Japanese shipping industry. Policy guidance and supervision for the Transportation Department came from the Eighth Section (Land and Sea Transport) of the Third Bureau (Communications and Transport) of the army general staff, and budgetary and logistical support came from the Economic Mobilization Bureau of the Ministry of War. The department received, in addition, the helpful attention of the army's liaison office with the navy general staff and the dominant First Bureau (Operations) of the army general staff. At the operational level the Army Transportation Department developed an elaborate system to provide waterborne transport services. Shipping engineer regiments were the most useful, with 1,200 officers and men in a headquarters and three companies [sempaku kōhei rentai] to operate landing craft and barges. The army also maintained debarkation units, trained and equipped to load or unload transports in port or over a beach, while shipping transport commands managed port facilities used by army vessels. The Army Trans-

---

[20] Unless otherwise noted, this analysis of Japanese amphibious warfare development is drawn primarily from two studies written from official Japanese archival sources by Capt. Masao Suekuni, JMSDF (Ret.), chapter 1, "Landing Operations of the Japanese Military," of the larger work, "Amphibious Operations in Military History and Their Background," [Senshi ni miru jōriku sakusen to sono urakata] published as a volume in the series Reference Materials in Naval History [Kaigun senshi sankō shiryō] (Toyko, 1982) and "The Evolution of Joint Landing Operations of the Japanese Military," Journal of Military History [Gunji Shigaku], 27, July 1991, pp. 55–56. I want to thank Prof. Hisao Iwashima of Iwate University for serving as my liaison with Captain Suekuni, and I am indebted to Professor David Evans of the University of Richmond for his translation of Captain Suekuni's pamphlet and to Mr. Walter Grunden for additional translations from the same text. Mr. Grunden also translated "The Evolution of Joint Landing Operations." I also acknowledge the contribution of Dr. Edward J. Drea, U.S. Army Center of Military History, also a Japanese linguist, for the use of his "The Development of Imperial Japanese Army Amphibious Warfare Doctrine," 18 February 1992, also based on Japanese archival documents and official histories. See also Saburo Hayashi and Alvin D. Coox, Kogun: The Japanese Army in the Pacific War (Quantico, VA, 1959), pp 1–28.

portation Department had both the money and supervisory staff to conduct research and development for amphibious operations without navy interference, especially in the design of landing craft, although in practice the two services worked closely together.

Although the army did not begin formal training and instruction in amphibious operations until 1921, one division (the 5th, also stationed at Hiroshima) conducted a minor landing exercise in 1918. Under general staff supervision, the army conducted a major exercise in 1920, observed by the chief of the general staff, General Uehara Masasaku, who then attached a high priority to development of a steel, armored, and self-propelled landing craft. Although it did not put the new landing craft into operation until 1925, the army conducted landing exercises with one division in 1921 and in 1922 with three divisions, out of a total force of seventeen divisions. In addition, it undertook command post exercises with amphibious operations scenarios, not only in its staff college but also in tactical armies and divisions. In 1926 the army general staff identified the 5th, 11th, and 12th Divisions as "especially designated" divisions for the Luzon landing and assigned them to the Army Transportation Department to develop ready forces and appropriate doctrine for amphibious assaults. The home stations of all three divisions placed them close to army waterborne transportation units and shipping engineer regiments.[21]

The army also soon supplanted the navy as the authoritative source of amphibious warfare doctrine, although navy participation in writing and testing doctrine, especially for naval gunfire support, did not end. In 1924 the army published its authoritative "Summary of Amphibious Operations and Operations Defending Against Amphibious Attacks" [*Jōriku oyobi joriku bōgyo sakusen kōyō*], which represented a major improvement of the navy's *Kaisen yōmurei*, the equivalent of the U.S. *Landing Party Manual* or guide for naval infantry units drawn from the fleet. Additional doctrinal pronouncements flowed from the army and navy and culminated in publication of "Outline of Amphibious Operations" [*Jōriku sakusen kōyō*] in 1932 after five years of collective drafting by both service staffs. This doctrinal manual carried the force of general staff endorsement and informal imperial sanction. The 1932 publication re-

---

[21] Suekuni, Sections 4 and 6, "Landing Operations"; Drea, "The Development of Imperial Japanese Army Amphibious Warfare Doctrine"; Suekuni, "The Evolution of Joint Landing Operations of the Japanese Military," *Gunji Shigaku*, previously cited; G-2, War Department General Staff, *Handbook on Japanese Military Forces*, 1 October 1944, TM E30-480 (reprint, Novato, CA, 1991), pp. 10–18, 49–50.

flected major advances in ship-to-shore movement, as tested in major joint landing exercises in 1929 and 1931.[22]

Developed in the hard school of the real ocean, which extracted payment in the form of wrecked landing craft and dead soldiers, Japanese amphibious warfare doctrine by 1932 showed a sound appreciation in both army and navy of the fundamental requirements for a successful opposed landing. First, the navy's covering forces would have to insure that an enemy air attack, submarine attack, or warship raid did not disrupt the amphibious force convoy or ship-to-shore movement. Moreover, the navy would provide sufficient naval gunfire to destroy beach defenses and suppress enemy mobile defense forces. Toward that end the navy conducted naval gunfire experiments with army participation beginning in 1926. The results were not heartening, but suggested that cruisers and destroyers could – with better range-finding and communications equipment – fire on targets identified by ground forces. The nature of naval ordnance was another limitation; shells designed to penetrate steel ships did not work well against field fortifications unless the structure was made of concrete and offered a vertical face toward the firing ship. In mounting an efficient ship-to-shore movement, the army, equipped with progressively larger numbers of motorized landing craft, showed increased ability to move troops and equipment rapidly from transports to landing craft – even at night – and then land them at the intended objective. After the Shanghai punitive landing (1932), when the navy landed soldiers without adequate ammunition and heavy weapons, the army planned to use only its own transports and landing craft for future operations.[23]

The army's domination of amphibious operations development, however, had two effects that reduced Japanese primacy in this operational specialty. The war in China after 1937 forced the army to release its three "amphibious divisions" for service on mainland Asia, and one of them, the 11th Division, suffered grievous casualties near Shanghai, which suggested that "especially designated" divisions had spent too much time in learning to land and too little in learning to fight. Therefore, the army made no special effort to make other divisions ready for amphibious landings on a routine basis. The Chinese campaign also demonstrated to the

[22] Suekuni, Section 3, "Landing Operations;" War Department General Staff, *Japanese Military Forces*, pp. 166–169; "Japanese Landing Operations and Equipment," May 1943, Office of Naval Intelligence, reprinted in A.D. Baker III, ed., *Japanese Naval Vessels of World War Two* (Annapolis, MD, 1987); Hans G. Von Lehman, "Japanese Landing Operations in World War II," in Lt. Col. Merrill L. Bartlett, ed., *Assault From the Sea: Essays on the History of Amphibious Warfare* (Annapolis, MD, 1983), pp. 195–201.

[23] Suekuni, Sections 8 and 9, "Landing Operations."

army that the old method of landing at undefended sites still worked; with no significant Chinese naval or air forces to disrupt operations, the Japanese could revert to landing with all deliberate speed and without any essential dependence on naval gunfire. No close air support was required. Although the army and navy could and did disagree about optimal landing sites, the Chinese campaign buttressed the traditional view in both services of amphibious operations, i.e., that landings at night by battalion-sized units, widely dispersed yet concentrated at the point of attack (Japanese landing craft tended to deploy in columns, not waves) brought startling success in terms of enemy confusion and demoralization. The emphasis upon preparing for an Asian Gallipoli disappeared in operations at Shanghai, Tsingtao, Canton, and Hainan Island. This form of "victory disease" was then reinforced by the army's preoccupation with a mechanized, artillery dominated war on its northern border with a revived Red Army.[24]

Although the use of naval landing parties to secure initial landing sites waned in theory and practice in the 1920s, the Imperial Japanese Navy maintained its own landing forces and its own programs, funded from its own budget. This naval program met no serious resistance from an army general staff absorbed with operations in China and training for a war with the Soviet Union. The navy's senior officers wanted to reduce "military" training for sailors (which closely followed rigorous army standards of physical fitness, weapons training, drill, and discipline) and to stress technical training, a fact that spurred development of specialized amphibious forces. More importantly, the use of sailors for landing parties reduced the readiness of a ship for fleet action to dangerous levels. However, if the navy still aimed to undertake initial landings, which it did, then it wanted better tactical performance than naval infantry showed in the Shanghai operation of 1932. Japanese naval infantry looked very smart in Shanghai but fought ineptly, and their performance accelerated the movement to establish permanent Special Naval Landing Forces [rikusentai] of reinforced battalion strength (1,069 officers and men) that included two infantry companies of six platoons and a heavy weapons unit armed with light artillery. The rikusentai formed at the four major naval bases at Sasebo, Kure, Yokosuka, and Maizuru and bore numbers related to their activation at each base, e.g., Sasebo 2nd Special Naval Landing Force. By 1941 the Special Naval Landing Force battalions looked and fought like the army battalions, distinguishable mainly by

---

[24] Ibid.; Hayashi and Coox, *Kogun*, pp. 6–27.

their use of navy rank and the anchor insignia rather than the army's star.[25]

The navy, however, saw no reason to create a second amphibious army. It did not buy or develop specialized transports, and its only specialized landing craft (as compared with the lifeboats and cutters found on any warship) was essentially the army's Type A landing craft. Instead, it planned to embark the Special Naval Landing Force battalions on warships, most often older destroyers, which carried naval guns and rapid-fire cannon that could support the landing force. This force, however, had a limited and highly specialized mission appropriate only to the central Pacific. Emphasizing surprise with night landings or the use of smoke-screens, it could enter the narrow passages in coral barrier reefs, navigate in shallow lagoons and put forces ashore on small atolls like Wake or poorly defended larger islands like Guam. It was not a substitute for the army divisions that would have to seize the Philippines, Hong Kong, and Malaya. Realizing the degree of risk they were accepting by assuming a rapid night landing, the Special Naval Landing Forces attempted to complicate the defender's problem by creating parachute-trained battalions, which land-based naval aviation might drop. In a sense the new Special Naval Landing Forces simply provided the navy with a capability to perform traditional limited missions with greater efficiency.

By December 1941, the Japanese Army and Navy had acquired a substantial body of theoretical expertise and operational wisdom about the conduct of amphibious operations. Although interservice rivalry in strategic planning complicated collaboration of the army and navy, it did not prevent the Japanese armed forces from conducting successful landings when and where they wanted after 7 December 1941. Their limitations came only after the American armed forces altered the conditions under which the Japanese had to operate, and that change did not occur until the counteroffensives on New Guinea and the southern Solomons in late 1942.

The United States

The American forces that eventually defeated the Japanese were already well developed by the time the Japanese Army began its China campaign in 1937, and in the four remaining years of peace the navy and the ma-

[25] War Department General Staff, *Japanese Military Forces*, pp. 76–79; David C. Evans, "The Japanese Navy in the Invasion of the Philippines," American Historical Association Meeting, 1991.

rine corps brought their understanding (if not their capability) of amphibious operations to a level unequaled in Britain or Japan. By 1940 the
U.S. Army had joined the annual fleet landing exercises and accepted
the naval services' published doctrinal guidance. The relative success of
the American armed forces in creating both doctrine and forces for amphibious operations stemmed from a complex interaction of strategic
guidance, service roles and missions, interservice and civil-military politics, and military-industrial collaboration.[26]

The challenge of solving the problems of an opposed amphibious landing fell to the United States Marine Corps. The marine corps was an organization so small and obscure before World War I that most political
leaders and the public did not know that it was an independent service
administered by the Navy Department, but not part of the navy itself. Association with the navy, however, had been close since the corps' founding in 1798, for marines served principally as ships' guards and naval base
security forces throughout the nineteenth century and then became the
cutting edge of the naval landing parties that went ashore with increasing
frequency and growing size throughout the Western Hemisphere and in
Pacific and Asian waters.

Marines also participated in land campaigns. During World War I,
a single marine brigade serving with the American Expeditionary Forces
in France made the corps the darling of the media and public and ce-
mented its already close ties with Congress. With the review of War Plan
ORANGE in 1919–1920, the marine corps recognized that it could fill an

---

[26] This section is drawn from the following sources, all of which represent extensive
research in official archives and private papers collections: Allan R. Millett, *Semper
Fidelis: The History of the United States Marine Corps* (New York, 1980, rev. 1991),
pp. 319–343; Jeter A. Isley and Philip A. Crowl, *The U.S. Marines and Amphibious War*
(Princeton, NJ, 1951), pp. 14–71; and Clifford, *Amphibious Warfare Development in
Britain and America*, previously cited. For a partisan, but "inside" account, see Gen. Holland M. Smith, USMC (Ret.), *The Development of Amphibious Tactics in the U.S. Navy*,
"Occasional Paper," History and Museums Division, HQMC, 1992, which is a reprint-
ing of General Smith's 1946–47 articles for the *Marine Corps Gazette*. Given the relationship of the U.S. Marine Corps and the function of amphibious warfare, it is unsur-
prising that the analysis is done from the marine corps' perspective, but these works, as
well as other sources in this section, represent navy perspectives as well. For an effort to
diminish the institutional role of the marine corps, see William F. Atwater, "United States
Army and Navy Development of Joint Landing Operations, 1898–1942," PhD Dissertation, Duke University, 1986. The most articulate and forceful presentations (other than
Smith's) of the Marine Corps perspective may be found in Lt. Gen. Victor H. Krulak,
First to Fight: An Inside View of the U.S. Marine Corps (Annapolis, MD, 1984) and Col.
Robert D. Heinl, Jr., *Soldiers of the Sea: The U.S. Marine Corps, 1775–1962* (Annapo-
lis, MD, 1962).

important void in its organizational existence. An amphibious assault specialty represented a focused wartime mission that would once and for all make it distinct from the army. The army, however, doubted the need for an independent marine corps, especially a large one with an important wartime function, but Congress routinely rebuffed army initiatives to absorb the corps. In truth, the interwar army had plenty of problems with modernization and aviation development, and its interest in amphibious operations of any kind remained limited, beyond negotiations over the language of *Joint Action of the Army and Navy.*

Four centers of study for amphibious operations existed in the Navy Department – the Office of the Chief of Naval Operations, the General Board of the Navy, the Naval War College, and the Commandant of the Marine Corps – but only the marine corps viewed the mission as critical. In 1920 the Commandant, Major General John A. Lejeune, grasped the importance of the amphibious assault mission and its contribution to War Plan ORANGE planning. He ordered a trusted staff officer, Major Earl H. Ellis, who had studied the defense and capture of advanced naval bases before World War I, to analyze the requirements of amphibious operations across the central Pacific. In seven months of manic work, Ellis wrote Operations Plan 712, "Advanced Base Force Operations in Micronesia," a detailed and prescient study that Lejeune endorsed on 23 July 1921 as the basis for future training and wartime mobilization planning in the marine corps.

Lejeune's commitment did not come from cosmic inspiration, but from a further refinement of his own and his predecessor's views that the marine corps ought to be the organization to take and hold advanced naval bases in any future naval campaign. Ellis's work, however, gave the commitment a valid geographic theater and a specific enemy as well as some refined, if speculative, thoughts about the exact equipment and training required for an opposed assault. Moreover, it represented a natural extension of the marines' "advanced base force" mission (established in 1900) and the existence of forces to fulfill that role, a mix of mobile and fixed defense battalions created before World War I and reconstituted after the war with a paper strength of sixty-five officers and 1,502 men. (Its actual numbers were less than fifty in 1920.)[27]

---

[27] Lt. Col. Merrill L. Bartlett, *Lejeune: A Marine's Life, 1867–1942* (Columbia, SC, 1991), pp. 190–202; Dirk A. Ballendorf, "Earl Hancock Ellis: The Man and His Mission," *U.S. Naval Institute Proceedings,* 109, November 1983, pp. 53–60; Major Gen. Cmdt. to CNO, September 30, 1919, File 223, and Major Gen. Cmdt. to SecNav (Operations), Quarterly Readiness Reports, October–March 1920, File 221 (1919), Confidential Cor-

In the extended and continuing debates over service functions in the interwar period, army and navy planners wrestled with the general problems of mutual cooperation in coast defense and overseas expeditions. The development of aviation in both services complicated the exchanges. Influenced by War Plan ORANGE, the members of the Joint Army-Navy Board did not exclude the army from amphibious operations. Historical precedents and good sense argued that the army might have to make an amphibious assault to retake Luzon, and it certainly needed a plan to embark overseas expeditions to the Philippines, Alaska, Hawaii, Puerto Rico, and the Panama Canal Zone, all of which it defended.

Only Guam and Samoa were sole navy responsibilities. In all the textual interpretation that accompanied revising *Joint Action* (1927), the marine corps still received the mission "to provide and maintain forces in support of the fleet for the initial seizure and defense of advanced bases and such limited auxiliary land operations as are essential to the prosecution of the naval campaign."[28] Nevertheless, the army and navy continued to discuss mutual cooperation in amphibious operations and set forth their general operational approach in *Joint Overseas Expeditions* (1933), which included concepts for making an opposed beach assault. The document, however, left open the question of which service's senior officer would command a joint expeditionary force, a nonexistent problem for the navy and marine corps because the senior navy officer afloat always held overall command. In any event, all the services recognized the requirement to test amphibious warfare theory in their classrooms and in field exercises.[29]

respondence, 1919–1926, RG 80, NA; both Col. B. H. Fuller to Major Gen. Cmdt., "Advanced Base Plans," August 1, 1921, File 2515, CMC General Correspondence, 1913–1932, RG 127, NA; Major Gen. Cmdt., memorandum for the General Board of the Navy, "Future Policy for the Marine Corps and as Influenced by the Conference on Limitation of Armament," February 11, 1922, File 432, General Board Records, Naval Historical Center, Washington, D.C.; Office to the Commandant, "History of Advanced Base Training in the Marine Corps," August 28, 1931, File 432, General Board of Records, Naval Historical Center, Washington, D.C.

[28] The Joint Board, *Joint Action of the Army and the Navy* (Washington, DC, 1927), p. 3. This publication superseded War Department and Navy Department, *Joint Army and Navy Action in Coast Defense* (Washington, DC, 1920).

[29] The Joint Board, *Joint Overseas Expeditions* (Washington, DC, 1933); Col. A.T. Mason, "Special Monograph on Amphibious Warfare," 1949–1950, mss. history, Command File World War II, Operational Archives, Naval Historical Center, Washington, D.C. The 1933 manual followed a complete review of the marine corps missions within the Department of the Navy and the validation of the requirement for an amphibious assault force under unambiguous naval control. This correspondence and set of records is in File 432, 1931–1933), General Board Records. The basic navy justification for the marine

Resembling their counterparts in Britain and Japan, the staff and war colleges of the American armed forces turned greater attention to instruction on amphibious operations in the interwar period. By 1930 the navy and army war colleges conducted annual exercises, largely drawn from Plan ORANGE. The Naval War College made its exercise a grand production that included navy and marine faculty and student officers from Newport and Quantico, site of the Marine Corps Schools. Although army instruction on joint operations came largely through lectures by naval officers, the Command and General Staff College at Ft. Leavenworth devoted approximately one week's work to an amphibious exercise in the mid-1930s. Nevertheless, the Marine Corps Schools (principally the senior course for majors and captains) made the greatest institutional commitment. From the mid-1920s through the mid-1930s the curriculum on amphibious operations increased from about 25 percent to 60 percent of the total hours of instruction. The professional interaction on amphibious operations drew additional strength from the fact that marine officers attended the Naval War College, the Army War College, and the Army Command and General Staff College, and these officers carried on missionary work for the specialization.[30]

Beyond school problems, officer instruction profited from the conduct of real amphibious exercises, however limited in scope. The navy and marines attempted their first offensive landing problem in 1924. Called the "Expeditionary Force" by Commandant Lejeune, 3,300 officers and men split into two reinforced regiments (one on the defense, the other in a beach assault) that battled each other at Culebra, Puerto Rico, and the Canal Zone. With the army taking the defensive role the following year, 2,500 marines assaulted Oahu, but the deployment of the two available marine brigades to Nicaragua and China in 1927 brought large-scale interservice exercises to a halt. In 1932, however, the navy, army, and ma-

corps is expressed in Chairman General Board to Secretary of the Navy, memo, "Examination of the Organization and Establishment of the U.S. Marine Corps," August 10, 1932, CMC Correspondence, 1913–1932, RG 127, NA.

30 Atwater, "United States Army and Navy Development of Joint Landing Operations," pp. 45–47; Michael Vlahos, *The Blue Sword: The Naval War College, 1980*; Anthony A. Frances, "History of the Marine Corps Schools," 1945, Breckinridge Library, Marine Corps University, Quantico, Va.; Research Section, Marine Corps Schools, "A Brief Historical Sketch of the Development of Amphibious Instruction and Doctrine at the Marine Corps Schools during the Years Prior to World War II," 1949, "Schools" File, Reference Section, Marine Corps Historical Center; Lt Col. Donald F. Bittner, "Curriculum Evolution Marine Command and Staff College 1920–1988," "Occasional Paper" (Washington, DC, 1988), pp. 1–30; Boyd Dastrup, *The U.S. Army Command and General Staff College: A Centennial History* (Ft. Leavenworth, KS, 1981).

rine corps resumed amphibious exercises in Hawaii. All of these operations revealed a basic problem: moving combat troops to landing boats
from transports was too slow and disorganized. The limitations of landing craft also meant that supporting artillery and light tanks could not get
ashore to help breach beach defenses, and assault engineering, close air
support, and naval gunfire were clearly inadequate. Compared to Japanese fleet exercises at the same time, American efforts did not represent a
meaningful current capability, but grist for doctrinal assessment.[31]

The void in amphibious exercises after 1925 helped clarify the general
interest in amphibious operations within the three services. Interest in the
navy was directly proportional to the availability of a landing force; unless scheduled as part of a general set of amphibious exercises, the fleet
showed little concern for developing naval gunfire techniques against land
targets. With transports of its own, operated by the Quartermaster Department, the army might have volunteered to fill the gap in landing forces
between 1925 and 1932, but, despite the navy's desire for greater participation, the War Department General Staff showed more interest in attacking the marine corps than in attacking defended beaches.

Alarmed by rumors of its own demise – a realistic fear, given the budget crisis of the Great Depression – the marine corps turned to writing
doctrine. What started as an effort to produce a textbook on amphibious
operations at the Marine Corps Schools in 1931 turned into an organizational "barn-raising" of doctrine-writing and resulted in the *Tentative
Manual for Landing Operations* in 1934, the most comprehensive and
detailed effort to think through the numerous problems involved in
amphibious operation. Spurred by navy and marine corps interest in the
Tentative Manual, the Marine Corps Schools continued the review and
revision process into 1935, when the *Manual* became the authoritative
guide within the Navy Department for future amphibious exercises and
for research and development. The writing of the *Tentative Manual* coincided with the withdrawal of the marine brigades from overseas, the re-
designation of the "Expeditionary Force" as the "Fleet Marine Force" in
1933, and the preparation for the resumption of fleet landing exercises.[32]

[31] General Board, "History of Advanced Base Training in the Marine Corps," 1931, previously cited; Mason, "Special Monograph on Amphibious Warfare," 1949–1950, previously cited; Col. Dion Williams, "The Winter Maneuvers of 1924," *Marine Corps
Gazette,* 9, March 1924, pp. 1–25; Brig. Gen. Dion Williams, "Blue Marine Expeditionary Force," *Marine Corps Gazette,* 10, September 1925, pp. 76–88.

[32] Clifford, *Amphibious Warfare Development in Britain and America,* pp. 92–108; Maj.
Gen. John H. Russell, "The Birth of the Fleet Marine Force," *U. S. Naval Institute Proceedings,* 72, January 1946, pp. 49–51; Maj. Gen. Cmdt., to CNO, August 17, 1933, and

With a landing force in hand – the 1st Marine Brigade on the East Coast and 2nd Marine Brigade on the West Coast – the navy renewed annual Fleet Landing Exercises (FLEXs) in 1935. The first two FLEXs in 1935 and 1936 proved to be limited undertakings with most of the landing force (two infantry battalions and a mixed artillery battalion) embarked on two old battleships or on the light cruisers and gunboats of the Special Service Squadron, the navy's nautical constabulary for Latin America. At the navy's initiation, FLEX-3, held off the Southern California coast in 1937, embraced all three services. The army provided an expeditionary brigade (built around one understrength infantry regiment and its own transports), and the navy embarked 2,500 marines, the full strength of the Fleet Marine Force. FLEX-4 ran from January to March 1938 in Puerto Rico and involved a similar number of marines, with an army expeditionary brigade participating again at the navy's request. This time the army's contribution climbed to three infantry regiments with supporting arms, and the brigade alternated in the part of landing force and shoreline defender. FLEX-5 in 1939 and FLEX-6 in 1940 reflected both increasing tension in world affairs and the first stirrings of rearmament and expansion in the armed forces. Not until February 1940, however, did the Fleet Marine Force's 1st Brigade approach anything close to a wartime expeditionary force in numbers and armament, and the navy had yet to provide a transport squadron for the marine brigade. Instead it used a group of old battleships and converted destroyers (APDs or fast attack transports). Neither FLEX-5 nor FLEX-6 involved army landing units.[33]

The fleet exercises provided additional experience to refine the concepts in the *Tentative Manual,* now raised to the august status of Fleet

CNO to Maj. Gen. Cmdt., September 12, 1933, File 1975-10, CMC General Correspondence, 1933–1938, RG 127, NA; Joint Board to Secretary of the Navy and Secretary of War, "Further Consideration of Joint Operations," April 5, 1927, File 350, Joint Board Records, RG 225, NA; Maj. Gen. J. T. Myers to Col. L. McC. Little, June 30, 1931, Maj. Gen. Louis McCarty Little Papers, Library of Congress.

[33] Lt. Col. Benjamin W. Gally, "A History of U.S. Fleet Landing Exercises," 1939, Discontinued Command File, Operational Archives, Naval Historical Center, Washington, DC; Mason, "Special Monograph of Amphibious Warfare," pp. 4–16; Headquarters, 1st Mar Brig, FMF, "Notes on the Organization and Activities of the Fleet Marine Force in Connection With Landing Operations," 1938, File 1975-10, CMC General Correspondence, 1933–1938, RG 127, NA; The Adjutant's General's Office, USA, "Notes on Fleet Landing Exercise No. 2," March 30, 1936, copy in Archives, U.S. Army Military History Institute; Major Gen. L. McC. Little to Maj. Gen. Cmdt. T. Holcomb, April 7, 1938 (review of FMF exercises), Little Papers; Commander in Chief U.S. Fleet, "Report on Fleet Landing Exercise No. 6," June 13, 1940, Atlantic Fleet Command Files, Operational Archives, NHC; HQ 1st Mar Brig, reports of FLEX 6 problems and critiques, File 1975-10, RG 127, NA.

Training Publication 167 (*Landing Operations Doctrine, USN,* 1938) and issued under the *imprimatur* of the chief of naval operations. The exercises also refined the mobilization concepts and training embodied in Marine Corps Contributory Plan, C-2, ORANGE, reviewed and revised each year in the 1930s. The marine corps in 1932 anticipated a minimal wartime requirement of one division and one brigade for amphibious assaults and one base defense force, in all 20,400 officers and men drawn from a wartime corps of 119,000. Clearly traditional naval duties still influenced planning. The total strength of the marine corps was only 17,000 in 1936 and 28,000 in 1940.

In the course of the FLEXs the navy and marine corps experimented with about every imaginable amphibious technique and tactical approach allowed by their equipment. They tried day and night landings, smokescreens, varieties of air and naval gunfire support, concentrated assaults and dispersed infiltrations, the firing of all sorts of weapons from landing craft, and an array of demonstrations, feints, subsidiary landings, and broad-front attacks. Although rich debate went on within the navy and marine corps over the "lessons learned" from the FLEXs, FTP-167 reflected the growing consensus that amphibious assaults might be possible, but not easy. Army opinion remained even more skeptical.

The amphibious force would have to isolate the objective area, then pound the defenders into a stupor with naval gunfire and close air support. The landing itself would require a violent assault by a combined arms team, probably over a broad front, perhaps a beach of a thousand yards' width or more. To secure the beachhead, the landing force would need rapid reinforcement, complete with artillery and tanks. The greatest threat to a landing was a disruptive air and naval attack, which might pull critical fleet units from the objective area, but a combined air and ground counterattack was the most immediate concern. A counterlanding might give the enemy a striking advantage because it would be difficult for a landing force to protect its supply line and logistics support areas as well as defend the perimeter of its enclave. An amphibious expeditionary force could not rely on guile for success, but would require local superiority in every element of air, naval, and ground combat power.[34]

---

[34] Office of Naval Operations, *Landing Operations Doctrine United States Navy* (FTP-167) (Washington, DC, 1938); Marine Corps Contributory plan, C-2, ORANGE, MCWP-2, October 1932, MCS Files, RG 127, NA; Isley and Crowl, *U.S. Marines and Amphibious War*, pp. 37–58. In naval gunfire, for example, opinion divided on two concepts, one borrowed from World War I artillery practice of saturation bombardment measured by the tonnage per area, the other conceived during testing of firing slowly with pinpoint fire at exact targets like pillboxes. At issue was the question of effective range

The debate over amphibious warfare in the United States did not have the same closed character that it had in Britain and Japan. Articles on the subject appeared with regularity in service journals and even occasionally in civilian magazines. More importantly, Congress followed the discussions in the annual reports of the service secretaries and in Congressional hearings. Whenever someone in or out of office, military or civilian, criticized the marine corps, it opened the issue of readiness for amphibious warfare. It required no championing from the chief of naval operations or secretary of the navy for the marine corps to keep alive its operational specialty, even if underfunded. Amphibious operations and the marine corps received a special boost when Franklin D. Roosevelt became president because he had spent seven years as assistant secretary of the navy in the Wilson Administration and viewed himself as an honorary marine. His personal friendship with Commandants John H. Russell and Thomas Holcomb could not have been more congenial. Yet Navy Department funding decisions, approved by Congress, still presented problems of materiel procurement that limited amphibious development.[35]

THE PROCUREMENT ISSUE: AMPHIBIOUS SHIPPING AND LANDING CRAFT

The materiel requirements of the amphibious assault – or less deadly types of landings – did not demand exciting or high-risk investments in new military technology but, rather, special adaptations of shipping, aircraft, vehicles, and weapons to maritime service and the conditions of amphibious combat. Such adaptations often involved trade-offs in capability. In a landing craft, for example, the designers needed to consider size, speed, degree of armored protection and armament (all weight builders), maneuverability, the ease of disembarking troops and equipment, engine noise, buoyancy and freeboard (a high profile invited accurate shore battery fire, but a low profile swamped in heavy seas), ease of retraction from the beach, transportability by carrier ships, and seaworthiness under a variety of conditions. As practical experience mounted in the 1920s, the materiel requirements of amphibious operations became increasingly clear to

and warship vulnerability to shore batteries. The arguments are summarized in Cmdr. C.G. Richardson, "Naval Gunfire Support of Landing Operations," MCS lecture, 1938–1939, and Lt. Cmdr. David L. Nutter, "Gunfire Support in Fleet Landing Exercises," September 1939, both Breckinridge Library, MCU.

[35] The observations on Chiefs of Naval Operations and Secretaries of the Navy are based on the essays on these men in Robert W. Love, Jr., ed., *The Chiefs of Naval Operations* (Annapolis, MD, 1980) and Paolo Coletta, ed., *American Secretaries of the Navy* (2 vols., Annapolis, MD, 1980).

British, Japanese, and American planners, who displayed some degree of similarity in their responses. All three groups, for example, saw the need to develop a special landing craft suitable for disembarking infantry, light artillery, vehicles, and even light tanks over a sand beach.

In some cases, the principal challenge was to identify army weapons, equipment, and vehicles most suitable for amphibious operations and then make modifications to protect the equipment from saltwater corrosion, flooding, foreign matter (mostly sand), and the amphibious troops themselves. For example, the Japanese Special Naval Landing Force assault battalions and U.S. Marine Corps' artillery battalions armed themselves with light pack howitzers designed for mountain warfare. In terms of weapons, the small specialized landing forces of Japan and the United States could hardly afford armaments not already designed and manufactured for army use. Yet in two areas – specialized ship and craft development and aviation development – the amphibious forces of the interwar period attacked their common material problems with results that had more to do with military politics and the procurement culture of each nation than with technology.

In Britain, the requirement for a motorized landing craft emerged out of the operational analysis and limited testing of the 1920s. The Royal Marines Mobile Naval Base Defense Organization project gave the development of such craft a modest sponsor, but landing craft development remained essentially an Admiralty responsibility and not one with a high priority. In theory, a joint Landing Craft Committee, composed of military and industrial representatives and eventually answerable to the Chiefs of Staff, should have developed landing craft programs, but landing craft development received little attention until the formation of the Inter-Services Training and Development Centre in 1938 under the patronage of the Deputy Chiefs of Staff. The Landing Craft Committee, however, let contracts that produced two prototype Motor Landing Craft (MLC) in 1926 and 1929. These MLCs were principally for transporting tanks and cargo as well as troops, carried armor and weapons, moved slowly despite jet-assisted engines, and weighed 36,000 and 45,200 pounds. Civilian passenger liners, which doubled as transports in wartime, could carry neither craft because their davits and cranes could not handle boats over 20,000 pounds. By 1939 eight MLCs had been built; the estimates at the time suggested that several hundred would be necessary to land just one division.[36]

---

[36] The development of British amphibious ships and craft are described in J.D. Ladd, *Assault from the Sea, 1939–1945* (New York, 1976); Fergusson, *The Watery Maze*, pp. 35–85; Clifford, *Amphibious Warfare Development*, pp. 72–82; Division of Naval

Under Captain Maund's direction, the Inter-Service Training and Development Centre (ISTDC) gave landing craft development its highest priority. Working with shipbuilders White of Cowes and Thornycraft, ISTDC sponsored the construction of a prototype Landing Craft Assault (LCA), which went through several modifications but remained under forty feet in length and ten tons in pre-loaded weight. The ISTDC also had Thornycraft produce a twenty-ton Landing Craft Mechanized, which might be carried by the sturdier cranes and davits of converted cargo ships. After World War II began, the British urgently accelerated work on LCA and LCM variants, but also began increasingly to concentrate on beaching ships and craft that did not require carrier ships and could sail from a friendly harbor to an unfriendly beach with embarked vehicles and supplies.

Spurred by the disaster of 1940 and a new prime minister, Winston Churchill, the Admiralty pushed a crash program to develop a fleet of beaching ships and craft that could carry heavy tanks, artillery, and pre-loaded supplies. From this program came the Landing Ship Tank (LST), a World War II workhorse in every maritime theater, and an associated family of smaller amphibious assault ships. These beaching ships featured bow ramps and a shallow draft forward, had at least one deck for loading (LSTs had two), and carried the weight of their engineering plants, living areas, and work spaces aft, allowing them to beach and retract under their own power.

British ship designers found a working prototype in commercial vessels designed to serve Latin American oil fields, but the "Maracaibo oilers" needed a great deal of reworking to strengthen and balance them for military loads, and their bow doors and ramps required fundamental re-engineering. Nevertheless, the British development of beaching ships and craft after 1940 produced a variety of vessels noteworthy for their durability and ease of construction. The British problem was timing, not ingenuity. Not until mid-1942 had their amphibious ship and craft inventory reached an operational level to support major raids, let alone a full amphibious assault. The British amphibious fleet also had short legs: it could not stay at sea for long periods and long distances, and it was vulnerable to air and naval attack. In operational terms it was a fleet designed for one purpose – an Allied amphibious return to northern Europe under the cover of land-based air power.

The Japanese attacked the problem of specialized shipping principally

Intelligence, *Allied Landing Craft of World War II* (1944 edition, reprinted Annapolis, MD, 1985).

through the shipping bureau of the Army Transportation Department, and their successes and failures stemmed directly from the army's definition of amphibious requirements. Neither the Japanese shipbuilding industry nor the navy had much impact until after Pearl Harbor. Less than a year after their devastating amphibious successes in the Philippines and Malaya, the Japanese armed forces had surrendered the strategic initiative and switched their interest from amphibious assaults to general logistical problems in the Pacific war in determining their shipping requirements.

The army's initial interest – as in Britain and the United States – centered on the development of self-powered landing craft that army transports or converted merchantmen could carry, which meant ships in the 10,000 ton range with limited convertability to accept davits and cranes strong enough to handle landing craft. By 1930 the Japanese Army had developed and adopted two basic landing craft, the Type A [*daihatsu*] craft, a forty-nine footer with a bow ramp designed to carry one hundred assault infantry, and the Type B [*shōhatsu*], a thirty-footer with no bow ramp for thirty troops. Both craft carried armor and machine guns. Although satisfied with the craft, the army's shipping engineers did not like the limitations of army transports or navy warships, also still used for landings. At army request the navy's design bureau began work on a specialized carrier ship, construction of which began in March 1933. In spring 1935, the army accepted the *Shinsū-maru* and initiated a new era in amphibious shipping.[37]

With the deployment of the 8,000-ton *Shinsū-maru* and a further refinement, the 9,000-ton *Akitsu-maru* (1941), the Japanese amphibious forces had in hand prototypes for all-purpose amphibious ships. Today the U.S. Navy and Marines use this fundamental concept to the exclusion of all others in their LHA and LHD class amphibious assault carriers. In 1937, British and American observers watched the *Shinsū-maru* at work off Shanghai and immediately recognized a significant development in amphibious warfare. The *Shinsū-maru* carried landing craft in a well deck

---

[37] The account of Japanese amphibious and ship and craft development comes from the following sources: Seukuni, Sections 11 and 12, "Landing Operations," previously cited; Administrative Division, Second Demobilization Bureau as compiled by Shizuo Fukui, ex-Lt. Cmdr. IJN, *Japanese Vessels at the End of the War*, issued 25 April 1947 for the American occupation authorities, copy in the author's possession; Hans Lengerer, Sumie Kobler-Edamatsu, and Tomoko Rehm-Takahara, "Special Fast Landing Ships of the Japanese Navy," in three parts in vol. 10 of Andrew Lambert, ed., *Warship* (Annapolis, MD, 1986); Ladd, *Assault from the Sea*, pp. 65–76; War Department General Staff, *Japanese Military Forces*.

that could be flooded, which allowed the landing craft to float free from an open stern gate. The ship could also hold additional craft on davits, but its next most impressive function was an ability to discharge vehicles from a deck-level parking garage directly onto a pier. It also carried two catapults for aircraft but did not embark operational seaplanes. It could, however, transport and unload aircraft if necessary, a capability further developed in the *Akitsu-maru*, which even had a short take-off flight deck.

The army, however, halted construction of this class of vessel and turned to another type, a beaching craft for tanks and other vehicles, which it designated a *koryu* or general purpose vehicle-carrying ship. Seeing the potential requirement for such a ship in Pacific operations, the navy persuaded the army to accept smaller ships (800 tons) that could be built more quickly, a proposal facilitated by naval control of the shipbuilding program and construction yards. The first Type 101 2nd Class Transport [*Nitō yusō-kan*] did not enter service until 1942; of the sixty-nine Type 101 transports built, twenty went directly under army control, and the navy obtained all the others.

The navy's modest interest in the Type 101 (the Japanese version of the LST) reflected its greater commitment to its own 1st Class Transport [*Ittō yusō-kan*], which had most of the operating characteristics and some of the armament of a destroyer. Its amphibious capability came from its troop and cargo compartments within the hull and from its aft deck configuration, stripped bare to carry four *daihatsu* or larger numbers of amphibian tanks or small craft. Almost 2,000 tons and capable of twenty-two knots, the 1st Class Transport, which entered service in 1943, had almost a 4,000-mile steaming radius. Its design reflected the navy's assumption that it would conduct emergency resupply missions to areas not necessarily under Japanese sea and air control and that it would depend upon its speed and anti-aircraft armament to survive. In the U.S. Navy inventory the 1st Class Transport would have received designation as a fast attack transport (APD), which in reality was a converted destroyer that could conduct amphibious raids and reconnaissance missions. Such a vessel did not have the capability to provide adequate lift for a major landing.

In many ways the American development of amphibious warfare ships and craft appears far less innovative than the prewar and early war efforts of Britain and Japan. One looks in vain for American equivalents of the LST or the *Shinsū-maru*. Yet the American effort had a certain simplicity that allowed it to create an adequate amphibious force in 1943 and a truly global force by 1944. Part of the wartime capability rested in the industrial means to mass produce British designs like the LST, and part of it came from the ingenuity to exploit the *Shinsū-maru* design and build a fleet of twenty-six Landing Ships Dock (LSD). (The British, for comparison, converted two Channel ferries to well-deck, stern-gate configurations and accepted one LSD from the United States.) Amphibian tractors were the one truly novel American development, a vehicle the marine corps sought with the determination of the quest for the Holy Grail. Nevertheless, the marine corps collaborated with the navy to create the foundations of an amphibious force before Pearl Harbor.[38]

The major barrier to building an amphibious force before 1941 lay in the tacit agreement between the navy and Congress to focus the most important ship-building programs on warships, not auxiliaries. Real capability for fleet support lagged badly behind the need; in 1934 the navy had *seven* support ships to service a combatant force of 149 warships, excluding submarines. Two transports served navy and marine corps bases world-wide and were unavailable for amphibious operations. Even when Congress approved a fleet expansion program in 1938 that would allow the navy to modernize and enlarge its numbers past treaty limitations, transports did not appear in the program. The same thing happened in 1939, in part because even pro-navy Congressmen like Carl Vinson thought that transports were not "defensive" enough for the times. Not until the "Two Ocean Navy" Act of 1940 – after the fall of France – did Congress approve the addition of twelve transports, all converted merchant vessels already in service, some for twenty years. By way of comparison, the transport fleet of the army numbered nine vessels in commission in 1940.

At the end of 1941, however, the navy had thirty-eight transports and a like number of cargo ships, but no specialized amphibious shipping. It commissioned its first LST in 1942, its first LSD in 1943. The navy followed the prewar assumption that it could create an amphibious transport force by converting merchantmen and liners to military service, including the installation of davits and cranes capable of handling landing

---

[38] This analysis is based on data from the four editions (1939, 1941, 1942, and 1945) of James C. Fahey, comp., *The Ships and Aircraft of the U.S. Fleet* (reprinted, Annapolis, MD, 1976); Calvin W. Enders, "The Vinson Navy," PhD dissertation, Michigan State University, 1970; DNI, *Allied Landing Craft of World War II*; Lt. W.F. Royall USN, "Landing Operations and Equipment," 1939 study with data and photographs, Breckinridge Library, MCU; Proceedings, Conference Concerning Various Types of Landing Craft, Their Capabilities and Limitations, 1943, Breckinridge Library, MCU; Michael Vlahos and Dale K. Pace, "War Experience and Force Requirements," *Naval War College Review*, 41, Autumn 1988, pp. 26–46; Thomas C. Hone, "The Navy, Industrial Recovery, and Mobilization Preparedness, 1933–1940," 1990, author's possession.

craft. Such a policy, however, demanded development of an optimum landing craft, not only for assault infantry, but for tanks and vehicles essential for establishing a beachhead. Marine officers assigned to amphibious development in the 1930s despaired that the Navy Bureau of Construction and Repair (later called the Bureau of Ships) would ever produce or accept an adequate landing craft. After a decade of experimentation with bureau boats, the Navy Department accepted a marine corps proposal to invite commercial competition, but this ploy produced nothing much better by 1938.

In the meantime, the small navy office for landing craft reopened stalled negotiations with Andrew Higgins, a Louisiana boat-builder who had offered one of his thirty-foot "Eureka" boats for testing in 1936. Although it did not have a bow ramp, the "Eureka" had excellent power for its size, even after being enlarged to thirty-six feet, and it had shallow-water handling capabilities. Its light draft forward allowed it to ground and retract without difficulty, an essential feature in the Gulf bayous. Tested by the navy and marines in 1939 the Higgins boat won universal praise, and in 1941 Higgins, prompted by pictures of the *daihatsu*, redesigned the boat for a bow ramp. At the same time, the navy worked with its own and Higgins's designs to produce a tank lighter, the Landing Craft Mechanized (LCM), which was also transport-borne.[39]

Confronted by coral reefs in the central Pacific, the marine corps pursued a special piece of equipment to supplement the navy's landing craft, an amphibian tractor. In 1923 the marine corps investigated an amphibian tank complete with propellers, designed by J. Walter Christie, the brilliant but stubborn inventor of armored vehicles. Tested by the marines in the landing exercises of 1924, the Christie amphibian, although rejected for mechanical fragility, whetted marine appetites for an amphibian tractor. They found another possibility in a tracked vehicle designed by Donald Roebling, a third-generation inventor and engineer then living in Florida. Roebling had built his prototype "alligator" for swamp rescues; the vehicle featured a light-weight buoyant aluminum hull and tracks that could propel it through water or across soft earth. It had to be strengthened and reengineered for military use, but from 1937 until its adoption in 1940, the Roebling tractor and its inventor developed a sound working relationship with the marine corps. Both the Japanese and British

---

[39] Clifford, *Amphibious Warfare Development*, pp. 108–117; Maj. John W. Mountcastle, "From Bayou to Beachhead: The Marines and Mr. Higgins," *Military Review*, 70, March 1980, pp. 20–29; Jerry E. Strahan, *Andrew Jackson Higgins and the Boats That Won World War II* (Baton Rouge, LA, 1994).

armies investigated amphibian tanks, but neither pursued the equipment with any sense of urgency, and neither had an operational amphibian in service in 1941 when the marine corps took delivery of the first one hundred of the 18,000 American-built amphibian tractors that served in World War II.[40]

The use of aviation for amphibious operations fell into two functional categories: 1) to gain air superiority over the land and sea space of the amphibious objective area, and 2) to conduct offensive air operations against enemy targets. While the development of naval aviation moved slowly in interwar Britain, Japan and the United States created carrier-based and land-based naval aviation units to protect or to attack fleets and naval bases. No responsible amphibious planner thought air superiority an insignificant requirement; the persistent challenge came in finding a way to move air units within range of an amphibious objective area, whether by establishing air bases nearby or deploying carriers. Understandably nervous about air attacks on transports laden with troops and equipment, naval aviation planners tended to think of amphibious operations in the same way they thought about fleet action. The naval aviators of Japan and the United States, for example, approached this problem with similar doctrinal conclusions. In contrast, Army Air Corps and Japanese Army aviators or those of the independent Royal Air Force gave hardly any thought to amphibious operations.

The aviation component of the U.S. Marine Corps – two small air groups in the 1930s – thought a great deal about amphibious operations because the Commandant told them to do so. Marine aviation had a unique status. Its personnel were marines, subject to corps policies on assignments, promotions, training, and operations, but its officers were designated "naval aviators" and certified as such by the navy's aviation training establishment. Aircraft were the same as in the navy, and money for them came from the navy budget, known in Washington as "blue dollars." The marine corps, for example, had a senior aviator stationed at Headquarters Marine Corps, to manage aviation policy, but he paid as much or more attention to the chief of the navy's Bureau of Aeronautics as to the Commandant. The navy's aviation leaders maintained fairly clear ex-

---

[40] Col. Victor J. Croizat, *Across the Reef: The Amphibious Tracked Vehicle at War* (London, 1989); Maj. Alfred D. Bailey, *Alligators, Buffaloes, and Bushmasters: The History of the LVT through World War II*, "Occasional Paper," (Washington, DC, 1986); George F. Hofmann, "The Marine Corps and J. Walter Christie: The Development of the Amphibian Tractor in the 1920s," paper presented to the Society for Military History, April 1992.

pectations for marine aviation squadrons: they were to carry the burden of aerial defense for established and advanced naval bases. This mission thus freed the navy's aviation squadrons for the role they most sought, service aboard fleet carriers and participation in great naval battles. Marine aviators received carrier training, too, but they did not expect to participate in fleet engagements, only to perform their "advanced base mission" of land-based air defense.

Even before 1914 some marine aviators began to recognize the contribution that air attacks could make to land campaigns, and they viewed World War I as an important example of the potential of tactical aviation in ground combat. These lessons, they thought, received validation in marine aviation operations in Haiti, the Dominican Republic, China, and Nicaragua. But marine aviators focused on their place in a naval campaign, and some of them concluded that they should create a unique role, close air support of amphibious assaults. This fire support role did not necessarily release marine squadrons from the performance of the general tasks of all naval aviation; it simply gave them a task in which only they specialized. Their aircraft were no different from the navy's, but their pilots had to serve in the ground marine corps before attending flight school, and their special interest was the attack upon beach defenses.

The problem of fire support for an opposed amphibious landing could not be wished away. Unless an enemy proved especially inept, poorly led, and deployed with carelessness, an established defense had inherent advantages over a landing force that only overwhelming supporting fire could negate. The artillery and small arms of the landing force, fired from boats or neighboring islands, was not likely to provide the edge. Naval gunfire – high velocity and flat trajectory fire whatever the weight of shell – could not destroy reverse slope defenses and might easily miss well-constructed fortifications on flat terrain like Pacific atolls. Dive-bombing by marine aviation provided the answer to targets that eluded naval gunfire.

The problem then became the proper identification of targets, which might require radio communications from ground observers to attack aircraft, and the coordination of ground tactical action with supporting air attacks. The navy solution to the problems of close air support reflected practices designed to control naval gunfire in fleet engagements: airborne spotters in seaplanes or carrier aircraft controlled the strikes. The marine corps did not believe that this method was adequate and thought instead of control from the ground, exercised by aviation experts attached to the ground forces. This system would only work, however, with marine-trained squadrons that could reach the amphibious objective area, and such squadrons were likely to be navy units based on carriers, not marines.[41]

THE TEST OF WAR: AMPHIBIOUS OPERATIONS, 1940–1942

The audit of war found all three of the great naval powers wanting in varying degrees, and their experience with amphibious warfare brought disappointments proportional to their prewar interest. With less time to adjust and a greater gap between requirements and forces, the British had the most unpleasant tutorial, but under the whip of Churchill they pioneered the creation of specialized techniques and equipment for landings, the creation of joint staffs, and the orchestration of naval forces and landing forces. Yet the British experience demonstrated that one cannot conjure specialized military capabilities into existence overnight, not just in scale but in the expertise and quality of operational management that comes only from time-consuming hard training and trial and error.

For the Japanese, their prewar investment paid off in the dazzling success of the Western Pacific campaign of 1941. Yet within that campaign the Japanese demonstrated liabilities that cost them the initiative in the South Pacific in 1942. For the United States the outbreak of war in 1939, especially the Allied collapse of 1940, galvanized rearmament and preparedness, which included the amphibious forces. The lack of Allied air and naval control in every theater until late 1942 meant that the first prerequisite for amphibious operations could not be met, thus sparing the premature commitment of American amphibious forces. Even their first use in the South Pacific and North Africa dramatized that much remained to be done to bring those forces to full efficiency.

Among its many defeats in 1940, Britain could attribute only one to its ineptness in joint operations, the Norwegian campaign of April–June 1940. Ironically the campaign paved the way for its author, Winston Churchill, to move from his position as First Lord of the Admiralty to prime minister. The Allied difficulties in taking and holding Narvik, Trondheim, Stavanger, and Bergen were more a result of the air and the naval aspects of the campaign, not its amphibious elements. Yet German

---

[41] Archibald D. Turnbull and Clifford L. Lord, *History of United States Naval Aviation* (New Haven, CT, 1949); Lt. Col. Edward C. Johnson with Graham A. Cosmas, *Marine Corps Aviation: The Early Years 1912–1940* (Washington, DC, 1977); Marine Corps Schools, *A Text on the Employment of Marine Corps Aviation*, 1935; Headquarters Marine Corps, *Marine Corps Aviation General – 1940* (Washington, DC, 1940).

aerial superiority represented the key to the campaign, and it rested on the daring and rapidity with which German seaborne and airborne ground forces seized and held Norwegian airfields. The Royal Navy had ravaged the surface forces of the *Kriegsmarine* before the Luftwaffe established itself ashore, and amphibious counterlandings might have slowed or even halted the German conquest of southern and central Norway. Moreover the Allies squandered their forces in the extended battle to seize and then defend Narvik, which they ultimately abandoned. Of all the campaigns in Western Europe in 1940, only the Norwegian campaign was inherently naval, and the British amphibious participation proved too little and too late. Although German naval losses were severe and the flight of the Norwegian merchant fleet was a plus for the Allies, the long-term impact of Norway's loss meant that German air and naval forces could operate with much greater freedom and effectiveness against Allied convoys in the North Atlantic and along the route to the Soviet Union. The seizure of Norway also insured the closing of the Baltic and the free transportation of Scandinavian raw materials to German factories.[42]

Stung by his Nordic Dardanelles, Churchill, ever given to unconventional thought and action, dramatically elevated amphibious operations within the British command hierarchy. He changed a Chiefs of Staff initiative to appoint an adviser for raiding and amphibious operations into a full-blown independent command, Combined Operations Headquarters, first under Admiral of the Fleet Sir Roger Keyes and then Admiral Lord Louis Mountbatten. With Churchill's firm support, the Combined Operations Headquarters developed an empire of construction programs, landing force development schemes for equipment and techniques, and training centers for commando raiding units and more conventional ground forces, including specialized armored units.

Although Churchill stressed the raiding mission early in the war, he expected Combined Operations Headquarters to prepare for a major invasion of France. In the meantime, Churchill and the bolder souls among his military family considered all sorts of amphibious raiding operations along the entire rim of the Axis – from Norway to Western Africa, from Vichy France to the Middle East, and from the Mediterranean to Madagascar. The most memorable of these enterprises, so worthy of Pitt in conception and of Falstaff in execution, was the amphibious raid on Dieppe

---

[42] Arthur J. Marder, "'Winston is Back': Churchill at the Admiralty, 1939–1940," in Arthur J. Marder, *From the Dardanelles to Oran: Studies of the Royal Navy in War and Peace, 1915–1940* (London, 1974), pp. 105–178. For a recent assessment of the Norwegian campaign, see H.P. Willmott, *The Great Crusade* (New York, 1989), pp. 70–80.

in August 1942, a learning experience to be sure, but an extremely cost-ly one to the Allied landing force, which lost more than half its 6,000 of-ficers and men. The vigor of the German defense froze amphibious tac-tics; direct assaults upon prepared defenses still looked suicidal, and the "commando concepts" of stealth, surprise, and small forces still domi-nated planning.[43]

The Japanese armed forces proved in their southern operations that they could 1) establish air and naval supremacy and 2) put landing forces ashore in speed and numbers that confounded Allied defenders. Whether the land-ing forces came from the Japanese Army, which put about one-third (eleven divisions) of its field force in the campaign, or from the Naval Special Landing Forces, the results were largely the same. Japanese aircraft, war-ships, and sometimes land-based artillery pummeled the objective area but not for long if surprise was a major factor. Then landing craft and barges brought the troops ashore at multiple, narrow landing sites distributed along long stretches of coastline. When possible, small craft brought troops inland by rivers and estuaries, and the Japanese launched parachute assaults in conjunction with the landings when the transports and troops were available. The grand tactics of infiltration and exploitation worked well, especially in confounding a larger Commonwealth army in Malaya. The Japanese continued to regard night operations as preferable unless they were confident of overwhelming fire superiority and adequate troop numbers. Japanese landing forces swept into Luzon, Hong Kong, Malaya, Guam, the Solomons, New Britain Island, and the Dutch East Indies and routed Allied defense forces without serious losses to their amphibious ex-peditionary forces and their supporting air and naval units.[44]

The Japanese, however, also proved that they could make errors, usu-ally resulting from their own hubris as warriors and their contempt for the soldierly qualities of their western opponents. They tended to attempt too much, with too little, and too quickly, and their operational plans ac-cepted a degree of risk and planned complexity that carried within them the seeds of interservice conflict and disaster. Amphibious operations were no exception, and in fact, official doctrine reinforced this ten-dency. The campaign against the Americans had its upsets, however tem-porary. The first landing attempt against Wake Island on 11 December 1941, resulted in disaster; U.S. Marine aircraft and coast defenses ruined

---

[43] Fergusson, *The Watery Maze*, pp. 70–185.

[44] Hayashi and Coox, *Kogun*, pp. 29–46; Suekuni, Section 13, "Landing Operations"; Will-mott, *Empires in Balance*, pp. 130–397; Paul S. Dull, *The Imperial Japanese Navy (1941–1945)* (Annapolis, MD, 1978), pp. 21–29.

the landing before the troops could even disembark from their transports. Even when the atoll finally fell, the marines on one island wiped out the Japanese landing force. During the Philippines campaign, the Japanese attempted to envelop the American defense line on Bataan in January–February 1942 with four different amphibious landings, all of them unsuccessful. In part because of the limitations of their amphibious ships and craft, the Japanese landed too few troops and failed to give them close artillery and armoured support, a problem the British, for example, fully appreciated.

In summer 1940, the United States had to face the twin possibilities of war with Japan and the spill-over of the war in Europe into the Western Hemisphere. American planners needed to consider the prospect that the victorious Nazis might jump the Atlantic to Vichy French bases or to ones deeded them by their Latin American admirers. The possibility of pre-emptive amphibious landings in the Caribbean or the Western Atlantic brought requirements for troops that the marine corps alone could not fill. By 1941 the Army-Navy Joint Board, which became the Joint Chiefs of Staff, had arranged for army divisions to undertake amphibious training as part of two joint training forces, one on each coast.

Eventually, four army infantry divisions participated in the training, and the Joint Training Force Headquarters, commanded by two marine major generals, evolved into amphibious corps headquarters for the Atlantic and Pacific Fleets. The training exercises revealed serious shortcomings in transports and landing craft, and the army, in fact, established its own amphibious training command in June 1942 and began to organize specialized engineer brigades for amphibious operations. The navy, in all truth, had its mind on its own inadequacies in gunnery, antisubmarine warfare, and aerial combat rather than amphibious landings. Senior marine officers like Major General Holland M. Smith had ample knowledge but little influence over fleet exercise priorities.[45]

After Pearl Harbor and six months of frustration, the Joint Chiefs of Staff accepted the idea of an African "Second Front" after prodding from Roosevelt and from the British and started planning for landings in Mo-

---

[45] Isley and Crowl, The U.S. Marines and Amphibious War, pp. 58–67; Atwater, "United States Army and Navy Development of Joint Landing Operations," pp. 118–165; Clifford, Amphibious Warfare Development, pp. 144–159; Norman V. Cooper, A Fighting General: Gen. Holland M. "Howlin' Mad" Smith (Quantico, VA, 1987), pp. 57–86; Joint Landing Force Board, "Study of the Conduct of Training Landing Forces for Joint Amphibious Operations during World War II," May 1953, Command File World War II, Operational Archives, Naval Historical Center; Operations Division Correspondence, 1942–1945, Records of the War Department General Staff, RG 165, NA.

rocco and Algeria. At the same time, Admiral Ernest J. King decided that the time was ripe after the Japanese defeat at the Battle of Midway in June 1942 to open a Pacific war "Second Front" along the eastern edge of the Malay Barrier. He planned to use navy units and a marine division to defend the logistical lifeline to Australia and New Zealand. This idea was also dear to the heart of General Douglas MacArthur, who longed to start his return to the Philippines from Australia. From August 1942 through the rest of the year, American expeditions explored the state of their combat readiness under fire and experienced a severe test from the Japanese but less so from the Vichy French forces that greeted the November landings in North Africa. In the South Pacific an Australian-U.S. Army task force of two divisions invested the Buna-Gona area on New Guinea's north shore and engaged the Japanese defense forces in a brawl reminiscent of World War I. Both forces suffered serious losses from disease and logistical hardships over the course of the offensive. The same conditions developed on Guadalcanal in the southern Solomons, invaded by a marine division on 7 August 1942, except this time Americans held the base area, the Henderson Field enclave. The Americans in both places proved they could move forces by sea and support them and, in the case of Guadalcanal and companion landings on Tulagi and Gavutu, stage an amphibious assault against opposition.[46]

The two Allied offensives in the South Pacific are important for revealing what the Japanese could *not* do. To relieve forces meagerly supplied and reduced by casualties and illness, they could *not* stage a major counterlanding, a concern for the local American commanders. The Japanese capability to introduce new troops to both battles was strictly a logistical effort and did not represent on operational concept that altered the fundamental battle of attrition. At Buna-Gona the Japanese reinforced their established defensive positions, and on Guadalcanal they deployed the better part of two army divisions to launch three separate attacks over three months at the 1st Marine Division's perimeter. In both campaigns the Japanese had or could have employed ample naval and air forces to cover an amphibious landing; in the Solomons, in fact, they badly mauled the U.S. Navy, especially in night actions. But the Japanese did not have an amphibious force that could make an opposed landing against determined troops. Had they possessed such a force they might have struck a decisive blow against the American landing forces. In the dark days of

---

[46] Richard B. Frank, *Guadalcanal* (New York, 1990); Jay Luvaas, "Buna, 19 November 1942 – 2 January 1943," in Lt. Col. Charles E. Heller and Brig. Gen. William A. Stofft, eds., *America's First Battles, 1776–1965* (Lawrence, KS, 1986), pp. 186–225.

1942 an Allied defeat in the South Pacific might have influenced the course of the Pacific war.

The landings in North Africa provided somewhat different lessons because the French mounted no serious air and naval challenge and waged only sporadic battle against the landing forces. British thinking about amphibious warfare influenced the landings, which focused on the ports of Casablanca, Oran, and Algiers. All three succeeded but not without significant problems. Night landings and airborne drops dispersed forces and injected unnecessary confusion. Artillery and tanks arrived slowly, and, two Dieppe-style *coups de main* ended in disaster in the inner harbors of Oran and Algiers. In Oran almost 500 American soldiers died or fell prisoner, and a similar size group of proto-rangers surrendered in Algiers. In actions elsewhere an errant shell from an American cruiser inflicted the only Allied casualties, while disarray among the troops and beach control groups allowed landed supplies to go astray.

The commanding general of the Moroccan force, Major General George S. Patton, Jr., landed on the beach at Fedala and immediately wished he were somewhere else, but he did report that "the performance of the navy . . . has been of the highest order. I am amazed at their efficiency, and I am delighted at the wholehearted spirit of cooperation they have evinced."[47] All told, the Allies lost 1,181 killed and missing in two days of fighting, about the same number of combat deaths suffered by the 1st Marine Division in four months on Guadalcanal. Clearly the Allies had some more training to do to improve their amphibious skills, which in any event were not needed for almost a year until the landings on Sicily. The marines and navy also had an additional learning experience ahead, the first atoll assault at Tarawa in November 1943. This bloodletting led to a revision of naval gunfire and air support doctrine as well as to the accelerated production of amphibian tractors.

The amphibious forces of Britain, Japan, and the United States all required changes in doctrine, organization, communications, weapons, equipment, training, and logistics as World War II dragged on, but the Allies, shifting to the offensive in 1942, had greater incentive and greater resources to make the necessary changes. Neither Germany nor Japan could be defeated without major landings in 1944, operations of a scale and

---

[47] Maj. Gen. George S. Patton, Jr., to Maj. Gen. A.D. Surles, 6 November 1942 and diary entries, 8 November 1942 and 9 November 1942 in Martin Blumenson, ed., *The Patton Papers* (2 vols., Boston, 1974), II, pp. 101–102, 103–106, 108. The landings are described in detail in George F. Howe, *Northwest Africa: Seizing the Initiative in the West* in the official history series *The U.S. Army in World War II* (Washington, DC, 1957).

scope unimaginable in 1942. Japan, by contrast, lost its amphibious capability in 1942 when it shifted to the strategic defensive. Japanese planners in both services apparently did not believe that amphibious forces had an offensive role to play within the general concept of a defensive naval campaign. Made by default, such a decision probably was ill-advised before 1944 because an amphibious counterstroke in the South Pacific in summer 1943 might have seriously disrupted Allied planning in that theater, a theater already plagued by inter-Allied and interservice disputes. In the end, British and American amphibious know-how provided an operational capability that was unmatched by the Axis and essential to the Allied victory.

CONCLUSION

The interwar development of amphibious warfare by Britain, Japan, and the United States reveals several common threads that explain each nation's limited success in developing the operational specialty. Each nation shared the same geopolitical characteristics: all avoided alliances until the eve of World War II, so they had to provide all of the armed forces they thought they required. As maritime nations all three exercised military power to a large degree through their navies; all three nations, however, had to maintain ground forces to occupy and defend territories away from their home territory. All three nations recognized the potential of military aviation; Britain's need for both offensive and defensive air forces, of course, became critical in the face of the Luftwaffe's challenge in the 1930s. Neither Japan nor the United States faced a comparable threat. Judged against the course of World War II, all three nations might have profited from a greater investment in amphibious forces, but such investment depended upon the relative balance of investment in air, naval, and ground forces, and it is difficult to fault the broad investment strategies adopted by the United States and Britain, both of which viewed air and naval forces as more important than a large ground force. Japan, on the other hand, entered World War II with too little navy for the protracted war that lay ahead. The quick war assumptions and "continentalism" of Japanese planners doomed the imperial armed forces to a high-risk strategy with limited forces, which produced dramatic operational swings from caution to impetuousness.

Yet given the nature of the resistance they faced in 1941–1942 and the success of their landing operations, the Japanese demonstrated they had more than adequate amphibious forces against the opposition they *then* *faced*. But their successes depended upon inept Allied air and naval defensive operations, which could be and were corrected with larger forces, better planning, and more skilled commanders. As Allied operations improved in late 1942, the Japanese found they were not fully prepared to conduct an aggressive strategic defense of their newly won islands. Time and again they suffered prohibitive losses because they could not seize or recapture bases for their air and naval forces. The ground forces they committed had little mobility and endured crippling logistical shortages. The Japanese had the novel experience of losing their amphibious assault capability after using it so well in the early stages of the Pacific war. This capability disappeared by institutional decisions, not enemy action. The landing forces became land-bound defensive garrisons, and the amphibious shipping shifted to general resupply missions. In a sense the Japanese amphibious forces committed collective *sepuku* after their greatest victories, largely for want of prudent strategic vision.

The Japanese experience reflected a factor that applied to the British and Americans, i.e., having staff proponency, doctrinal publications, and a place in the officer educational system does not ensure that a developing operational specialization will reach fruition. There must be a foundation in institutional commitment, and a major organizational embrace of a new mission. The British military establishment had many splendid ideas about amphibious operations, but the thinkers had no firm organizational foundation and funding. Only the Inter-Services Training and Development Centre met this requirement, and its modest mandate produced equally modest accomplishments in landing craft design. The Japanese problem was different and the same; its amphibious development followed service lines and service conceptions of strategic requirements. The Japanese Army had its own amphibious force, the navy another. Only in the opening campaign of the Pacific war did they work in concert, and even then these forces worked simultaneously but most often separately in accordance with their missions and operational capabilities. Neither the army nor the navy viewed amphibious forces as anything other than forces of limited utility and peripheral to their services' real functions, which were war on land and sea, not the projection of military power from the sea to the land. The United States, on the other hand, had a politically potent independent service – the U. S. Marine Corps – that had a clear wartime mission after 1921, which was to seize advanced naval and air bases in a war with Japan across the Pacific Ocean. Even with the foreign intervention disruptions of 1927–1933 the marine corps did its best to train and equip itself for amphibious assaults, and the senior leadership of the U.S. Navy (at least some of it) cooperated because the admirals understood that base-seizure was a real mission. Few admirals or generals in Britain and Japan showed equal clarity of vision. Only American planners fully appreciated that if amphibious forces prepared to capture and defend objectives of critical military utility (air and naval bases), they could also conduct amphibious operations of a less demanding nature.

3

STRATEGIC BOMBING
The British, American, and German experiences

WILLIAMSON MURRAY[1]

Throughout the twentieth century, fundamental changes in how wars between nations are fought have been influenced by a variety of political, technological, geostrategic, and cultural factors that have been largely external to military services and organizations. For instance, technological changes affecting the weapons and other tools employed in military operations have usually been part of the external environment in which peacetime military innovation occurred, not indigenous to military organizations themselves.

Of all the military innovations during the years 1918–1939, strategic bombing was particularly influenced by these kinds of external factors. There are a number of reasons why this should be so; among other factors the rapid development of aviation technology as well as the catastrophic impact of World War I on Western civilization exercised a crucial impact on the minds of airmen, politicians, and ordinary citizens alike. Such societal and intellectual influences in turn influenced development of the doctrines and conceptions with which military organizations approached the question of strategic bombing.

Much of air power's appeal to military and political leaders lay in its potential to combine physical destruction with the reach and speed to overfly intervening oceans, plains, rivers, and mountains and focus that destructive power against the vital centers of the enemy nation in a matter of hours. Above all, this sort of strategic use of air power seemed to offer an escape from another terrible war of attrition on the ground. The-

---

[1] I am indebted to Barry Watts of the Northrop Corporation for his patient and incisive criticisms of the several drafts of this chapter and to Richard Muller of the Air Command and Staff College for his help with the German sections of this essay.

oretically, it would allow those nations that possessed strategic bombing capability to attack the enemy's economic and social base from the opening moments of conflict. The primary instrument of this potentially new form of warfare was, of course, the long-range or heavy bomber. To strategic bombing proponents it opened the possibility of pursuing "objects which were beyond, not only the range, but also the strategy of armies and navies."[2]

The possibilities of powered flight first appeared in the late nineteenth century, but the first powered flights had occurred early in the twentieth. Despite considerable public and military interest,[3] aircraft design had not advanced greatly by 1914. But under the pressure of war, the opposing sides accelerated technological development. By 1918, air forces had experimented with virtually every mission of air power familiar to us today.

Strategic bombing appeared early in the war with Zeppelin raids on southern England. By 1917 the Germans were using aircraft to attack the British Isles, and by the conflict's last year the British were replying in kind with attacks on the Rhineland.[4] Moreover, by 1918 the British were investing heavily in the creation of a strategic bombing force for strikes deep into Germany. Nevertheless, whatever the initial attempts at strategic bombing, both the extent of such attacks and their results left room for considerable debate as to its potential effects on future warfare.

The second influence on development of strategic bombing during the interwar period had to do with the romantic conceptions with which the public and even the military viewed flight. Throughout the interwar period, popular views accorded aircraft, airmen, and flight itself with attributes of almost mythological proportions.[5] Even the Great War's air-to-air combat received an aura of adventure that stood in stark contrast to the brutality and horrors of the fighting in the trenches. Ironically the re-

---

[2] Sir Charles Webster and Noble Frankland, *The Strategic Air Offensive Against Germany 1939–1945*, vol. 1, *Preparation* (London, 1961), p. 7.

[3] Of course, not all senior officers were perceptive on the possibilities that powered flight might offer armies and navies. In 1910 Ferdinand Foch commented on aircraft: "That's good sport, but for the army the airplane is of no use." Robert Debs Heinl, Jr., *Dictionary of Military and Naval Quotations* (Annapolis, MD, 1966), p. 22.

[4] For a general discussion of German raids against Britain in World War I see Francis K. Mason, *Battle over Britain: A History of German Air Assaults on Great Britain, 1917–1918 and July–December 1940, and of the Development of Britain's Air Defenses between the World Wars* (New York, 1969).

[5] For the romanticization of flight in the post-World War I period in Germany see Peter Fritzsche, *A Nation of Fliers: German Aviation and the Popular Imagination* (Cambridge, MA, 1992).

ality of aerial combat was anything but romantic.[6] For every Ball, Rickenbacker, Immelman, Fonck, or Richtofen – and most of the great aces were dead by 1918 – there were thousands of young men, killed or maimed in their first combat missions. But reality has rarely gotten in the way of myth and romance; to most in the 1920s and 1930s, civilian as well as military, flight represented a romantic avenue to the future.

The third element that shaped strategic bombing's development between 1918 and 1939 was technology. For most of the 1920s change was gradual, but in the 1930s the pace of technological change in aviation began to accelerate. Admittedly, this situation reflected increases in defense budgets as the arms race heated up. But also involved was solid developmental and experimental work that had occurred in the 1920s. Among such changes altering air power's potential were drastic improvement in air frames as metal replaced wood-and-canvas construction, significant redesigns of wings, and improvements in power plants – all of which brought great increases in speed, payload, and range. By early September 1939 the Germans were flying the first jet aircraft. If such dizzying technological change were not enough, the environment of air combat was changing with introduction of radar and coordinated air defense systems. Such an era of revolutionary change was not conducive to coherent innovation, particularly after a long period of minimal funding. Nevertheless, air forces innovated with considerable success despite the difficulties. If strategic air war looked quite different from prewar concepts in reality, that was largely due to its complexity and the relative paucity of prewar experience.

THE STRATEGIC AND POLITICAL ENVIRONMENT

The nature of air war

The same broad factors that govern war on land and sea work on those who conduct war in the air. Friction, ambiguity, and uncertainty all affect air operations. Nor are airmen freed from the demands of training and logistics. Yet there are aspects of war in the air that make analysis and understanding of air operations substantially more difficult than those on land or at sea. There is a temporal and terrestrial framework to operations on land and at sea that is not evident in air operations. The loss or

[6] For the most recent and thorough discussion of the war in the air see John Morrow, *The Great War in the Air, Military Aviation from 1909 to 1921* (Washington, DC, 1993). For a general discussions of British losses on the Western Front during the war see Malcolm Cooper, *The Birth of Independent Air Power, British Air Policy in the First World War* (London, 1986).

occupation of territory on the ground, for instance, has no obvious correlate during an air battle.

Aerial combat often dissolves to the smallest common denominator, while an inherent chaos, speed, and lack of landmarks make it difficult for even the participants in any action to discern what has happened. And much of air power's effects deal with intangibles. In air war survivors return to their bases, but what they have achieved, at least for the immediate present, is in most cases only calculable in changes in percentages of aircraft destroyed and pilots lost.

Peacetime conditions only increase such ambiguities and uncertainties. Can bomber formations actually defend themselves in combat? Can fighters break up bomber formations? Under combat conditions will pilots and bombardiers be able to hit their targets accurately? Will bombers even be able to find their targets? Such questions bedeviled those who wished to think seriously about the employment of air power in the interwar period. The imponderables of peacetime practice and exercises only magnified the uncertain lessons of World War I. Consequently those who innovated in the interwar period had to confront large numbers of ambiguities and uncertainties.

The nature of strategic bombing

Before we look at innovation and strategic bombing in the interwar period, it is useful to examine the complexities of the Combined Bomber Offensive as the RAF and the U.S. Army Air Forces were waging it in 1944. By January 1944 Bomber Command was into its fifth year of flying over Germany and its fourth year of major raids against German targets. In that month, the command possessed 1,224 bombers, of which over 1,000 were Lancasters, Mosquitoes, or Halifaxes. Over the course of the year Bomber Command lost 3,220 aircraft.[7] Yet the command dispatched approximately 9,000 aircraft per month throughout the year against targets in Germany and elsewhere on the continent.[8]

Bomber Command's crew losses were, at best, depressing; the night bombing offensive against German cities and transportation targets between 3 September 1943 and 2 September 1944 cost the command 17,479 flying personnel killed in action or dead of wounds.[9] Despite this terrible

---

[7] PRO AIR 14/3489, War Room Manual of Bomber Command Operations, 1939–1945, Air Ministry War Room (Statistical Section).

[8] Sir Charles Webster and Noble Frankland, *The Strategic Air Offensive Against Germany, 1939–1945*, vol. IV, *Annexes and Appendices* (London, 1961), appendix 40, p. 433.

[9] Ibid., appendix 41, "Bomber Command Casualties," pp. 440–443.

rate of aircrew casualties and aircraft losses, the command remained a functioning and effective military force capable at times of dealing terrible damage to its targets.

Eighth Air Force, the American strategic bombing force based in England, was carrying an equally heavy load by early 1944 despite the fact that it had begun major operations less than a year earlier. In February 1944, Eighth disposed of 1,481 bombers in its squadrons, of which 1,046 were fully operational. In addition, its fighter units possessed 883 fighters, with 678 fully operational.[10] For the next four months Eighth Air Force engaged the Luftwaffe in a battle of attrition for daylight air superiority that ultimately hinged on each side's ability to replace its aircrews. That battle cost Eighth Air Force between 20 and 25 percent of its bombers and nearly 30 percent of its crew strength each month. The three months of February, March, and April 1944 saw the command lose 1,057 bombers. Yet despite this heavy attrition, Eighth's base strength continued to grow, eventually reaching 2,427 "on hand" bombers and 2,007 "available" bomber crews in July 1944.[11] By contrast, during the first five months of 1944, the Luftwaffe lost some 2,260 fighter pilots out of an average strength of 2,280, including many of its best squadron, *Gruppe*, and *Geschwader* commanders.[12] Not only was the German fighter force unable to match Eighth's growth in numbers during this period but the overall quality of its pilots dropped dramatically.

The payload capacity that the Anglo-American bomber forces eventually achieved is suggested by the fact that Bomber Command dropped 281,708 tons of bombs on German targets, and Eighth Air Force 192,330 tons (474,038 total tons) during the year.[13] These totals do not even include those for Fifteenth Air Force, which struck at targets in Austria and the Balkans throughout the year.

The above figures partially suggest the massive economic mobilization of industrial capacity and manpower that was necessary for the Anglo-American strategic bombing effort. What they do not illuminate is the enormously complex task of adapting and innovating under the pressures of combat to achieve effective results against the enemy's industrial ca-

---

[10] "Statistical Summary of Eighth Air Force Operations, European Theater, 17 August 1942–8 May 1945," Air Force Historical Research Center.

[11] "Statistical Summary of Eight Air Force Operations: European Theater, 17 August 1942 – 8 May 1945," p. 14

[12] Williamson Murray, *Luftwaffe* (Baltimore, MD, 1985) Table LII.

[13] Webster and Frankland, *The Strategic Air Offensive against Germany, 1939–1945,* vol. IV, appendix 44, "Tonnages Dropped by the R.A.F. Bomber Command and the U.S. Eighth Air Force, 1939–1945," p. 456.

pacity and military forces. By 1944 the American and British bomber forces were relying on a host of complex technological and scientific devices to accomplish their missions. Radio navigation aides, such as "Gee," now gave some hope of arriving in the general vicinity of a target, while radar allowed bombers to drop with some degree of accuracy at night or in bad weather. Long-range escort fighters protected the daylight bomber formations.

The whole effort rested on sophisticated intelligence agencies to estimate not only the bomb damage of air attacks but the enemy's complex economic system and its vulnerable points. Intelligence effort represented much more than an effort to assess aerial pictures of the bomb damage. As we now know, decryption work based at Bletchly Park allowed bomber commanders to look directly into German efforts to repair the damage to the Reich's oil industry occasioned by Allied attacks.[14]

Virtually none of this complexity or the mobilization of resources it required was in the writings or preparations of airmen in the 1920s and 1930s. Given how little experience was available from the air battles of World War I, we should not find this particularly surprising. But it is an instructive point. What should interest us, however, is how realistically airmen innovated in preparation for the next war in the interwar period, given *what was foreseeable at the time*. And secondly, we need also to focus on how realistically they adapted their concepts of strategic bombing to the conditions of war after 1939. Above all we should remember that the bomber forces and efforts of 1944 resulted from the harsh experiences of war and strategic choices made during the conflict; consequently, the thinking and preparations of the interwar period reflected only dimly through war's harsh school.

The British

In relative terms, the British suffered less from World War I than any of the other European combatants. Not until summer 1916 did their manpower become fully involved in the battles of the Western Front; the war

---

[14] In fact the work of Bletchly Park may have been more useful to Allied air commanders in estimating the effectiveness of their raids. Bomb damage to targets could be misleading; Schweinfurt is a good example; the view from the air gave the impression of a target flattened by the bombing. In fact the damage largely consisted of collapsed roofs and walls, while the machine tools sustained relatively little damage. "Ultra" information in 1944 made clear to American airmen how effective their raids were against oil targets and when they needed to reattack such sites due to German repair efforts. See Murray, *Luftwaffe*, pp. 257–261.

never touched British territory; and while British civilians suffered some privation, their sufferings paled in comparison to those of Germany, Austria-Hungary, France, and Italy; Britain's population remained relatively well-fed and tended to the end of the war.[15]

Nevertheless, the British public reacted with intensity to the war; the mass of antiwar literature that deluged the reading public in the late 1920s reflected a national rejection of the policies that had involved the country in the last war. Such attitudes also placed the British services, the Royal Air Force (RAF), the navy, and especially the army in the position of wayward children, rejected by much of the society that they were to protect.[16] The RAF, however, was the least affected by the postwar anger. At the least, it escaped blame for the Western Front or failure to win another "Glorious First of June" in the North Sea. Moreover, the advent of air power had changed Britain's strategic situation: for the first time since Napoleon, the British Isles were vulnerable to a direct attack by an enemy. There was great fear among British civilians that Britain was vulnerable to a "knockout blow." Reinforcing such attitudes was a generation of civilian propagandists who wrote voluminously about the air threat.[17]

The politicians reinforced such fears in public pronouncements. Stanley Baldwin warned the British people in 1932 (before Hitler came to power) that "I think it is well for the man in the street to realize that no power on earth can protect him from being bombed. Whatever people may tell him, the bomber will always get through."[18]

The appearance of the Luftwaffe only served to reinforce such fears. In April 1938 Samuel Hoare commented to his colleagues in the Cabinet on an army report that urged commitment of British troops on the continent:

> The impression made [on him] . . . by the report was that it did not envisage the kind of war that seemed probable. In a war against Germany our own home defenses would be the critical problem. . . .The problem

[15] For the best general study of the impact of World War I on Britain see Sir Llewellyn Woodward, *Great Britain and the War of 1914–1918* (Boston, 1967).

[16] See the essay in this study on the development of armored doctrine for a more detailed elaboration of this point.

[17] Among a whole host of publications that predicted a terrifying end of western civilization through air war see P.R.C. Groves, *Our Future in the Air: A Survey of the Vital Question of British Air Power* (London, 1922), and *Behind the Smoke Screen* (London, 1934); J. Griffin, *Glass Houses and Modern War* (London, 1938); and L.E.O. Charlton, *War from the Air* (London, 1935), *War over England* (London, 1936), and *War from the Air* (London, 1938). All of these authors presented future air war in terms of smashed-up cities, rioting civilians, and the collapse of government.

[18] Keith Middlemas and J. Barnes, *Baldwin: A Biography* (London, 1969), pp. 731–736.

was to win the war over London! . . . We should need in the initial stages all our available troops to assist in the defense of this country [to keep the population from collapsing in the face of German bombing attacks].[19]

The RAF regarded such currents with some ambivalence.[20] On one hand senior officers were delighted to receive public support in interservice arguments. On the other, in the early 1930s the RAF confronted a political establishment interested in achieving international accord at the Geneva Disarmament Conference. Such agreements would have banned the bomber and eventually disestablished the RAF – at least until the next war.[21]

But the most serious problem the RAF confronted lay in underfunding of the British defense establishment. The lack of financial support had the greatest impact on the RAF's ability to innovate with the technological changes that cascaded in the last half of the 1930s.[22] By the onset of rearmament, Britain's aircraft manufacture had degenerated into a cottage industry with neither the technological nor production expertise to produce modern aircraft or increase production. Increased funding in 1939 eventually allowed production of enough modern fighters – although just barely – to fight the Battle of Britain to a successful conclusion. But by limiting RAF spending in 1937 on bombers, Chamberlain insured that until 1943 Bomber Command failed to receive technologically up-to-date aircraft needed to wage a strategic bombing campaign.[23]

Underfunding combined with other substantial problems to exacerbate

---

[19] PRO CAB 24/276, CP 94(38), "Staff Conversations with France and Belgium, Annex II," CID, Extract from Draft Minutes of the 319th Meeting, 11.4.38., p. 157.

[20] It is worth noting that the RAF at times did feed such fears among the politicians. As the air staff warned in a general strategic survey of the situation on the continent produced in spring 1938: "In point of fact, the scale of attack which Germany could direct upon us in April of this year (1938) is unlikely to be less than 400 tons [of bombs] per day, but, so far from our defense preparations being similarly advanced, they are, in fact, very far from complete. The danger to the country must, therefore, be considered as correspondingly greater than we thought it would be when we recorded the above opinion [for 1939]." PRO CAB 53/37, COS 698 (Revise), Committee of Imperial Defense, COS Sub-Committee, "Military Implications of German Aggression against Czechoslovakia," 28.3.38., pp. 145–146.

[21] See the discussion of the British policy during the Geneva disarmament conference in Malcolm Smith, *British Air Strategy Between the Wars* (Oxford, 1984), pp. 112–121.

[22] For the underfunding of the RAF see ibid., pp. 111–112.

[23] The impact of this decision on the flying squadrons of Bomber Command is graphically described by Max Hastings, *Bomber Command* (London, 1979). See also Webster and Frankland, *The Strategic Air Offensive against Germany*, vol. 1, *Preparation*.

the RAF's difficulties in the late 1930s. Having drastically underfunded the RAF, the British government reversed course and stuffed accelerating resources and demands for expansion down the throat of the air staff. As rearmament gathered steam, the result was a steady decline in operational capabilities. Arthur Harris, future commander of Bomber Command, remarked in 1937: "We cannot run a highly complicated and technical business on the floating population of a casualty ward."[24]

The RAF gained its independence in 1918 – due to public outcry over German bombing attacks on London.[25] Thus, the tenuous support that the services received from British governments through to March 1939 placed the new service in a vulnerable political position. With the minuscule defense spending of the 1930s, the army and navy looked hungrily at the funding that the RAF was receiving. Both happily argued they could do the job more effectively and cheaply, if they possessed their own air support. If the RAF missions were only to support the army on the ground and the navy at sea, then why should Britain allocate resources to a service that had no independent mission?[26]

The danger seemed considerable to the RAF; at times the senior services more than lived up to RAF paranoia that they had designs not only on its budget, but its existence as an independent service. The response by the first postwar chief of staff of the RAF, Sir Hugh Trenchard, was to push strategic bombing as the RAF's central mission – at least in interservice arguments. There was some irony in the stridency of his arguments, because the government itself appears to have been convinced of the RAF's legitimacy by its successful and inexpensive suppression of colonial troubles in Iraq.[27]

Trenchard's greatest contribution to British history lay in saving the RAF from dissolution during a period (the 1920s) when there appeared to be few strategic, operational, or technological reasons why Britain should maintain the only independent air force in the world. Equally important was the fact that Trenchard actively identified and pushed the careers of airmen who provided the RAF's leadership in World War II. Sig-

---

[24] Quoted in Smith, *British Air Strategy Between the Wars*, p. 276.

[25] For the sharpest discussion of the birth of the RAF see Cooper, *The Birth of Independent Air Power*.

[26] For an interesting discussion of the squabbles among the services see Smith, *British Air Strategy between the Wars*, pp. 22–28.

[27] In fact many of the most outstanding airmen in the RAF during World War II won their spurs in the efforts to suppress the unruly Iraqi tribes: among others Tedder, Coningham, and Harris.

nificantly, there was no pattern to his choices, except that he picked officers of strong, independent turns of mind; his choices ranged from Arthur Harris to Hugh Dowding, Arthur Tedder, and John Slessor.[28]

By the 1930s interservice relations were even worse than during the 1920s. The placement of Harris, with his penchant for coining wounding phrases, as Director of Plans on the air staff exacerbated bad feelings. But one also senses that bad blood, built up over a decade of underfunding, deeply poisoned interservice relations.

Finally, in understanding the environment within which the RAF operated, one must recognize the heavy contribution that geography – influenced by technology – made to British conceptions of strategic bombing. As an island Britain has always had a certain detachment from continental affairs. Thus, the British have never regarded European battles as the final arbiter of their fate. They could always return to fight another day, however badly their ground forces performed. Air power, however, had changed the strategic equation, and in a fashion not entirely clear to British leaders. Nevertheless, the air situation left little room for complacency. As Baldwin suggested in 1934 to the House of Commons: "Let us not forget this: since the day of the air, the old frontiers are gone. When you think of the defense of England, you no longer think of the White Cliffs of Dover; you think of the Rhine."[29]

The RAF proved of two minds on the issue of a continental commitment. Most senior officers recognized that bases in the Low Countries and Northern France were indispensable for bombing Germany. Moreover, if the Germans seized this region, they would acquire bases that would improve their prospects for a strategic bombing campaign against the British Isles. But as aircraft capabilities improved in the late 1930s, some on the air staff pulled away from a belief that Belgian and Dutch bases would be necessary for bombing the Reich. Not surprisingly, Harris was a chief advocate of this line of reasoning; he argued that army claims about the Low Countries' strategic importance represented an opening wedge for a wholesale continental commitment.[30] Here air staff views found a recep-

---

[28] Trenchard had a reputation for sacking squadron and wing commanders who displayed any qualms about pushing their crews into the breach throughout the course of World War I. Nevertheless, Dowding, because of the catastrophic losses that his squadron had suffered in the air battles over the Somme, had gone to Trenchard and asked that his unit come out of the line. Neither then nor later did Dowding's career suffer.

[29] *Hansard*, vol. 292, 30 July 1934, column 2339.

[30] Smith, *British Air Strategy between the Wars*, pp. 87–88.

tive audience from the appeasers; until February 1939 Chamberlain refused to fund the army at the levels required by a continental commitment.[31]

In the end, Britain's strategic position was conducive to thinking about,
but not necessarily through, the innovations necessary to create a strategic bombing capability. The British confronted a general malaise, as well
as severe economic and strategic problems that offered of no easy solutions. The German air threat came late in the game and made coherent
and thoughtful innovation difficult. Nevertheless, the strategic environment does not fully explain either British failures or successes.

The Americans

In many ways the American environment mirrored much of the framework within which British airmen developed their concepts of air power.
Despite the fact that the United States entered World War I at its end and
suffered few casualties, American society rejected military commitments
and disparaged the military even more deeply than did the British.[32] The
return to "normalcy" also reflected the traditional American view of the
international arena, a view largely influenced by the removal of the United States from world affairs by two great oceans. The American polity
found it difficult to recognize dangers in what Neville Chamberlain would
characterize at the height of the Czech crisis as the "quarrels in a far away
country between people of whom we know nothing."

America's isolation also made it difficult to justify strategic bombing.
Unless the United States were to wage a protracted campaign against either Mexico or Canada, either scenario almost inconceivable, it was difficult for Americans to see a role for bombers beyond that of coast defense.[33] Admittedly the Japanese appeared to be a possible enemy, but it
was easier to see how naval power could play in a war with Japan than
air power, except in the immediate defense of the Philippines. Only as
technology provided significant increases in payload and range did the
possibility of bombing Japan enter the consciousness of military analysts.

---

[31] For a discussion of British attitudes towards the continental commitment see: Michael
Howard, *The Continental Commitment* (London, 1972).

[32] American literature displayed a high level of fury over what it regarded as the waste of
the war and generally echoed the themes of Britain's great antiwar literature. The novels
of Hemingway, Faulkner, and Dos Passos all underline this point.

[33] Allan R. Millett and Peter Maslowski, *For the Common Defense: A Military History of
the United States of America* (New York, 1984), p. 383.

In the early 1930s America's geographic situation led the army air corps to assume part of the mission of coastal protection, which pushed it to support development of long-range aircraft that could reach out into the oceanic expanses. While such a requirement forced technology in terms of weight and distance, it also encouraged the belief that bombers would remain invulnerable to fighter attack, provided they were flown in self-defending formations.[34]

Nevertheless, as in Britain and Germany, American airmen found considerable enthusiasm for air power on the part of their countrymen. Almost from his return from Europe, Billy Mitchell found himself the darling of the media. But no matter how attractive Mitchell or his ideas on air power, Americans saw little reason to devote substantial resources to defense spending until the late 1930s. One of the myths of the interwar period depicts the visionaries of air power struggling "manfully" against a tide of service ignorance and entrenched conservatism – army as well as navy. There is little truth in the picture; the army courtmartialed Mitchell because of his outrageous statements about the "criminal" disregard of air power on the part of the U.S. government, not because of his advocacy of air power.[35]

Moreover, the services were receptive to the more realistic claims of air power advocates. Navy reformers, led by William A. Moffett, found Mitchell a useful foil in pushing the navy's leadership towards serious investment in naval air power. Whatever prejudice existed in the army certainly was not reflected in the enthusiasm with which George C. Marshall viewed air power; after all, the basic plan on which the army air forces fought World War II, AWPD-1, saw life under his authority as army chief of staff. By the late 1930s officers with aviation experience were achieving flag rank in the navy, while Marshall recognized General "Hap" Arnold, chief of the army air forces, as almost his co-equal. Such developments hardly suggest that service conservatism impaired air power innovation or concepts and preparations for strategic bombing.

A great strength of the American environment lay in the development of civil aviation; airlines not only spanned the North American continent, but eventually reached out to South America and across the Pacific and Atlantic. Civil aviation in Europe represented a matter of national prestige; given the enormous distances in the United States, civil aviation was a necessity, particularly as technology advanced in the 1930s. Key was

---

[34] Ibid., p. 384.
[35] For the best discussion of Mitchell's career, see Alfred F. Hurley, *Billy Mitchell* (New York, 1964).

government support for aviation in its widest application. That base of civil aviation in turn provided the military with solid industrial potential as well as technological developments in navigation and air frame construction. Moreover, American industry was particularly well-tuned to the concepts of mass production, of enormous advantage in the contest of industrial production during the coming war.

All things considered, the U.S. environment may have been the most conducive to innovation in strategic bombing. Orson Welles could terrify Americans with tales of alien invaders, but American leaders never came close to the national psychosis that characterized British fears. In the end Germany's European opponents provided sufficient time for America to mobilize its latent economic strength for the great air battles over Central Europe.

The Germans

German innovation in air power was substantially different from the British and American cases. Not only did two substantially different governments (the Weimar Republic and Hitler's Nazi regime) control policy during the interwar period, but Germany had to deal with the strictures of the Treaty of Versailles which forbade possession of military aircraft. In the end Versailles both hindered and aided concepts of strategic bombing in Germany. But at least through the early 1930s the treaty prevented German development of military air power and aircraft; such a situation, however, was not entirely disadvantageous. It prevented Germany from acquiring large inventories of obsolete aircraft, as the Italians and French did. On the other hand, the Germans were to have great difficulty in matching Anglo-American power plants suitable for strategic bombers, a major hindrance in designing and producing a strategic bomber.[36]

German geography, history, and military culture combined to push development of strategic bombing in significantly different directions. Most Luftwaffe officers found the idea of strategic bombing attractive, but Luftwaffe officers also had to recognize Germany's geographic and strategic vulnerabilities. No matter how successful the Luftwaffe might prove in attacking enemy cities or industries – and here the Germans initially thought in terms of Paris, Prague, and Warsaw – if enemy ground forces captured the Ruhr or Silesia, Germany would lose the war. British airmen could discount the loss of Western Europe in 1940, not only because such a loss

---

[36] Edward L. Homze, "The Luftwaffe's Failure to Develop a Heavy Bomber before World War II," *Aerospace Historian*, March 1977.

involved someone else's territory, but because it allowed them to concentrate on a strategic bombing campaign. The Germans confronted a different situation; since their country lay in the center of Europe, the Luftwaffe developed a more broadly based conception of air power, one that included air superiority, interdiction, and even close air support for the army in addition to strategic bombing.[37]

Nevertheless, geography did have a negative impact on aviation theory in Germany. Throughout the interwar period the Germans thought in terms of Central European distances, and the resulting mental framework remained in place until 1941.[38] Moreover, the shadow of catastrophic defeat at Jena/Auerstadt in 1806 hung over all German military culture. That memory gave an operational focus to everything that the German military did and helps explain why the Germans would perform so well on the battlefield in two world wars. But the narrow focus of German military thought on the problems of war in Central Europe helps explain the extraordinary weaknesses the Germans displayed in strategy, intelligence, and logistics in those conflicts.[39]

German society was enormously enthusiastic about flight. Where many British feared the military ramifications of air power, the Germans delighted in its military potential. Denied an air force and even civilian aircraft, at least immediately after the war, the Germans looked eagerly to the rebirth of the Reich's air power. In fact Versailles' restrictions made flight a symbol of Germany's anger over the peace.[40] The rejection of Ver-

---

[37] For a fuller discussion of these issues, see Williamson Murray, "The Luftwaffe before the Second World War: A Mission, A Strategy?" "The Luftwaffe and Close Air Support," and "The Luftwaffe against Poland and the West," in *German Military Effectiveness* (Baltimore, MD, 1992).

[38] That fact helps to explain the difficulty that postwar historians have had in understanding the interests in strategic bombing that many in the Luftwaffe had before the war. At least through 1940, the two-engine bombers that the Luftwaffe possessed seemed more than adequate for strategic bombing attacks against targets like Paris, Warsaw, or London. When, however, the war expanded to the wider dimensions of continental Europe and the Mediterranean, then the Germans found themselves in a situation with which they were technologically unprepared to cope. And that situation became ever more difficult as the war progressed.

[39] Throughout its short history, the Luftwaffe displayed the same disregard of strategic issues that the German army and navy displayed throughout this period. It did no better in the realms of intelligence and logistics. The Luftwaffe's basic doctrinal manual for air war, *Die Luftkriegführung (The Conduct of the Air War)* concludes with the note that the last two chapters on logistics and intelligence remained to be written. In fact, the Luftwaffe's general staff never finished these two sections – a clear indication of its interests and priorities. *Die Luftkriegführung* (Berlin, 1935).

[40] See particularly the wonderful photograph of the Hamburg monument commemorating the impact of Versailles on German aviation in Fritzsche, *A Nation of Fliers*, p. 105.

sailles found its most avid symbol in the national gliding movement, particularly on the Wasserkuppe in central Germany. There thousands of enthusiasts watched soaring gliders in air exhibition with both patriotic and military overtones. A great war memorial overlooked the scene with an inscription that linked military past through depressing present to a glowing future:

> We dead fliers
> remain victors
> by our own efforts
> Volk fly again
> and you will become a victor
> by your own effort.[41]

The Nazi seizure of power in January 1933 signalled a drastic turn in the fortunes of Germany's military. Hitler demanded a massive buildup of the Reich's military forces whatever the political and diplomatic risks.[42] At the same time he set in motion creation of a massive air force under Hermann Göring. Either one of these situations, creation of a new service or a rapid buildup of air power capabilities, would have caused great difficulties; together they represented an extraordinary challenge, but one that the Germans showed considerable ingenuity in mastering.

The Luftwaffe's growth to 15,000 officers and 370,000 men within six years represented an enormous success in any terms.[43] But this expansion created problems at all levels of the officer corps, even more so than with the army, because of the technological demands of the new service. Moreover, Nazification of the Luftwaffe's officer corps occurred to an even greater degree than with the army or navy. While this situation had a limited impact on the Luftwaffe's ability to innovate before the war, it exercised a catastrophic impact on its assessment of the threat during the war. Symptomatic of the influence of ideology on military judgment was an incident that occurred in late 1938. Hitler had just demanded the Luftwaffe expand its operational forces by a factor of five by 1942, an expansion that would have cost the equivalent of the whole defense budget between 1933 and 1939, bankrupted the nation, and required 85 percent of the world's production of aviation gasoline. Senior Luftwaffe officers concluded that there was no prospect of realizing such a plan. However, Hans

---

[41] Ibid., p. 111.

[42] See Hitler's depiction of his goals and the role of the military in the new Reich in "Aufzeichnung Liebmann," *Vierteljahrshefte für Zeitgeschichte* 2, no. 4, October 1954.

[43] Wilhelm Deist, et al., *Das Deutsche Reich und der Zweite Weltkrieg*, vol. I, *Ursachen und Voraussetzung der deutschen Kriegspolitik* (Stuttgart, 1979), p. 479.

Jeschonnek, soon to be the Luftwaffe's chief of staff, commented: "Gentlemen, in my view it is our duty to support the Führer and not work against him."[44]

Perhaps the most perplexing problem the Luftwaffe confronted in its expansion was the fact that disarmament and depression had reduced Germany's aircraft industry to a small group of companies living a hand to mouth existence with few research and development capabilities. The size of the aircraft industry in 1933 – 4,000 workers – suggests the extent of the problem in building up its production base.[45] As late as 1932, only Junkers and Heinkel – successful builders of airliners and seaplanes respectively – possessed the limited capacity for the series production of aircraft. Other companies worked according to the "craftsman ethic" – a wasteful and time-consuming process of constructing "hand-built" aircraft.[46] Particularly serious was the lag in engine development. The vast majority of the engines powering German aircraft in the 1920s and early 1930s were foreign designs – hardly a sound basis on which to create an independent air arm.[47] Even the Ju 87 and the Bf 109, two types on which the Luftwaffe depended during the first years of World War II, flew in their early variants with Rolls-Royce power plants.

Generally then, the Luftwaffe confronted similar problems to Anglo-American air forces. Nevertheless, Germany's geographic, cultural, and political framework was substantially different than was the case with its future opponents. Consequently, German innovation in strategic bombing would move the Luftwaffe in very different directions in its approach to air power and strategic bombing.

THE OPERATIONAL AND TACTICAL DEVELOPMENTS

Three main factors drove the development of land-based air power in Britain and the United States throughout the interwar period: a desire to escape the horrific slaughter of the trenches; the institutional imperatives and yearnings that arose, in the British case, from protecting air force autonomy, or, in the American case, from seeking it; and, the promise of technological advances that would make air power's potential in future conflicts almost limitless. To British and American strategic-bombard-

---

[44] Edward Homze, *Arming the Luftwaffe, The Reich Air Ministry and the German Aircraft Industry, 1919–1939* (Lincoln, NE, 1976), pp. 223–224.
[45] Deist, *et al.*, *Das Deutsche Reich und der Zweite Weltkieg*, vol. 1, pp. 480–481.
[46] Homze, *Arming the Luftwaffe*, p. 26.   [47] Ibid., p. 27.

ment enthusiasts between the wars, the heavy bomber held out the possibility of winning future conflicts independent of armies or navies. It was
this vision of air power striking directly at the heart of the enemy nation
– either by attacking its population centers or economic structure – that
became the focus of both British and American innovations in land-based
air power. What was *not* accorded sufficient attention in such an approach was the evidence from the previous war as to the complexities, difficulties in employment, and the limitations of air power. Among other
things, American and British proponents of strategic bombing prior to
1939 substantially overestimated the accuracy and efficacy of such an approach and generally disregarded the need to gain control of the air over
enemy territory *before* their air forces could exploit the full potential of
friendly air power. Even then, as the official historians of the U.S. Army
Air Forces in World War II note, the effects of strategic bombing "were
gradual, cumulative, and during the course of the campaign rarely measurable with any degree of assurance."[48]

By comparison German air power took a substantially different direction between 1918 and 1939. The more traditional, combined-arms context within which German thinking about air power evolved during the
interwar period reflected not only the Reich's strategic position but
the German military's thorough examination and institutionalization of
the tactical and operational lessons of World War I air operations. The resulting German view was not to ignore strategic bombing, as has repeatedly been asserted,[49] but to place that particular use of air power within

[48] Wesley F. Craven and James L. Cate, *The Army Air Forces in World War II*, vol. II, *Europe: Torch to Pointblank, August 1942 to December 1943* (Chicago, 1949), p. ix.
[49] In the view of the American Eighth Air Force in 1945, the German high command "saw
an air force as a weapon primarily for use in obtaining decisive results in ground battle"
and lacked the "imagination" to "envision that the decisive battle itself might take place
in the air" (Eighth Air Force and Army Air Forces Evaluation Board [European Theater
of Operations], *Eighth Air Force Tactical Development: August 1942 – May 1945* [En-
gland, July 1945], p. 84). Sir Charles Webster and Noble Frankland offered similar views
in their official history of the strategic air offensive against Germany (*The Strategic Air
Offensive against Germany*, vol. 1, *Preparations* [London, 1961], p. 125). The fact that
the Luftwaffe never managed to field heavy bombers comparable to the American B-17
or the British Lancaster has undoubtedly been one of the reasons for such views. See, for
example, Corum's *The Roots of the Blitzkrieg*, pp. 167–168. In fact, the German Air
Ministry cancelled development of the four-engine Dornier Do 19 and Junkers Ju 89 in
1936 because the German aircraft industry had not been able to produce sufficiently
powerful engines, not because Luftwaffe leaders like Walther Wever or Helmut Wilberg
lacked imagination about the potential of strategic bombing or ignored it as a mission
(Williamson Murray, *Strategy for Defeat: The Luftwaffe 1933–1945* [Maxwell Air Force
Base, Alabama, 1983], pp. 8–9. However, in 1937, the year after Wever's death in a plane

a broader, combined-arms context. While the Germans neglected neither air superiority nor air support of ground forces, by 1940 the Luftwaffe also had the largest bomber fleet in the world and had done more to cope with the difficulties of locating and hitting targets in bad weather or at night than any other air force.[50] As a result, the Luftwaffe was able to deal out levels of damage through independent bombing operations such as the 1940 *Blitz* against London that neither the British nor American air forces were able to match until 1943. At the same time, German air power was generally more effective than its British or American counterparts in influencing the battlefield during 1939–1942. Nonetheless, the Luftwaffe's failure to achieve daylight air superiority over England during the 1940 Battle of Britain ended whatever hopes the Germans may have had of invading the British Isles and probably reinforced the inclinations of many within the German military to believe that air power was most efficient and effective when employed in combination with other forces.[51]

Senior British and American airmen, of course, drew different lessons from the Battle of Britain. Unlike the Germans, they realized in its wake how important the full mobilization of national industrial capacity to pro-

---

crash, the Germans began development of the He 177, which they hoped would enable them to bypass the need to develop the high-powered engines required for four-engined heavy bombers by mounting two less-powerful engines in a single nacelle. German engineers were never able to make this design work (ibid., p. 9). In retrospect, aircraft-engine technology may well have been one area in which the Versailles treaty succeeded in constraining German military developments.

[50] For example, Luftwaffe scientists began experimenting in 1939 with radio-direction systems for navigation and as an answer to the problem of bombing targets at night or in bad weather. The Knickebein blind-bombing system, first used during the Battle of Britain, was a direct result and preceded by about two years the fielding of a similar system by the Royal Air Force (Murray, *Strategy for Defeat*, pp. 16 and 20). The Germans used a version of this system (fine X-beams or X-Gerät) on the evening of 14 November 1940 to mark the city of Coventry for a night attack that killed 554 people and seriously injured 865 (Jones, *The Wizard War*, pp. 141 and 147–151). Basically these systems consisted of radio-direction beams broadcast from two separate locations that crossed over the target. The German bombers that attacked Coventry flew to the city down the western beams emanating around Cherbourg and used the eastern crossing beams emanating from Calais to provide timing signals for bomb release (Jones, *The Wizard War*, pp. 135–145; also, Cajus Bekker [pseudonym for Hans Dieter Berenbrok], *The Luftwaffe War Diaries* [Garden City, NY, 1968], pp. 179–180).

[51] R.J. Overy, *The Air War: 1939–1945* (New York, 1980), pp. 36–37. The Bf 109's limited range meant that German twin-engine bombers could only attack southern England, if they were to have adequate fighter escort to keep losses to acceptable levels. Consequently, in the Battle of Britain, the Luftwaffe could only impose on Fighter Command a rate of attrition that its commanders were willing to accept, and the Germans were never in a position to attack the Royal Air Force over the full length and breadth of its domain (Murray, *Strategy for Defeat*, p. 46).

vide the aircraft necessary for success in air would be to final victory.[52] But far from giving up on independent bombing in light of the Luftwaffe's failure in the Battle of Britain, bomber advocates on both sides of the Atlantic promoted such an approach as the only way to strike directly at Hitler and perhaps win the war without a costly land battle. Still, it would take time, enormous industrial resources, and the lives of thousands of airmen on both sides for the British and American air forces to realize the potential of the heavy bomber, whether operating independently or in support of land operations. In the American case, it was not until spring 1944 that U.S. strategic air forces in Europe developed the wherewithal, including long-range escort fighters and a growing capacity to replace heavy losses, to wrest daylight air superiority from the Luftwaffe over Europe, thereby making the Normandy invasion and the sustained daylight bombardment of the German heartland possible.[53] And even in early 1945, after American heavy bombers had, at last, been able to concentrate in daylight on industrial targets, the results seemed less decisive than American airmen like Arnold, then commanding general of the army air forces, had hoped.[54]

The divergent approaches to continental air power taken by the Germans, as contrasted with the British and Americans, suggest how extraordinarily difficult it was to foresee with clarity or detail the proper direction in which to develop air power during the interwar period. To understand

---

[52] Murray, *Strategy for Defeat*, p. 56; also, Geoffrey Perret, *Winged Victory: The Army Air Forces in World War II* (New York, 1993), p. 42. In the American case it was Robert Lovett, by April 1942 the assistant secretary of war for air, who eventually tied production to strategy (*Winged Victory*, p. 141).

[53] Perret, *Winged Victory*, pp. 295–296. U.S. Eighth Air Force heavy bombers began combat operations against targets in German-occupied France on 17 August 1942; Eighth's heavies first bombed targets in Germany in late January 1943 (Roger A. Freeman, with Alan Crouchman and Vic Maslen, *Mighty Eighth War Diary* [New York, 1981], pp. 9 and 35). During the first two years of Eighth Air Force heavy bomber operations there was "nothing resembling the repeated heavy attacks on vital industrial targets that the Maxwell theorists and authors of AWPD-1 had envisaged" (Perret, *Winged Victory*, p. 325).

[54] As Arnold wrote feelingly to his air senior commander in Europe, General Carl "Tooey" Spaatz, on 14 January 1945: "We have a superiority of at least 5 to 1 now against Germany and yet, in spite of all our hopes, anticipations, dreams, and plans, we have as yet not been able to capitalize to the extent which we should. We may not be able to force capitulation of the Germans by air attacks, but on the other hand, with this tremendous striking power, it would seem to me that we should get much better and much more decisive results than we are getting now." (Quoted in Wesley F. Craven and James L. Cate, *The Army Air Forces in World War II*, vol. 3, *Europe: Argument to V-E Day, January 1944 to May 1945* [Chicago, 1951], p. 716).

the choices that airmen confronted in the interwar period, one must consider the "lessons" that air power's role in World War I might have suggested to dispassionate observers. The four years between 1914 and 1918 saw vast technological changes that transformed the airplane into a crucial component of land warfare. Moreover, beyond rapid technological change and industrial mobilization, the war suggested a number of tactical and operational lessons that subsequent experience would confirm.

First was the absolute requirement for aerial superiority before air forces could embark on *any other* air operations without suffering unacceptable losses. Second was the general inaccuracy of the bombing both by day and night. In September 1917 Lieutenant Commander Lord Tiverton of the Royal Naval Air Service reported that "experience has shown that it is quite easy for five squadrons to set out [at night] to bomb a particular target and for only one of those five ever to reach the objectives; while the other four, in the honest belief that they had done so, have bombed four different villages which bore little, if any resemblance to the one they desired to attack."[55] The third, and perhaps most important, lesson had to do with the role of the airplane in ground operations. By 1918 the great offensives returned movement to the battlefield, but without the help of aircraft, particularly reconnaissance and close air support, the new concepts underlying the German breakthrough and exploitation tactics would not have worked.

At the outset of war in 1914, the use of air power had approximated the intelligence gathering duties of cavalry. What intelligence aircraft provided was largely idiosyncratic and murky in nature. A typical aerial reconnaissance report from the early days of the war might mention thousands of Germans moving down the road to Brussels without providing further details. By 1915, though, aircraft were providing aerial photographs which were crucial in developing artillery capabilities during the remainder of the war. By 1917, air reconnaissance had become so important that the failure of French fighters to protect their reconnaissance aircraft was a major contributor to the catastrophic defeat of the Nivelle offensive – a defeat that almost resulted in the collapse of France.[56] As early as 1917 the Germans had liaison officers in the front-line trenches to coordinate air strikes against enemy positions by use of radio.[57] By

---

[55] Quoted by Group Captain R. A. Mason in "The British Dimension," *Airpower and Warfare*, ed. Alfred F. Hurley and Robert C. Ehrhard (Washington, DC, 1979), p. 32.

[56] Lupfer, *The Dynamics of Doctrine*, p. 66, note 115.

[57] General der Kavallerie von Hoeppner, *Deutschlands Krieg in der Luft* (Leipzig, 1921), pp. 114–118 and 149–150.

1918, close air support aircraft were supporting in a major fashion the breakthrough battles that changed the nature of war in the twentieth century.[58]

What was different about the German postwar approach was the fact that German airmen carried out a searching examination of air operations in the last war paralleling and reinforcing the exhaustive analysis of ground operations directed by Seeckt – an effort to find out what had *actually* happened in the air war.[59] Consequently, when the Nazi regime began rearming Germany in 1933, the German military possessed careful, thorough studies of the actual experience of the last war. German doctrinal statements on air power – particularly those in the aerial-reconnaissance and air-support sections in Part 1 of the 1921/1923 *Leadership and Battle with Combined Arms* and in the Luftwaffe's 1935 doctrinal manual *Die Luftkriegsführung (Conduct of the Air War)* – rested on a solid historical foundation.

This foundation produced a reasonably balanced approach to air power. The Germans promoted no one application of air power excessively at the expense of others. Instead, German thinking, grounded firmly on the experiences of 1914–1918, recognized the importance of achieving air superiority by destroying the enemy's air force, underlined the crucial interdependence between ground and air forces on the modern battlefield, and suggested that emphasis or priorities among air power's various tasks in future campaigns would depend on local circumstances, which could vary widely from one campaign to another.

The British

No air force placed strategic bombing at the heart of its existence more than the RAF. Even in World War I the British had proven enthusiastic supporters of the concept of strategic bombing. By summer 1918, British representatives on the Inter-Allied Aviation Committee had pushed strongly for a coordinated bombing campaign against Germany by Allied air forces. They suggested that a "special long-range striking force" could

---

[58] The best work on the doctrinal and tactical changes during 1917–1918 remains Lupfer's *The Dynamics of Doctrine*. Nevertheless, Lupfer underestimates the contribution of air power to these developments.

[59] The German Air Service (Luftsreitkräfte) initiated its program for studying the lessons of World War I about two weeks before Seeckt issued his 1 December 1919 directive (Corum, *The Roots of Blitzkrieg*, p. 144). The Air Service's 1919–1920 examination of World War I air operations involved eighty-three officers working on twenty-one subcommittees (ibid.).

disrupt "the enemy's key industries" and have a significant impact on the enemy's morale.[60]

What particularly marks the RAF's approach to strategic bombing was a willingness to dispense with uncomfortable evidence. Much of this had to do with a military culture that rejected history – and even evidentiary efforts to make a reasonable cut at honest "lessons learned" analyses – in coming to terms with air power's potential. What much of the RAF's leadership did in the interwar period to prepare the service for the next war can only be explained in recognizing that the organization's culture believed that the past had *no* relevance to the future and was comfortable with manipulating the evidence to support conclusions that were in line with current doctrine.

Even before the Great War had ended, the RAF had begun cooking the books.[61] The RAF's official history of the war in the air was a masterpiece of propaganda to justify its continued existence rather than a realistic appraisal of the cold, harsh realities of the air war over the trenches. In fact the air staff made its dismissal of history explicit as early as 1924; in a memorandum its officers argued that the forces employed in attacking an enemy nation

> can either bomb military objectives in populated areas from the beginning of the war, with the objective of obtaining a decision by moral[e] effect which such attacks will produce, and by the dislocation of the country, or, alternatively, they can be used in the first instance to attack enemy aerodromes with a view to gaining some measure of air superiority, and when this has been gained, can be changed over to the direct attack on the nation. The latter alternative is the method which the lessons of military history seem to recommend, but the Air Staff are convinced that the former is the correct one.[62]

As suggested above, the crucial importance of air superiority to bombing operations and the problems of target identification had been central to the conduct of the air war throughout World War I. Nothing in the evidence suggests that the RAF's leadership ever addressed these problems

---

[60] Supreme War Council, Annexure to Process Verbal, Third Session 1-a, Aviation Committee. "Remarks by the British Representative for the Third Session of the Inter-Allied Aviation Committee Held at Versailles, 21st and 22nd July 1918." Trenchard Papers, RAF Staff College, Bracknell.

[61] The deduction of RAF studies in 1918 on the effect of bombing attacks on German towns was that the enemy's fighting capacity decreased "as the number of raids increased. . . . Though material damage is as yet slight when compared with the moral[e]."

[62] PRO AIR 20/40, Air Staff Memorandum No. 11A, March 1924.

in depth throughout the interwar period. Target identification and bombing accuracy remained major problems until the beginning of the next
war. In May 1938 the assistant chief of air staff admitted:

> it remains true, however, that in the home defence exercise last year,
> bombing accuracy was very poor indeed. Investigation into this matter
> indicates that this was probably due very largely to failure to identify
> targets rather than to fatigue.[63]

While the RAF's doctrine for strategic bombing called for trained aircrews
to precede the bomber force and mark targets for following aircraft, there
was no organization within either Bomber Command or the air staff devoted to developing technology to support such an approach until the late
1930s. When asked how "trained" aircrew would find their targets, Tedder ironically replied, "You tell me!"[64] What is astonishing is that only
gradually in the late 1930s did the extent of the RAF's navigational and
bombing accuracy problems emerge. The air staff made a beginning in
1937 by creating the Committee for the Scientific Survey of Air Offense
under Henry Tizard. Its hope was that this committee would do for
Bomber Command what the Committee for Scientific Survey of Air Defense had done for Fighter Command. But the latter body worked directly with and for Dowding, a man of great vision, leadership, and technological sophistication. The new committee on bombing, however, worked
for the Air Ministry, which filtered out much of its advice before it reached
a command not exactly enthralled about receiving civilian input.[65]

At the same time Bomber Command established its own internal committee to look into the question of bombing accuracy, but the evidence at
that committee's disposal was scanty indeed. As the official historians indicate:

> Thus, the Bombing Committee had to rely on the trials at the Arma-
> ment Training Camps and theoretical reasoning. But the trials provid-
> ed no tests for the identification of a target. They were often made at
> levels which would be impossible in wartime against defended targets.
> They took place in daylight and good weather. There were hardly any
> tests as to what could be done at night or in cloudy weather. Under these
> conditions some squadrons were able in practice to produce a high de-
> gree of accuracy. But in the large-scale exercises which approached
> more closely to war conditions, their deficiencies were exposed. When

[63] PRO AIR 2/2598, Air Ministry File #541137 (1938).
[64] Guy Hartcup, *The Challenge of War* (London, 1967), p. 126.
[65] Webster and Frankland, *The Strategic Air Offensive against Germany*, vol. 1, p. 114.

remedies were proposed, and the relative merits of different forms of bombing were discussed, it was realized that there were not sufficient data on which to found reliable conclusions. . . .Nevertheless, the Manual of Air Tactics contained minute instructions on the various kinds of bombing, special attention being given to high level bombing in daylight. Most of this was necessarily based on theoretical reasoning since there had been so little practical experiment.[66]

Navigation in bad weather or night may have been the greatest weakness in the RAF's bombing force. Bomber crews themselves described their navigational methods as "by guess and by God."[67] At least the leaders of Bomber Command recognized that they had a serious problem. In May 1939 the commander of 3 Group reported that by using dead-reckoning his crews could at best bring their aircraft within fifty miles of a target.[68]

But here funding problems did raise difficulties. The lack of resources in the 1920s and 1930s resulted in unlighted airfields, a lack of radio beacons, little navigational training, and even a dearth of mechanical devices such as sextants and bomb-aiming devices. The commander of Bomber Command, Air Marshal Sir Edgar Ludlow-Hewitt, noted in the late 1930s: "little real progress can be made in training bomber units to operate under the conditions in which they must operate in war until adequate provision is made for the reasonable security of the aircraft and crews undertaking this training."[69] Under such conditions, squadrons hardly trained at night or in bad weather for fear of increasing already depressingly high accident rates that resulted from the RAF's rapid expansion in the late 1930s. That in turn helped hide the extent of the problem.[70] Nevertheless, the evidence clearly indicates that there was plenty of warning about the inability to find or hit targets – whatever the difficulties in funding. What is almost inexplicable is the fact that the leadership persisted through the first two years of bombing operations in its prewar belief that the bomber would find its target – and with the added benefit of having watched the Luftwaffe flounder in *its* night attacks despite the fact that German crews possessed technological aids that would not be available to British crews until 1942.

Funding difficulties also affected aircraft procurement. In 1937 when

---

[66] Ibid., pp. 117–118.   [67] Ibid., p. 112.   [68] Ibid., p. 112.

[69] Quoted in John Terraine, *The Right of the Line, The Royal Air Force in the European War, 1939–1945* (London, 1985), p. 83.

[70] As the official historians note: "The training of crews in night flying was, indeed, obviously a difficult problem and it was never really faced in the pre-war period." Webster and Frankland, *The Strategic Air Offensive against Germany*, vol. 1, p. 113.

the RAF was preparing a new generation of bombers, the Chamberlain government stepped in and refused to support a bomber program for the RAF. Instead, it decided to support Fighter Command and a complex air defense system that depended on radar as well as a new generation of fighters.[71] This decision rested on the cost of a bomber force versus that of an air defense system, but in retrospect it was the right decision, one of the few correct defense choices that Chamberlain's government made in the late 1930s. At the outbreak of the war Bomber Command aircraft were less than satisfactory. Production and design innovation simply failed to match the standards of Fighter Command. The "Wellington" was a satisfactory bridge to the new generation of bombers, but the "Whitley" and "Hampton" were unsatisfactory. To let anyone fly the "Battle" in combat was just short of criminal. Moreover, there were also problems with the follow-on bombers; the Short "Stirling" was little more than a sitting duck. The "Halifax" proved an excellent workhorse, but the "Manchester" represented a misguided effort – similar to the He 177 – to couple two engines in a single nacelle. Nevertheless, private efforts at design innovation suggest that the RAF could have achieved more with its bomber designs than it did. The "Mosquito", developed by De Havilland with no government support, was one of the finest aircraft of World War II; not surprisingly, when presented with the aircraft in 1940, the Air Ministry had not a clue what to do with an aircraft made of wood.

Part of the problem was due to British resource allocations. Nevertheless, the contrast between Fighter Command and Bomber Command is indeed striking. The former under Dowding's prodding, pushed the limits of technological development in aircraft, radar support, and communication links needed to protect the British Isles. Bomber Command remained well behind in every area.

Development of a long-range escort fighter underlines the extent of the difference between the two commands. In March 1940, Dowding suggested to the air staff that the technology was available to create a long-range escort fighter. As with his work in creating Fighter Command, Dowding aimed to push technological innovation to support operational concepts that would guide the air war against Germany. His proposal received a curt dismissal:

> It must, generally speaking, be regarded as axiomatic that the long-range fighter must be inferior in performance to the short-range fighter. . . . The question had been considered many times, and the discus-

[71] Ibid., p. 51.

sion had always tended to go in circles. . . . The conclusion had been reached that the escort fighter was a myth. A fighter performing escort functions would, in reality, have to be a high performance and heavily armed bomber.[72]

A year later Churchill asked the same question and received the same answer from the chief of air staff; Churchill's response to Sir Charles Portal's gloomy conclusion was that such a view "closed many doors."[73]

How to explain the extraordinary differences in innovation between Fighter Command and Bomber Command despite the fact that both were within the same service? Admittedly, Dowding and his command received support at the critical moment when air defense became a viable possibility. But Dowding and those who supported him also displayed imagination and willingness to push technological innovation as far as it would go in support of the mission.

The opposite obtained in the case of Bomber Command. The extent of the problem was apparent by the late 1930s: the command could not find its targets, could not bomb them accurately if it found them, and could not carry sufficient loads to damage such targets significantly even if it were to hit them. One can almost go so far as to suggest that most in Bomber Command displayed a willful desire to dismiss the dismal results that were already apparent in 1938 and 1939. For the first year of its strategic bombing campaign against Germany, Bomber Command struck at precision targets at a time when only one of three of its bombers could drop its bombs within a target with an area of seventy-five square miles. Only in summer 1941 when the irrefutable evidence of the Butt report indicated that British bombers were doing more damage to cows and trees than to Germany's cities or industries, did Bomber Command become interested in addressing its problems.[74] And one suspects that only the threat that the politicians might divert bomber production from a failed effort finally forced the command to make the necessary changes in attitude.

At the heart of the problem was a belief in the Air Ministry, the air staff, and Bomber Command that the bomber *had to do substantial damage to an opponent*. Thus, there was much willingness to dismiss any evidence to the contrary. Until there was a recognition of what was actually happening, then the preconditions required for innovation were simply not in

---

[72] PRO CAB AIR 16/1024, Minutes of the 20th Meeting of the Air Fighting Committee, held at the Air Ministry, Whitehall, 12.3.40.
[73] Webster and Frankland, *The Strategic Air Offensive against Germany*, vol. 1, p. 177.
[74] Ibid., vol. 4, p. 205.

place. What is extraordinary in the British case is the fact that it took near-ly two years of war to prove to Bomber Command what was obvious be-fore the war began: its aircrews needed significant technological help and until they received that help they were not going to accomplish their mis-sion.

The Americans

American thinking about air power during the interwar years faced sev-eral interlocking problems. First, unlike their British counterparts, Amer-ican army airmen did not emerge from World War I as an independent service. Instead, the Army Reorganization Act of 1920 established the Air Service as a combat arm of the U.S. Army.[75] Hence, in no small part a growing desire for independence drove an emphasis on strategic bombing at the expense of other missions that increasingly characterized the think-ing among army aviators after 1926 (the year in which the air service be-came the army air corps).[76]

In retrospect, this desire was not venal. American airmen sincerely be-lieved that independence was necessary to realize air power's full potential, and combat operations in North Africa established, in early 1943, that com-plete subordination of assets to ground commanders was a recipe for disas-ter.[77] The airmen also had unique technological needs. Still, there is a cer-tain irony in the fact that this desire led them to emphasize strategic bombardment. Whereas British airmen under Trenchard's leadership fo-cused single-mindedly on strategic bombing to *defend* the RAF's autonomy, American airmen, chaffing under the absence of "air mindedness" displayed by their nonflying colleagues, became more and more enthusiastic over the technological potential of the heavy bomber as a means to autonomy.

The second problem that confronted army airmen was the difficulty in-volved in holding on to the lessons suggested by their limited experience with air power during World War I. The United States did not enter the

---

[75] Lieutenant Colonel Thomas A. Fabyanic, "Strategic Air Attack in the United States Air Force: A Case Study," Air War College Report No. 5899, April 1976, p. 15.

[76] "The quest for autonomy led to the advocacy of strategic bombardment, which led, in turn, to the depreciation of not only defensive pursuit aircraft but all pursuit aircraft. Bombardment and autonomy were so inextricably bound together that the questioning of bombardments by an Air Corps officer was not only impolitic but unwise." Perry McCoy Smith, *The Air Force Plans for Peace: 1943–1945* (Baltimore and London, 1970), p. 34.

[77] Perret, *Winged Victory*, pp. 188 and 198; Craven and Cate, *The Army Air Forces in World War II*, vol. 2, p. 164.

conflict in strength until the war's end, when German military power was collapsing. Consequently, the American operational experience was not nearly as extensive as that of the British or Germans. Nonetheless, in the early 1920s the concepts taught at the fliers' "graduate" school showed reasonable balance between pursuit, attack, and bombardment aviation.[78] Accurately reflecting the collective experience of Allied and German airmen, the school taught that "the primary role of the air arm was the defeat of the enemy air force; secondarily, it would destroy military and industrial targets in the enemy's homeland."[79]

Billy Mitchell, however, set an unfortunate tone to the debate over air power in the United States. Nevertheless, alone among air power theorists, he saw air war in classical terms. In his first book, he suggested that the proper ratio among types of aircraft in a postwar air corps should be 60 percent pursuit (fighter), 20 percent bombardment, and 20 percent observation.[80] To realize their potential, he believed that air forces would first have to destroy their opponents in the air before moving against other targets.

But Mitchell's arguments that air superiority was a *sine qua non* faded over time. By the late 1920s the emphasis in the air corps had shifted to a belief that the bomber represented the beginning and end of modern air power.[81] As early as 1926 its training manuals were suggesting that an enemy's economic system would possess "vital parts" or "sensitive points", the destruction of which would bring his economic life to a halt.[82] It did not take a great jump to argue that destruction of a vital por-

---

[78] The creation of the air service in 1920 brought about the establishment of a graduate school for its officers. The school, founded by Colonel Thomas DeWitt Milling, was originally known as the Field Officers' School and located at Langley, Virginia (Perret, *Winged Victory*, p. 24). In 1922 it was renamed the Air Service Tactical School (ibid.). When the air service became the air corps in 1926, the school underwent its final name change and became the Air Corps Tactical School (Fabyanic, "Strategic Air Attack in the United States Air Force," p. 15). Four years later, in 1931, the school moved to Maxwell Field, Alabama (ibid.).

[79] Fabyanic, "Strategic Air Attack in the United States Air Force" p. 20.

[80] Hurley, *Billy Mitchell*, pp. 63–83.

[81] There are a number of important studies on the development of air power doctrine in the army air corps and later the army air forces. See in particular: Robert T. Finney, *History of the Air Corps Tactical School, 1920–1940* (Washington, DC, 1992); Thomas A. Fabyanic, "A Critique of United States Air War Planning, 1941–1944," PhD Dissertation, St. Louis University, 1973; Thomas H. Greer, *The Development of Air Doctrine in the Army Air Arm, 1917–1941* (Montgomery, AL, 1955); Robert F. Futrell, *Ideas, Concepts, Doctrine: A History of Basic Thinking in the United States Air Force, 1907–1964* (Montgomery, AL, 1971).

[82] "Bombardment Text," Langley Field, Virginia, 1925–1926, Air Force Historical Research Center, Montgomery, Alabama.

tion of the enemy's industrial potential would cause the collapse of his economic structure, and therefore, his will to continue.

The implicit character of this doctrinal shift, both in substance and in terms of its promulgation, is important to appreciate. During the interwar years, many key officers of the U.S. Army Air Forces (USAAF) in World War II went through the Tactical School; a number of them also served as faculty members.[83] Over time, first attack and then pursuit declined in significance in the teachings of the Air Corps Tactical School. That said, the strategic bombing enthusiasts who gained ascendancy at the school by the mid-1930s won the debate over the proper direction for American air power fairly:

> Their's wasn't the only subject taught at Maxwell. And anyone at the Tactical School was free to make any argument he wanted. Even after [the fighter advocate Claire] Chennault left there were other fighter advocates, such as Earle Partridge and Hoyt Vandenberg, on the faculty. Each side put its case to the students, who were free to decide which, if either, they wanted to believe.[84]

Nonetheless, it was the bomber advocates who came to dominate the thinking of American airmen by the late 1930s. Innovations in doctrine and theory by the early 1930s were exercising important effects on the growth and force structure of the air corps. In turn these developments pushed technology in certain, discrete directions. For obvious reasons, daylight operations would simplify problems of navigation and bombing accuracy; advances in design pushed bomber capabilities faster than those of fighters. As a result, new bombers in the mid-1930s possessed range and performance almost the equal of the fighters then in service.

Along with enhanced capabilities came increases in the defensive armament of bombers. Since it was difficult to work out the implications, air corps thinkers slid into assumptions about the relative invulnerability of daylight bomber formations. As early as 1930, one text at the Air Corps Tactical School suggested

> Bombardment formations may suffer defeat at the hands of hostile pursuit; but with a properly constituted formation, efficiently flown, these

---

[83] Ira C. Eaker, Carl Spaatz, Curtis E. LeMay, Haywood S. Hansell, Jr., and Claire E. Chennault all went through the Tactical School; Hansell the bomber advocate and Chennault the fighter advocate both taught at the school as well.

[84] Perret, *Winged Victory*, p. 27. Attack fell by the wayside at the Tactical School with George Kenney's departure to attend the Army War College in 1931; the neglect of pursuit is often dated from Chennault's forced retirement due to chronic bronchitis in 1936 (ibid., pp. 26 and 27).

defeats will be the exception rather than the rule. Losses must be expected, but these losses will be minimized by proper defensive tactics.[85]

The combination of easier operating conditions, the assumed defensive capabilities of bomber formations, and the relatively insignificant differences between the speeds of bombers and fighters pushed doctrinal emphasis toward the belief that daylight unescorted bomber formations could fight their way through enemy skies without suffering prohibitive losses. As one instructor put it in the early 1930s: "A well-planned and well-conducted bombardment attack once launched, cannot be stopped."[86] Concurrent with the belief that bomber formations could fight their way through enemy defenses went the assumption, stated in the bombardment text for 1935, that "escorted fighters will neither be provided nor requested unless experience proves that bombardment is unable to penetrate such resistance alone."[87] By the late 1930s army air corps doctrinal innovation had crystallized into a set of assumptions that presented a tightly focused conception of air war. It had developed the operational concept of high altitude, level flight, daylight, precision bombing.

Hand in glove with this confidence in the defensive capabilities of U.S. heavy bombers went a faith in the bomber's effectiveness against targets. The chapter on bombing accuracy and probabilities in Maxwell's 1939 tentative manual, "Delivery of Fire from Aircraft," argued that one could quantify the question of how large a bomber force would be necessary to destroy a given target "with a reasonable assurance of success" while avoiding overkill.[88] Such assurance rested on the application of probability and statistics to peacetime experience on bombing ranges. Necessary and even ground-breaking as this sort of analytic work was, subse-

[85] "Bombardment Text," Air Corps Tactical School, Langley Field Virginia, 1930, p. 109, Air Force Historical Research Center.

[86] Haywood S. Hansell, Jr., *The Air Plan that Defeated Hitler* (Atlanta, 1972), p. 15.

[87] Quoted in Fabyanic, "Strategic Air Attack in the United States Air Force," p. 31.

[88] Air Corps Board, "Delivery of Fire from Aircraft," Part One, "Precision Bombing," chapter 4, "Bombing Accuracy and Probabilities," 10 June 1939, Alfred F. Simpson Historical Research Center, 167.86–4, Maxwell AFB, Alabama, p. 1. This manual contains, among other things, a clear definition of "Circular Probable Error" (CEP) and charts for determining the number of *independently aimed* bombs that would need to be dropped to have various probabilities of hitting the target. The manual also states that "a large proportion of individual bombardment objectives will be destroyed if hit by one properly selected size of bomb" (ibid., p. 2). As Fabyanic has pointed out, the Tactical School's assumption that proper bomber formations could permit unescorted penetration was incompatible with the assumption that each bomb would be independently aimed ("Strategic Air Attack in the United States Air Force," pp. 37–38).

quent events indicate that it had the unfortunate consequence of encouraging the belief that, through statistical methods and probability theory, the important aspects of bombing could be quantified and their effects against the enemy predicted.

The heady mixture of faith in air power's future, confidence in the technical superiority of American bombers, and "mathematical assurance" as to the efficacy of bombing created an atmosphere that gave disproportionate weight to vision and "scientific" calculations over firsthand combat experience. Instead of a careful study of foreign wars, which might have questioned the plausibility of doctrinal tenets such as unescorted, daylight bombing in airspace defended by enemy fighters and anti-aircraft artillery (AAA) – or the ease with which air attacks could destroy target systems – bomber proponents either disregarded combat experience in Europe or interpreted it in light of American doctrinal preferences.

This assessment may seem harsh. But it is difficult to reach any other judgment in light of later events. When American airmen, some of whom were to lead the army air forces after Pearl Harbor, went to England in 1940 to study European air war at first hand, their reports doubted that British or German experiences applied to U.S. forces:

> for all its ferocity, the Battle of Britain could not duplicate the sort of air battle that the American air planners had in mind. As a result, the concrete "lessons" simply did not materialize. True, both German and British bombers proved vulnerable to fighters, but then they were medium bombers, poorly armed and flying at relatively low altitude. . . . American bombers were much better armed and they would be flying at high altitude.[89]

A further point that the British airmen pressed on American observers was that, in any event, a long-range, high-performance fighter was technically impossible.[90] As a result, when the Air War Plans Division revised its August 1941 estimate of the air requirements for defeating Germany in August 1942, it concluded that it was *"perfectly feasible to conduct pre-*

[89] Major General Haywood S. Hansell, Jr., *The Air Plan That Defeated Hitler* (Atlanta, 1972), pp. 53–54.
[90] James Parton, *"Air Force Spoken Here": General Ira Eaker and the Command of the Air* (Bethesda, MD, 1986), pp. 123–124. Also see William R. Emerson, "Operation POINT-BLANK: A Tale of Bombers and Fighters," *The Harmon Memorial Lectures in Military History, 1959–1987*, ed. Harry R. Borowski (Washington, DC, 1988), pp. 452–453; and Fabyanic, "Strategic Air Attack in the United States Air Force," pp. 65 and 69.

cise bombing operations against selected precision targets, from altitudes of 20,000 to 25,000 feet, in the face of anti-aircraft artillery and fighter defenses [italics in original]."[91]

By 1940, then, the Americans had developed a coherent conception of strategic bombing. For the eventual outcome of World War II this represented a crucial innovation; it provided the *raison d'être* for AWPD-1, the basis of the great bomber fleets that attacked Nazi Germany in 1943 and 1944. But that *raison d'être*, which allowed airmen to advocate strategic bombing with such enthusiasm, carried with it seeds that almost resulted in defeat in 1943. The result of air corps doctrine, which became dogma to some, was faith in a series of interlocking assumptions: that vital targets in the enemy's war economy existed, could be identified, and were vulnerable to precision bombing; that bomber fleets unescorted by fighters could fight their way to those targets through German air defenses without suffering unacceptable losses; that the bombers, once there, could achieve enough accuracy to destroy the targets; and that the bomber force could achieve sufficient intensity of attack against entire target systems vital to German war production that the Germans could not avoid their collapse or find alternatives. The most fundamental difficulty in this doctrine was the interlocking nature of these assumptions.

> [The American air] planners failed to recognize the critical importance of collective risk, a failing that perhaps was a natural consequence of trying to simplify complex matters while maintaining a firm belief in an as yet unproven doctrine. They thus tended to discount many factors of potential significance, leaving a more or less clean slate for consideration of the next problem. Yet by not allowing the potential difficulties to accumulate, they fostered the development of erroneous assumptions of what American air power could do . . . All of the assumptions contained inherent weaknesses, which, for the most part, were not serious if taken independently. But, collectively, the shortcomings were mutually exclusive and thus made the entire concept a tenuous one.[92]

---

[91] As quoted in Fabyanic, "Strategic Air Attack in the United States Air Force," p. 65. For purposes of estimating force or munitions requirements, the use of probability and statistics to estimate the numbers of bombers required to destroy various targets was unavoidable. But extending these kinds of calculations to cover the defensive viability of bomber formations against a determined enemy was a different matter altogether. However valuable the reliance on numbers that grew up at the ACTS may have been for force planning, its extension to combat outcomes against a reactive adversary proved a mistake.

[92] Fabyanic, "Strategic Air Attack in the United States Air Force," pp. 162–3 and 164.

The Germans

A British intelligence officer, summing up the Luftwaffe's strategic bombing in World War II, concluded: "Unfortunately for the success of its efforts, the German air force was gradually forced to give up its integrity as an independent fighting force capable of waging large scale warfare on industries and communications, and became more and more the tool of the army, or, to use the [Luftwaffe] general staff's own expression, it became the army's fire brigade."[93] This assessment is true as far as it goes. What most commentators, however, have failed to convey is that strategic bombing theory and preparation remained an integral thread in Luftwaffe and German thinking through to the war's end.[94]

Indeed, the Luftwaffe went to war as the only air force capable of launching effective "strategic" attacks on its opponents. In the short run, it had solved the problems associated with folding strategic bombing into the capabilities of a versatile, effective air force. On one level, the German case is an excellent example of realistic and practical innovation, especially in institutional flexibility. But the Germans largely excelled in technical innovations that produced equipment or scientific support for aircraft. They proved less able to place their approach to air war within a larger framework of national strategy.[95]

As with the army after 1918, the Reichswehr's leadership attempted to learn the operational and tactical lessons of the war in the air in World War I. As early as November 1919, groups of former air service officers (as were their ground comrades) were examining the organizational, tactical, and technological questions of the last conflict.[96] Predictably, most of the resulting Reichswehr studies emphasized tactical support missions.[97]

---

[93] A.I. 12/USAFE/M. 74, USAFE/ Air Ministry Intelligence Party (OKL), Intelligence Branch Report No. 74, "Strategic Bombing in the GAF," USAFHRA 512.63512M-74, 3 Dec 1945, p. 2.

[94] Both the V-1 and V-2 represented German replies to Allied strategic bombing attacks and an attempt to get around the training and fuel problems that Nazi Germany confronted in 1944. From the German perspective such weapons – just as cruise and ballistic missiles are considered "aerospace" power – represented "strategic" air power, though in a pilotless form.

[95] Since both army and navy suffered from the same weaknesses, we are dealing here with a problem of national culture that extends beyond the ideological idiosyncrasy of the Nazi state.

[96] James S. Corum, The Roots of Blitzkrieg, Hans von Seeckt and German Military Reform (Lawrence, KA, 1992), p. 144.

[97] James S. Corum, "The Luftwaffe's Army Support Doctrine, 1934–1941," paper presented to the Annual Meeting of the Society for Military History, Quantico, VA, 1992.

Interestingly, Hans von Seeckt, father of the postwar army, strongly supported the idea of an independent air force and argued that nations possessed what Clausewitz had termed "centers of gravity" which "have become much more vulnerable" with the advent of offensive aviation.[98] Although the concept of strategic bombing remained an acknowledged, if unimportant aspect of *Reichswehr* doctrine, it received less attention in Germany during the 1920s than in Britain and the U.S. The reasons are obvious: the lack of an air force to push the concept, defense planning that concentrated on clandestine rearmament, retrenchment, and preparations for a possible war with Poland.

It was not until the Nazis came to power in January 1933 that air power (and by extension strategic bombing) came to the forefront in German defense policy. The Nazi seizure of power, the new regime's overt and covert goals, and the accelerated pace of rearmament served as catalysts for German air power innovation. Unlike the various forms of tactical aviation, which followed a roughly linear path from World War I to 1939 in German military thought, strategic bombing development was primarily a post-1933 phenomenon. The "strategic" aspects of German air power theory "were interdependent with and influenced by the aggressive policy of the Nazi regime."[99]

Certainly, Hitler waxed enthusiastic about what the military technology of aircraft might allow him to do. The fact that the number two man in the Nazi state, Hermann Göring, headed the new service did not hurt the Luftwaffe's pretensions or ambitions. As early as May 1933, the first members of the new Air Ministry were arguing that the Third Reich should immediately embark on creating a strategic bombing capability to deter those European powers tempted to put a stop to German rearmament. Dr. Robert Knaus argued beyond mere deterrence, however. He suggested that in the future "terror bombing of enemy capitals or an air offensive against industrial areas will produce moral collapse so much earlier, the weaker the national cohesion is, the more the urban masses follow materialistic interests [i.e., Marxist or liberal-democratic proclivities] and the more the masses are divided by sociopolitical tensions and political controversy."[100]

---

[98] Hans von Seeckt, *Gedanken eines Soldaten* (Berlin, 1929), p. 92.

[99] Klaus A. Maier, "Total War and German Air Doctrine before the Second World War," in Wilhelm Deist, ed., *The German Military in the Age of Total War* (Warwickshire, 1985), p. 212; see also Klaus A. Maier, *et al.*, *Das Deutsche Reich und der Zweite Weltkrieg*, vol. II, *Die Errichtung der Hegemonie auf dem europäischen Kontinent* (Stuttgart, 1979), pp. 44ff.

[100] Bernard Heimann and Joachim Schunke, "Eine geheime Denkschrift zur Luftkrieg-

The realities, however, of the situation were such that the State Secretary of the Air Ministry, the future Field Marshal Erhard Milch, scuttled the Knaus proposal on the sensible grounds that German industry was as yet incapable of producing advanced bombers. Nevertheless, Knaus's career as an air theorist and writer prospered; he contributed to the Luftwaffe's "death from the skies" genre of literature with a study entitled *Luftkrieg 1936* and later took over as commandant of the Air War Academy at Berlin-Gatow, which throughout its short history laid a heavy emphasis on strategic bombing.[101]

A more serious and far-reaching attempt to devise an air power doctrine for the Luftwaffe appeared shortly after its unveiling in March 1935. This was the seminal doctrine manual, *Die Luftkriegführung (Conduct of the Air War)*, written by a committee headed by the Luftwaffe's first chief of staff, Walther Wever. This team of officers consisted of many of those who had remained in the *Reichswehr*'s disguised general staff. If they displayed favoritism towards the army, it was only in their attempt to fit Luftwaffe doctrine into a wider framework.

As with most German doctrine, "Conduct of the Air War" presented a clear, concise formulation of the problems that employment of air power would confront in the coming war.[102] Within the framework of national strategy, the Luftwaffe's main missions, it argued, would be the attainment and maintenance of air superiority, support for the army and navy, attacks on enemy industrial power, and interdiction of the battlefield from the enemy homeland. Thus, strategic bombing had an important role within Luftwaffe conceptions, but it represented only one of several roles. In stark contrast to other current theories of air power, "Conduct of the Air War" emphasized that strong cooperation between the services was essential. Moreover, "strategic" attacks against the enemy's industrial base would occur only 1) when there was an opportunity to affect a war quickly and decisively, 2) when ground and naval operations had prepared the way, 3) during a stalemate, or 4) when that approach was the only means to strike the enemy's economy. "Conduct of the Air War" provided the Germans with a realistic appreciation of air war.

skonzeption Hitler-Deutschlands vom Mai 1933," *Zeitschrift für Militärgeschichte*, vol. III, 1964, pp. 72–86.

[101] Air Ministry, *The Rise and Fall of the German Air Ministry (1933–1945)* issued by the Air Ministry (ACAS) (London, 1948), p. 48.

[102] What is indeed astonishing, at least in the view of the author, is the fact that *Die Luftkriegführung* represents a more useful and intelligent discussion of air doctrine than any of the innumerable iterations of the basic doctrinal manual of the U.S. Air Force, AFM 1-1.

As with the British and Americans, indigenous thought about strategic bombing proved more important than foreign influences. Germany produced a number of air power advocates who were enthusiastic supporters of strategic bombing. These officers were primarily responsible for developing the concept known as *operative Luftkrieg*, or "operational air war." Although the term had a number of connotations and embraced a wider conception of air power than in Britain or the United States, it referred to the use of air power in a systematic campaign of attacks on an enemy nation and against "enemy production centers and their supply of raw materials."[103]

Yet most Germans argued that the activities of the three services must be synergistic in the next war, especially after the Luftwaffe gained air superiority. Typical prewar scenarios suggested that bomber units would smash enemy war industries that sustained its forces, power stations supplying those industries, and communications systems linking the enemy's homeland and military. But German thinking also emphasized that concerted pressure by air attack and combat losses in the ground war, working together, would bring about the collapse and defeat of the enemy.[104]

Despite active interest in strategic bombing, the Luftwaffe in 1938 and 1939 was far from ready to execute such a campaign. While reasonably able to attack the Reich's immediate neighbors, the Luftwaffe possessed little capability to strike Britain or exert the "intercontinental effect" that many German air power theorists espoused.[105] The explanation lies in a number of factors, all of which limited development of a full-blown strategic bombing capability. Most important was the impossibility before the war of estimating the force structure or capabilities required by a strategic bombing campaign. None of the air forces in this period got it right, a reflection of the intangibles that made up the potential equation.

Luftwaffe planners, keenly aware of Germany's continental position, recognized that pursuit of an air strategy divorced from ground operations represented a luxury the Reich could not contemplate. Even Rohden, one of the foremost proponents of "operational air war," admitted that

---

[103] Hans Detlef Herhudt von Rohden, *Vom Luftkrieg* (Berlin, 1938), p. 10.

[104] Horst Boog, *Die deutsche Luftwaffenführung, 1933–1945* (Stuttgart, 1982), pp. 151–164. This is, of course, how the Combined Bomber Offensive and land operations from the east and west eventually broke Germany in World War II.

[105] Murray, *German Military Effectiveness*, pp 63–64. As the commander of Second Air Force reported to the Luftwaffe high command in September 1938 during the height of the Munich crisis: "Given the means at his disposal, a war of destruction against England seems to be excluded." L. W. Gr. Kdo., Führungsabteilung, Nr. 210/38, g.Kdos., 22.9.38.

Germany was poorly served by its geographical position with potentially hostile neighbors in both east and west.[106] As a result, German air operations, particularly in the opening weeks of war, would have to support the army. Strategic bombing, a long-term investment, would avail Germany little in such a campaign.

The weaknesses of Germany's aircraft industry further reinforced the fact that the Luftwaffe was better suited to conducting tactical rather than strategic missions. By the late 1930s Germans were producing excellent two-engine bombers, but all had significant deficiencies. None possessed sufficient range, bombload, or other performance characteristics.[107] In addition, low munitions production pushed the Germans to emphasize dive bombing capabilities as a means of achieving greater accuracy; for single-engine aircraft such a requirement made sense, but for four-engine bombers it represented insanity.[108]

These elements combined to influence development of the German heavy bomber. Under Wever, the air staff energetically pushed technological innovation in search of a suitable four- engine bomber to provide striking capability outside of Central Europe. The first four-engine prototypes were ready in 1936, but there were serious problems with both designs. The most serious was the inadequacy of power plants. Lack of resources as well as inadequate long-range planning rendered the German aircraft industry incapable of sustaining a serious heavy bomber program. Kesselring, Wever's successor as chief of the general staff, pointed out that "with the prevailing shortage of raw materials, production of strategic bombers in any adequate numbers could only [be] achieved at the expense of other types."[109] As a result the Air Ministry cancelled both prototypes and almost concurrently issued specifications for a follow-on heavy bomber of significantly greater capabilities. The new aircraft, "Bomber A," or *Projekt* 1041, later became the Heinkel 177.[110]

The Luftwaffe gambled on the new design being ready by 1942, but serious design problems with the "coupled" engines along with other design flaws ensured that the He 177 never reached the Luftwaffe in significant numbers. Hitler later derided the craft as "obviously the worst junk that has ever been manufactured."[111] His comments were unfair; in many

[106] Rohden, *Vom Luftkrieg*, p. 4.   [107] Homze, *Arming the Luftwaffe*, p. 120.

[108] Murray, *Luftwaffe*, p. 16.   [109] Bekker, *The Luftwaffe War Diaries*, p. 554.

[110] Edward L. Homze, "The Luftwaffe's Failure to Develop a Heavy Bomber before the Second World War," *Aerospace Historian*, March 1977, 21.

[111] David Irving, *The Rise and Fall of the Luftwaffe: The Life of Field Marshal Erhard Milch* (Boston, 1973), p. 262.

respects the He 177 was an outstanding aircraft. But the persistence of Luftwaffe and Heinkel engineers in forcing technical innovation went well beyond reasoned analysis and explains much about failure of the project.

Despite factors inhibiting innovation, the Luftwaffe did make significant strides in the technical and technological capabilities necessary for strategic bombing. In 1939 its basic bomber designs, the He 111 and Dornier 17, were the most capable medium bombers in the world. Robust, fast, and capable of extensive refinement and design, these aircraft were the backbone of Luftwaffe bomber wings. The Junkers 88, which arrived in operational squadrons at the end of 1939, was one of the most versatile aircraft of the war. It would have been an even more formidable aircraft were it not for the 250,000 design changes reputedly required by the air staff – design changes that increased its weight from seven to twelve tons and reduced its speed from 500 kmph to 300 kmph.[112] By way of comparison, comparable RAF bombers, like the Hampden and Whitley were markedly inferior.[113]

With a recognition of the importance of air superiority, the Germans did try to provide long-range escort fighters to bomber formations deep in enemy territory. They expressly designed the Bf 110 for the long-range escort mission. Fortunately for the British, that aircraft proved markedly inferior to both the Spitfire and Hurricane; in Spain, the Germans also experimented with equipping Bf 109s with drop tanks to extend range to support bombers. Inexplicably, they failed to use such tanks during the Battle of Britain, most probably because they believed that the Bf 110 could successfully perform the mission. When they discovered that it could not, it was too late.

The Germans also made significant progress in developing navigational and blind bombing aids, a technology which the British failed to develop in the prewar period. In 1938, the Luftwaffe began operational testing of the so-called "X-apparatus," a sophisticated system of interlocking navigational beams that enabled German bombers, under the right conditions, to bomb within 300 yards of designated bombing points.[114] At the same time, the Telefunken Company developed a simpler system, *Knickebein* ("bent leg"), so easy to use that it did not require specially trained crews to be effective.[115] The existence of these and later devices

---

[112] Deist, et al., *Das Deutsche Reich und der Zweite Weltkrieg*, vol. 1, p. 490.

[113] One Whitley pilot noted with traditional British understatement that his aircraft "was not the sort of vehicle in which one should go pursue the King's enemies." Alfred Price, *Battle over the Reich* (New York, 1973), p. 12.

[114] Boog, "The Luftwaffe and Indiscriminate Bombing," p. 383.

[115] Alfred Price, *Instruments of Darkness* (New York, 1978), p. 22.

insured that the Luftwaffe was the only air force in the world capable of blind bombing with some degree of accuracy for the first three years of the war. As with tactical air power, the Spanish Civil War exercised an important role in German conceptions about strategic bombing. Despite the fact that the Luftwaffe committed relatively few aircraft to the conflict,[116] the Germans learned useful lessons. Spain underlined the difficulties involved in navigation and blind bombings; the development of "X-apparatus" and *Knickebein* were direct results.

In discussing German strategic bombing one also needs to discuss the role of "morale" or "terror" bombing in Luftwaffe doctrine. Obviously, the Nazi leadership sought to develop an air force as the most effective means of cowing its opponents. Nevertheless, "Conduct of the Air War" failed to emphasize air power's psychological role. While it suggested that air power might have important effects on enemy governmental and population centers, it ruled out attacks for the sole purpose of generating "terror among the population."[117] Nevertheless, the Luftwaffe's presumed capability was a primary factor in the abject surrender of the Western Powers at Munich. In fact the Luftwaffe's potential capability at the time was largely illusory – but it was perception that counted.[118]

However, creation of an effective strategic bombing capability required more than doctrinal and industrial developments. It also demanded intensive intelligence preparation, particularly regarding the industrial make-up of potential adversaries. The Germans never mounted such an effort in the prewar period and only began an amateurish effort in the middle of the war. Luftwaffe intelligence remained the most neglected and undermanned branch of the service.[119] As a result, it lacked targeting information essential to conduct a sustained strategic bombing campaign. Its appreciation of the British aircraft industry, for example, rested almost entirely on a standard reference book on British industry, obtained from a London bookstore, the book order written on Reich Air Ministry stationery.[120]

German disdain towards intelligence and analysis seriously inhibited

---

[116] At its height in autumn 1938 the Condor Legion only totaled forty He 111s, five Do 17s, three Ju 87s, forty-five Bf 109s, and four He 45s. Matthew Cooper, *The German Air Force, 1933–1945, An Anatomy of Failure* (New York, 1981), p. 59.

[117] *Die Luftkriegführung*, paragraph 186.

[118] Murray, *German Military Effectiveness*, p. 64.

[119] Horst Boog, "German Intelligence in World War II," *Aerospace Historian* 33, June 1986, pp. 121–129.

[120] Derek Wood and Derek Dempster, *The Narrow Margin* (New York, 1961), p. 103.

development of any coherent targeting doctrine. Manuals referred in general terms to attacks on key industries, railways, gas, electricity, and so forth, but only on a general level.[121] Some German air power advocates sought to identify specific target groups to provide far-reaching effects; electrical power generation and, in the case of the Soviet Union, petroleum and transportation, were frequent recommendations.[122]

Yet the Luftwaffe neglected the sort of sustained "nodal" or "chokepoint" analysis that characterized preparations for the American strategic bombing effort. The first serious attempts along these lines occurred only in 1943 when the Soviet war economy came under sustained (and belated) scrutiny.[123] Admittedly American airmen did not get into the business of serious economic analysis until 1939 with creation of a small office of economic analysis in that year.[124] But the crucial point is that the Americans and British recognized the importance of intelligence from the start of the war and devoted significant resources to building up their capabilities. The Germans did not and rapidly fell behind.

When the war came in 1939, the Germans possessed more broadly based conceptions of air power, and hence the Luftwaffe intervened in military operations more effectively during the first two years. Even in strategic bombing the Germans possessed capabilities that were far in advance of those possessed by their Anglo-American opponents. Admittedly, the innovations in the Luftwaffe were not sufficient to win the Battle of Britain, but its strategic bombing capabilities had a significant impact on the Dutch – the bombing of Rotterdam caused the collapse of Dutch resistance – and the Yugoslavs – the bombing of Belgrade resulted in the collapse of Yugoslav resistance. But in terms of executing a strategic bombing campaign over great distances the Germans held a marginal lead, a lead that they would soon squander.

---

[121] *Die Luftkriegführung*, pp. 49ff.

[122] Oberst Frhr. von Bülow, "Die Grundlagen neuzeitlicher Luftstreitkräfte," *Militärwissenschaftliche Rundschau* 1 (1936): 83; Major Macht, "Engpässe der russischen Wehrwirtschaft," *Die Luftwaffe* 2 (Heft 3, 1937), p. 53.

[123] RdL u ObdL, Az 8 n 10 02 Nr 2161/43 g. Kdos (LD Ag III 10), "Vorschläge für Luftangriff auf die Stormversorgung des Bereiches Moskau-Obere Wolga," 31.7.43, OKL/485, National Archives T321/104/109.

[124] Wesley Frank Craven and James Lea Cate, *The Army Air Forces in World War II*, vol. I, *Plans and Early Operations, January 1939 to August 1942* (Washington, DC, 1983), p. 53. Not until early 1941 did the RAF significantly improve the position of intelligence within the air staff to support its own offensive against Germany. R.V. Jones, *The Wizard War, British Scientific Intelligence, 1939–1945* (New York, 1978), p. 182.

Innovation and war

The primary focus in this essay has been on innovation before the war, but innovations in strategic bombing that occurred during the conflict raise important issues. The Germans waged the first strategic bombing campaign of the war against Great Britain and in many respects they came closer than historians have admitted to victory. In the summer's air offensive they put the RAF in an extraordinarily dangerous position. Only the German decision to switch from a counter-force to a counter-value strategy in early September (the decision to take out London) took the pressure off Fighter Command at a critical moment in the battle.[125] That decision reflected Hitler's and Göring's conception of air war – one very much within the Douhetian approach. It also received strong support from Field Marshal Albert Kesselring. Moreover, the Germans were well aware of the strong fears in British appeasement circles about the threat of air attack, while Churchill's political position in late spring and summer, current mythology notwithstanding, remained tenuous, given substantial support in the Cabinet for a deal with Hitler.[126]

The defeat of the daylight offensive brought on a sustained night campaign against Britain's defenseless cities. But British scientific intelligence had uncovered the nature of German blind bombing and navigation devices and developed means to counter those bombing aids before the night-time offensive moved into high gear. As R.V. Jones points out:

> By February 1941 the Battle of the Beams was as good as won. We had another three months of bombing to endure, but all three German systems . . . were defeated. Many bombs therefore went astray, often attracted by the decoy fires that were now part of the countermeasure programme. . . . With the last major raids of April and May 1941, the Luftwaffe was therefore not only tending to miss its targets, but it was beginning to encounter losses on a potentially prohibitive scale. . . .

> What would have happened [had we not discovered the German systems]? We should certainly have been slower with radio countermeasures, and the bombing of inland towns must have been worse, perhaps much worse. With our night fighters and guns powerless, radio countermeasures were our only means of defense. Not only could there have

---

[125] For the most complete discussion of the Battle of Britain see Mason, *Battle over Britain*.

[126] John Costello, *Ten Days to Destiny, the Secret Story of the Hess Peace Initiative and British Efforts to Strike a Deal with Hitler* (New York, 1991).

been more Coventrys, but Milch's aim of knocking out our aero-engine factories might have been achieved.[127]

And it is worth mentioning that despite RAF countermeasures German bombing did have a significant impact on British morale.[128]

In June 1941 the Germans moved against the Soviet Union in a campaign that the army was supposed to finish by fall. In July 1941 on the basis of this assumption, German production priorities switched to aircraft and submarines to destroy Britain's strategic position. The collapse of Barbarossa in December and the declaration of war on the United States that same month, however, confronted the Third Reich with a strategic situation that was no longer within the capacity of Germany's military or production base to solve. In the east, Germany faced an intractable land opponent; in the west, the Reich confronted the growing threats of RAF and U.S. Army Air Force strategic bombing, while the *Kriegsmarine* conducted its own massive campaign against Allied seaborne commerce.

By fighting all three battles the Germans insured their defeat. But the response to the air threat raises interesting issues. Instead of meeting the Combined Bomber Offensive with an effective air defense, the Germans, particularly Hitler and Göring, demanded a Douhetian approach: through early 1944 the Germans stressed the production of bombers to retaliate against the British.[129] Then in 1944 they attempted to overcome problems in training, fuel, and production by creating a strategic bombing capability to retaliate against the attacks on Germany's cities and industries – one that would not require piloted vehicles. The V-1 and V-2 weapons failed, but both represented impressive technological innovation; there was never a counter to the V-2. But it was in terms of their strategic effects that these weapons represented fundamentally flawed, technological innovations. The resources expended on these weapons during the last eighteen months of the war equaled production of 24,000 fighter aircraft.[130] The German response to strategic bombing reflected a

---

[127] Jones, *The Wizard War*, pp. 179–180.

[128] As Harold Nicholson noted in his memoirs on 10 May 1941: "Herbert Morrison is worried about the effect of the provincial raids on morale. He keeps on underlining the fact that the people can not stand this intensive bombing indefinitely and that sooner or later the morale of other towns will go, even as Plymouth's has gone." Quoted in Angus Calder, *The People's War, Britain, 1939–1945* (New York, 1969), p. 248.

[129] For German air defense policies see Murray, *German Military Effectiveness*, chapter 4.

[130] U.S. Strategic Bombing Survey, "V-Weapons (Crossbow) Campaign," Report #60 (January 1947). A recent estimate for the economic resources consumed by the V-weapon programs estimates that it may have cost the Reich an amount equal to a quarter of the

human emphasis on retaliation instead of defensive capabilities. But as with most German military history in the twentieth century, the Germans proved able innovators in developing techniques and technical capabilities; but a strategy that confronted a single German nation with multiple threats made such innovations irrelevant to the outcome.

The British and American responses to the challenge that the war posed to their concepts of strategic bombing had a number of similarities. Both had the opportunity to observe the Luftwaffe's mistakes in the Battle of Britain. Nevertheless, both seem to have disregarded virtually all the lessons.

The Americans ignored such danger signals as the fact that without fighter support German bombers were not able to survive in the face of British fighter opposition. Neither the British nor the Americans worried much about the difficulties the Germans had in achieving significant levels of damage to industries or even cities. What they did get right was that much greater levels of production and force structure were going to be needed to support a strategically effective bombing campaign. The enormous production totals of British and American industry in 1943 and 1944 thus formed the base on which success in the Combined Bomber Offensive rested.

But the conduct of the campaign by both air forces encountered considerable difficulties. In August 1941 the British discovered that they were not hitting much of anything.[131] Only then (and not after the *Blitz*) did Bomber Command's senior leaders become interested in technology to support their efforts. But even at the height of its successes in 1943, Bomber Command ran a terrifying race with German night defenses; and only after defeat in the Battle of Berlin in winter 1943–1944 had brought Bomber Command's raids into Germany to a halt, did its leader interest himself in having night-time, long- range escort fighter support.[132] In September 1944, with the reconquest of France, Harris resumed control of his bombers; yet despite substantially improved accuracy, he persisted in a single-minded campaign to wreck German cities – one suspects largely to prove prewar doctrine correct.

---

cost of the U.S. Manhatten Project that built the first atomic bombs and, relative to the size of the German economy, imposed "a burden on the Third Reich roughly equivalent to that of the Manhatten Project on the United States." Michael J. Neufeld, *The Rocket and the Third Reich: Peenemünde and the Coming of the Ballistic Missile Era* (New York, 1995), p. 173.

[131] Webster and Frankland, *Strategic Air Offensive against Germany*, vol. 4, appendix 13, p. 205.

[132] Ibid., vol. 2, *Endeavour*, p. 193.

The American effort pursued a similar path. The buildup of strategic bombers began in 1942; but not until spring 1943 did the bomber force reach sufficient strength to raid German targets and only in August 1943 did Eighth Air Force have the forces to attack targets in central Germany. The August and October raids against Schweinfurt are the most famous attacks in a summer and fall of major efforts that nearly broke the command.

Eighth Air Forces's difficulties in 1943 underline a mismatch between doctrine and conceptions on one hand and estimates of the enemy and his industrial capabilities on the other. The issue is not that there were substantial problems with American concepts at the outset. Rather they have to do with the persistence in following a particular path in the face of contrary evidence. Not until U.S. airmen came to grips with that mismatch were they able to utilize air power to break the Luftwaffe and damage the German economy's ability to support the war.[133]

The problems that the RAF and U.S. Army Air Forces had in adapting to the real conditions of war are part of a broader pattern in the military history of the twentieth century. Military organizations often take mistaken conceptions into war. This should not be surprising. *What is surprising* is the tenacity with which they have often maintained these mistaken conceptions in the face of overwhelming, contrary evidence and at the cost of large numbers of young men called upon to fight their battles. This is as true of the airmen in the RAF and U.S. airmen as it is of the generals responsible for British troubles on the Western Front in World War I or those responsible for the debacle in Vietnam.[134]

CONCLUSION

There are a number of perspectives on innovation that this case study suggests. The military organizations of all three nations found the concept of strategic bombing worthy of serious study during the interwar period. How effectively they actually innovated depended on a number of vari-

---

[133] For the contribution of the Combined Bomber Offensive to the winning of the war see Williamson Murray, "Reflections on the Combined Bomber Offensive," *Militärgeschichtliche Mitteilungen*, 51 (1992), Heft 1.

[134] Two books appeared at approximately the same time and closely examined the persistence of prewar conceptions in the conduct of operations in these two examples. The patterns match with those of the airmen responsible for the American and British air efforts. See Timothy Travers, *The Killing Ground, The British Army, The Western Front, and the Emergence of Modern War, 1900–1918* (London, 1987); and Andrew F. Krepinevich, Jr., *The Army and Vietnam* (Baltimore, 1986).

ables, some of which the different environments influenced in subtle but important ways.

Perhaps the greatest problem that confronted airmen and innovators was the extent and range of assumptions they had to make in thinking about and preparing for strategic bombing. Virtually everything was up in the air. There were no clear answers to a number of crucial questions: how many aircraft would be needed? Could bomber formations protect themselves? What kind of intelligence would one need and how would one obtain it? How could one measure the effectiveness of bombing attacks? How accurate would bombing raids be under the pressures of combat? Where was technology going in a period of rapid change? What options were open to enemy defensive forces? The list of such questions was almost endless and only the war eventually provided a clearer path. What this suggests for modern planners in a period of rapid innovation and technological change is the extent to which they and their organizations must challenge basic assumptions.

By the late 1930s all three powers had access to the technology on which strategic bombing in World War II would rest. But again there were subtle differences in how they utilized the technological marketplace of the late 1930s. The Germans faced the most difficulties. Because of the short period of time available before the outbreak of war the Luftwaffe failed to solve the developmental problems associated with creating power plants that four-engine bombers required. They attempted to short-circuit the deficiency by designing an aircraft that coupled two power plants within a single nacelle. Had the design worked, the He 177 might have been the best heavy bomber of World War II; but here the best was the enemy of the good. What was not praiseworthy was the persistence with which the Germans stuck with a flawed approach.

The Americans possessed the advantage that the distances over which they operated forced development of civilian aircraft and engine technology that matched the requirements of military innovation. By 1935 they had a prototype of a four-engine bomber available, one that with improvements provided the workhorse of the daylight strategic bombing campaign against Germany in 1943 and 1944. The British had less success in this area which underfunding only exacerbated. But one suspects, particularly given the innovations in Fighter Command, that Bomber Command contained too many unimaginative officers accepting of assumptions that the bomber would always get through and do significant damage to the enemy whatever the difficulties.

On the electronic side of innovation, the Germans provided the lead in their support of navigation, blind flying, and blind bombing. Admittedly,

the Americans did ground-breaking work on blind landing and navigation technologies, but the assumption on which American doctrine rested – namely that strategic bombing would occur in daylight, visual flight
conditions – limited the potential that civilian and even air corps technological innovations might have opened.

Several factors pushed the Germans towards recognition that they
needed blind-bombing technology to support strategic bombing. First,
they had a clear sense they would be the ones to unleash the war; consequently, they had the clearest grasp of what the next war would look like
– at least as it unfolded in Central Europe. Admittedly, once they got beyond the framework of that war, they proved inept at solving the problems which they themselves had raised. Experience gained in Spain also
helped to underline the problems involved in finding and hitting targets.

When the Germans confronted the problem of launching a strategic
bombing campaign against the British Isles in summer 1940 they possessed technological innovations that their opponents would not possess
for the next two years. These technological aides brought them perilous-
ly close to success in the *Blitz*. But British scientific intelligence, something
the Germans never came close to matching, thwarted the German effort.

We might at this point turn to questions of innovation within an organizational context. As in armored warfare, the Germans suggest a number of interesting lessons. Since the Luftwaffe's culture drew heavily from
the army this should not be surprising. What made the Germans effective
innovators was their willingness to recognize their problems. If their
bombers had difficulty in hitting targets, then the senior leadership recognized it had a problem. And the system attempted to provide solutions.
Where the Germans failed lay in their incapacity to fit strategic bombing
within a coherent national strategy. One might suggest, of course, that a
coherent national strategy would have resulted in the Germans laying
down their arms in 1943.

On the Anglo-American side, Allied air forces became prisoners of their
own doctrine. Consequently, both air forces confined innovation within
a narrow framework. The clearest example is the disdain that British and
American airmen exhibited in considering escort fighters. But the British
belief that "the bomber would always get through" resulted in a minimization of electronic innovations through to the Butt report of 1941.
Ironically, when Bomber Command created enormous innovations in
both technology and tactics in 1943 and 1944 that opened up a wide
range of possibilities, Bomber Command persisted in its single-minded
path of turning over the rubble of Germany's cities again and again.
Similarly in the United States, airmen cast a doctrinal framework that

blinded them to virtually all the warning signs that the Battle of Britain should have raised. It took Eighth Air Force's commander until October 1943 to recognize that the long-range escort fighter was an *essential* element in the survival of his bombers, not a luxury.[135]

We have suggested in this case study that the German approach to land-based air power in the interwar period largely rested on previous combat experience with air power (World War I), the actual prospects that German armed forces confronted in the late 1930s, and lessons learned from foreign conflicts such as the Spanish Civil War.[136] We also have argued that German airmen placed air innovation within a broader context of the operational goals of the army as well as the Luftwaffe, which resulted in a more balanced approach than the Anglo-American emphasis on strategic bombing.[137] It is possible at this juncture to push this line of analysis further and observe that, in the German and American cases, the differences in their approaches reflected different assumptions about the very nature of combat processes. The 1933 *Truppenführung* expressed a decidedly Clausewitzian view of the essential nature of combat interactions. Consider this passage from the *Truppenführung*'s introduction:

> Situations in war are of unlimited variety. They change often and suddenly and only rarely are from the first discernible. Incalculable elements are often of great influence. The independent will of the enemy is pitted against ours. Friction and mistakes are of everyday occurrence.[138]

In much the same vein, the Luftwaffe's basic doctrinal manual underscored that "the nature of the enemy, the time of year, the structure of the [enemy] nation, the character of the [enemy] people, as well as one's own

---

[135] It is worth noting that not all American airmen were so blind. General Jimmy Doolittle in the Mediterranean was writing Arnold in May 1943 "intruders would greatly increase the effectiveness of our strategic operations." Letter from Doolittle, Subject: Escort Fighters; To: Commanding General, US Army Air Forces; Thru: Commanding General, Northwest African Air Forces, 22.5.43.

[136] The basic lessons that the Germans drew from the Spanish Civil War were: first, that "strategic" bombing was going to be more difficult than many thought; second, that bombing accuracy was also going to be a major problem; and, third, that close air support would often have an important role to play in ground operations.

[137] However, interservice cooperation between the Luftwaffe and the German Navy remained virtually nonexistent throughout the history of the Third Reich. Thus, the RAF's postwar assessment that the Luftwaffe's "main role" during World War II had been "to annihilate enemy air power and then to give the army maximum support" was reasonably close to the mark (*The Rise and Fall of the German Air Force: 1933–1945*, p. 49).

[138] *Truppenführung*, trans. of Part 1, U.S. Army Report No. 14,507, 16 March 1936, p. 1. The explicit reference in this passage to friction and mistakes being a constant feature of military operations obviously reflects a Clausewitzian outlook.

military capabilities" would determine air power's role in a future war.[139] These German views about the underlying nature of war are decidedly at odds with those manifest in Air Corps Tactical School bombardment theory or AWPD-1.

The contrast between German and Anglo-American efforts to develop land-based air power also highlights how resistant doctrinal beliefs – especially visionary ones – can be to empirical data, including actual combat experience. If the experience of U.S. airmen before and during World War II is any guide, doctrinal beliefs are almost impervious to countervailing evidence when those beliefs bear on institutional autonomy or existence, however justified those ends might be.

In turn, this insight raises the issue of how important the intellectual atmosphere in which peacetime military innovations occur is to success. Without some institutional process or consensus on the importance of subjecting doctrinal tenets, theoretical conclusions, or quantitative effectiveness calculations to *honest* evidentiary tests, it appears all too easy for military organizations to follow their hopes and dreams into catastrophe. A further implication of the different directions in which German and Anglo-American air power developed between the wars is that none of the three air forces were fully, or even adequately, prepared for the full gamut of prospective missions at the outset. The Germans had certainly prepared more thoroughly in 1939–1940 than their adversaries to employ air power to influence the outcome of land operations. The Luftwaffe's bombing of French forces in and around Sedan on the afternoon of 13 May 1940 offers a good example of the flexibility that the Luftwaffe had built into its bomber and dive-bomber forces.[140]

In the end, as the beginning of this paper suggested, the massive fleets of bombers and fighters and the almost seemingly endless battle of attrition resembled none of the prewar conceptions of airmen. The prewar innovations of the Germans on the ground and in strategic bombing brought them close to victory. But their incapacity to adapt to a different war resulted in their ultimate defeat. Their opponents eventually did innovate from the technological and tactical level to the strategic in a fashion that eventually brought victory. The cost of that victory, however, should give small room for complacency.

---

[139] *Die Luftkriegführung*, paragraph 11.

[140] Doughty notes that the Luftwaffe's aerial assault on 13 May 1940 sapped the will of French defenders by disrupting the fire of numerous artillery batteries and providing immediate fire support (*The Breaking Point*, p. 137). Besides accompanying escort fighters, II Air Corps put 310 bomber and 200 dive-bomber sorties into this effort; these attack sorties were also supplemented by StG 77 from VIII Air Corps (Bekker, *The Luftwaffe War Diaries*, pp. 118–119).

4

CLOSE AIR SUPPORT
The German, British, and American experiences, 1918–1941

RICHARD R. MULLER

The history of air power in Germany, Britain, and the United States during the interwar period provides excellent examples of successful military innovation. To some degree, each air force combined the lessons learned in World War I, postwar theoretical development, and increasingly advanced aviation technology into viable doctrines, weapons systems, and combat formations. The creation of a long-range heavy day bombardment force by the United States Army Air Corps and the establishment of an integrated air defense system in Britain are only the most well known of these developments. Although in most cases capability fell far short of the "ideal" suggested by army air corps, Luftwaffe, or RAF doctrine, the leadership of each service had thought through the problems of, and implemented the means of, executing the missions of conventional bombardment, pursuit, and observation aviation.

In contrast, the ability of the major air powers to carry out the task of close air support remained minimal until the final two years of World War II, with the qualified exception of Germany.[1] The air forces of the interwar era even experienced difficulty creating an accepted vocabulary for this form of air attack. "Close air support" (CAS) in its modern usage refers to "air action against hostile targets which are in close proximity to friendly forces and which require detailed integration of each air mission with the fire and movement of these forces" in order to reduce the danger from friendly fire.[2]

---

[1] Two recent works provide excellent narrative and detail on the evolution of close air support. They are: B. F. Cooling, ed., *Case Studies in the Development of Close Air Support* (Washington, DC, 1990), and Richard Hallion, *Strike from the Sky: The History of Battlefield Air Attack, 1911–1945* (Washington, DC, 1989).

[2] USAF Manual 1–1, *Basic Aerospace Doctrine of the United States Air Force*, vol. 2 (Washington, DC, 1992), p. 276.

During the interwar period, air action in direct support of ground forces occupied an ill-defined position within the air doctrines of the great powers. In the army air corps, this mission came under (but was only one component of) "attack aviation." In Britain, "trench strafing," "ground strafing," "direct air support" and "close support," were the most commonly employed terms, although their meanings frequently overlapped. The first two terms connoted an auxiliary mission against a target of opportunity.[3] In the German air service, the task was described initially as the "battle aviation mission" (*Schlachtfliegereinsatz*), and subsequently as "close combat aviation" or "direct army support," (which included tactical interdiction) before the Luftwaffe officially resurrected the World War I terminology in 1943.[4] All three air forces recognized the need to develop a capability for intervening in ground actions, closely coordinated with surface forces; yet, for a variety of reasons the development of the doctrine, training methods, and specialized aircraft and equipment required for the mission received a low level of attention.

This essay will examine the factors that facilitated or, more usually, inhibited the process of developing close air support techniques within three major air forces. Peter Paret has framed the general problem as follows: "How do military institutions adjust to new realities; what forces carry innovation forward, and what obstacles stand in its way?"[5] This chapter will also explore the institutional, technological, political, and doctrinal factors that influenced the interwar development of army support aviation. Finally, it will attempt to illuminate common obstacles to the process, as well as some common elements in the successful innovation that did occur.

WORLD WAR I

Aviation experiences in World War I were rich and diverse. The combatants developed every modern air power role and mission in the conflict,

---

3 Air Ministry, *The Second World War, 1939–1945*, Royal Air Force, Air Support (London, 1955), United States Air Force Historical Research Agency (USAFHRA) K512.041-3235, p. 7.

4 For evolution of the terms, see Horst Boog, *Die deutsche Luftwaffenführung 1935–1945* (Stuttgart, 1982), ch. 3, and Generalstab 8. Abteilung, "The Development of the German Ground Attack Arm and Principles Governing its Operations up to the End of 1944," 1.12.44, Air Historical Branch Translation No. VII/14, USAFHRA 512.621 VII/14, p. 2.

5 Peter Paret, "Innovation and Reform in Warfare," *The Harmon Memorial Lectures in Military History*, Number Eight (Colorado Springs, 1966), p. 2.

some to a remarkable degree of sophistication.[6] Although the German Zeppelin, Gotha, and R-plane strategic bombing raids on Britain and the activities of Trenchard's Independent Air Force provided much of the grist for the mill of postwar air power theory, the potential of army support aviation was much clearer in 1918 than the promise of strategic bombardment. Close support of ground forces played a major, if perhaps not decisive, role in the land campaign on the Western Front in the final year of the war. Yet the trend of rapid innovation and development that characterized close air support in World War I did not carry over into the interwar period. Like much of the legacy of the Great War, these were lessons that would need relearning.

Imperial Germany

Although some authorities credit the British Royal Flying Corps with inventing the concept of close air support or "ground strafing,"[7] the Imperial German air service (*Luftstreitkräfte*) developed army support aviation into a viable military instrument. The German emphasis on mastering the operational level of warfare, coupled with realistic training and a productive tradition of inter-arm cooperation, ensured this outcome.

Early in World War I, the nascent German air service developed observation and infantry contact aircraft to cooperate with the ground forces as auxiliaries. The duties of these aircraft were not primarily offensive: they were to gather information and pass it to the concerned ground formations. The other powers also developed aircraft to perform this function. German regulations for "infantry aircraft" called for their occasional use against enemy ground troops, attacking with fixed or flexible-mounted machine guns or, in some cases, with light bombs.[8] In order to facilitate this "contact patrol" work, the German staffs developed simple visual recognition devices and procedures to delineate front line positions and to ease liaison. Although "infantry aircraft" gained individual successes against ground targets during the Battle of the Somme in summer 1916, "organizational and tactical deductions" did not emerge from this experience until the following year's campaign in Flanders.[9]

The use of observation aircraft as makeshift close air support machines

---

[6] Lee Kennett, *The First Air War, 1914–1918* (New York, 1990).

[7] Hallion, *Strike from the Sky*, p. 20.

[8] "The Infantry Aeroplane and the Infantry Balloon," British translation of a captured German document dated 1st September 1917, Army War College, February 1918, p. 9.

[9] "German Ground Attack Arm," p. 1.

was common. Had the *Luftstreitkräfte* pursued this line of development it probably would not have achieved noteworthy success. The German solution to the problem of close support promised to bring larger dividends. The battle for air supremacy over the Western Front compelled the Germans to develop specialized escort aircraft for their infantry contact machines. These escort aircraft, organized into *Schutzstaffeln* ("protective flights") were fast, rugged two-seaters, with long loiter times, powerful armament, and speed and altitude performance approximating that of contemporary fighter aircraft.[10]

Experience gained during summer and fall 1917 inspired the Germans to experiment with employment of *Schutzstaffeln* as ground attack formations. The results of early combat trials were so promising that Ernst von Höppner, commanding general of the air service (*Kogenluft*), elected to form these aircraft into specialized ground attack units, eventually known as *Schlachtstaffeln* (literally "battle flights," usually translated as "ground attack squadrons").[11] Although the aircraft themselves, such as the Halberstadt CL II and IV, and the Hannover CL III, did not represent technical breakthroughs *per se*,[12] their method of employment, or operational doctrine, was revolutionary.

The German search for the ideal close support aircraft (CAS) brought to light two opposing design philosophies that have since dominated the debate on CAS aircraft. The findings of a fall 1917 conference of ground attack pilots strongly favored two-seater fighters.[13] Indeed, the *Amerika Programm*, a massive mobilization of the air armament industry to offset the entry of the United States into the conflict, called for increased production of these proven types.[14] Additionally, the Germans pursued a parallel line of research and development which emphasized all-metal, heavily armored machines, such as the Junkers J 1 "furniture van," designed solely for close support missions. Although poorly equipped to engage swifter Allied fighters, the Junkers aircraft were well-protected against small-arms fire from the ground. The appeal of the armored aircraft had a discernible impact upon later German and American practice, even

---

[10] Hans Arndt, "Die Fliegerwaffe," in Friedrich Seesselberg, ed., *Der Stellungskrieg, 1914–1918* (Berlin, 1926), pp. 344–346.

[11] Hans Arndt, "Der Luftkrieg," in M. Schwarte, ed., *Der Große Krieg, 1914–1918*, vol 4 (Leipzig, 1922), p. 599.

[12] Georg Paul Neumann, *Die deutschen Luftstreitkräfte im Weltkriege* (Berlin, 1920), pp. 90–93.

[13] "German Ground Attack Arm," p. 1.

[14] John H. Morrow Jr., *German Air Power in World War I* (Lincoln, NE, 1982), pp. 115–116.

though the more conventional two-seater types predominated until the end of the war.

While German technological advances in close support aviation were noteworthy, the doctrinal changes proved more significant. The German command successfully integrated the new weapon into the overall plans for breaking the trench stalemate on the Western Front. Aircraft were to be used in conjunction with Colonel Georg Bruchmüller's new methods for employing artillery and the new infantry infiltration or "stormtroop" tactics.[15] Preparations for the March 1918 "Michael" offensive, which would combine the new techniques on a grand scale, were meticulous. They featured realistic war games and maneuvers, and included live-fire exercises involving assault troops and aircraft.[16]

The procedures for employing close support aviation that grew out of the maneuvers were simple but effective. The attacking Germans would use aircraft *en masse*, with up to six *Schlachtstaffeln* employed together against a decisive point. The aircraft of each squadron would cross the enemy's trenches in line abreast, with subsequent waves in trail. They would then attack the enemy front line positions and artillery emplacements from the rear, at low altitude, with gunfire, grenades, and light fragmentation bombs.[17] The German army general staff's official memorandum on the subject maintained, "The object of battle flights is to shatter the enemy's nerve by repeated attacks in close formation and thus to obtain a decisive influence on the course of the fighting."[18]

No fewer than thirty-eight ground attack squadrons massed on the Western Front for the offensive; their numbers far exceeded the number of conventional bombers in the German air service's inventory.[19] Most importantly, the aircraft operated in conjunction with assault troops thoroughly grounded in the principles of air-ground cooperation. Employment of the garishly-painted Halberstadts and Hannovers played a significant part in the breakthrough battles of spring 1918, and the German command came to rely upon these weapons – despite the inevitably heavy

---

[15] John R. Cuneo, "Preparations of German Attack Aviation for the Offensive of March 1918," *Military Affairs*, 7, Summer 1943, pp. 69–70. This admittedly dated source remains an unsurpassed discussion of the subject.

[16] James S. Corum, "The Old Eagle as Phoenix: The Luftstreitkräfte Creates an Operational Air War Doctrine," *Air Power History*, 14, Spring 1992; Cuneo, "Preparation of German Attack Aviation," p. 70.

[17] Arndt, "Fliegerwaffe," p. 345.

[18] "Employment of Battle Flights," February 1918, in R. A. Jones, *The War in the Air*, vol 4 (Oxford, 1934), p. 434.

[19] Arndt, "Fliegerwaffe," p. 352.

losses sustained – because of their mobility and "shock effect" during the defensive battles of the final months of the war.

Close support was not without its problems. Apart from the alarming casualty rate, the German air and ground commanders expressed dissatisfaction with the available methods of command, control, and communications. While the ground attack formations, based at frontline airfields and served by an efficient communications network, could be dispatched with remarkable speed to a static front, support of a fast-moving advance was another matter entirely. Höppner recalled that maintaining reliable communications between the advancing troops and their supporting aircraft was the greatest technical problem plaguing close support operations.[20]

The German experience with the *Schlachtstaffeln* in World War I established the pattern for the further development of close support aviation in Germany. The issue remained a tactical and operational one that representatives of the army and air service commands involved would work out with only general guidance from higher headquarters. The Germans solicited, received, and subjected to rigorous evaluation suggestions from the lowest levels of command before developing written doctrine. Yet German close support aviation would not regain the prominence it held in 1918 until late 1943, after a later air war had degenerated into something quite unexpected by planners.

Britain

Developments in Britain represented another of the intellectual currents regarding close air support. Although the British deployed small numbers of aircraft designed primarily for the "ground strafing" role late in the war, the standard practice of the Royal Flying Corps (RFC) – later the Royal Air Force (RAF) – saw aircraft designed for other missions, most notably pursuit and observation, employed for ground support as the situation demanded.[21] Although this method produced dividends, such unsystematic employment hampered thoughtful consideration, and thus development, of army support aviation.

In common with the Germans, British air and ground integration began with the employment of artillery and infantry observation and contact aircraft. During the Aubers Ridge attack in 1915, the first recorded use of cloth signalling panels for air to ground communication took

---

[20] Ernst von Höppner, *Deutschlands Krieg in der Luft* (Leipzig, 1921), p. 149.
[21] Hallion, *Strike from the Sky*, p. 20.

place.[22] The British borrowed freely from French practice and based their infantry contact patrol doctrine on a memo issued by General Henri Joffre in April 1916.[23] They then put the new methods to the test in a series of training exercises; provisional instructions for air and ground cooperation appeared the following month. Then, as now, the difficulty in distinguishing friendly troops rapidly was the major problem in such operations. Accordingly, much of the early British cooperation work concerned simple identification. The instructions pointed out, "The infantry were to indicate their progress by lighting flares on the initiative of the company or section commander; in reply to the question from the aeroplane, 'Where are you?'; at certain hours previously specified in orders; and on reaching pre-arranged positions."[24]

The premeditated use of aircraft for direct ground attack did not occur until the Passchendaele battle of summer 1917. At the same time that the Germans were systematizing their employment of two-seaters, the British began attacking ground targets with small twenty-five pound Cooper bombs, carried in makeshift racks beneath pursuit aircraft. The official historian noted, "As pilots of some of the fighting squadrons were up most of the night before the attack, fitting the improvised bombing racks to their aeroplanes, it may be assumed that the idea was adopted in a hurry."[25]

Indeed, Trenchard lamented the resulting lack of coordination: "In the attacks on troops on the ground . . . the actual objectives were left to individuals concerned and no organized attack was made in connection with any particular operation on the ground."[26] Even the official historians admit, "Not much of [the close support] was of the kind that had been carefully prepared beforehand, but it represented a determined effort, dictated by the special conditions, to give the infantry a helping hand."[27] Nevertheless, the assessment of one RFC pilot is telling. He noted in late 1917, "these low-flying attacks that we had to make, for which most of my young pilots were quite untrained, were a wretched and dangerous business, and also pretty useless."[28]

Despite unimpressive early results, by mid-1918 the RAF had learned to deliver effective close air support to the army. As with the German air service, the success of the RAF did not depend primarily upon the intro-

[22] R.A. Jones, *The War in the Air*, vol. 2 (Oxford, 1922), pp. 109–110.
[23] Ibid., p. 179.   [24] Ibid., p. 180.   [25] Ibid., vol. 4, p. 166.
[26] Brereton Greenhous, "Evolution of a Close Ground-Support Role for Aircraft in World War I," *Military Affairs*, 39, February 1975, p. 23.
[27] Jones, *War in the Air*, vol. 4, p. 166.
[28] Greenhous, "Close Ground-Support Role," p. 25.

duction of new weaponry. The British did develop the Sopwith TF 2 "Salamander," a specialized trench strafer with downward-firing guns, but standard pursuit types such as the Sopwith Camel or the magnificent Bristol F2B two-seat fighter performed most of the close support work.[29]

The British successes were the product of rethinking the methods for employing their existing aircraft. After perusing after-action reports, Trenchard noted, "Reports which came in from the battle-field indicated the demoralizing effect of the low-flying attacks on the German troops, but made clear also that these attacks should not be left to the discretion of individual pilots, but must be properly organized in coordination with ground operations."[30]

From the Battle of Cambrai onwards, British ground attack missions took place with increasing frequency; the RFC, however, had proved to be "rather slow learners."[31] Above all, the fact that ground attack was not the primary mission of any of the British units made it all too easy to ignore the dangerous and unpleasant task. Nevertheless, the Cambrai operation proved to be a developmental watershed for British close air support. It was conceived as a limited offensive, employing tanks in their first significant appearance on the Western Front, and the manageable size of the operation allowed for more careful and thorough training than in previous offensives.[32] While the training and exercises were not as ambitious as their German equivalents during the winter of 1917–1918, the British made considerable strides towards integrating air, infantry, and armor. Although no codified mechanisms for such joint operations existed, the close cooperation between commanders went a long way towards insuring the (admittedly transitory) success of the Cambrai attack.

The British gleaned a number of lessons from Cambrai, as well as from subsequent close support work, which peaked in August 1918. Analysts concluded that the "influence" of close support aircraft was "beyond dispute." Proponents of army support aviation frequently referred to General Erich Ludendorff's memoirs, in which the German commander, recounting the Amiens breakthrough on 8 August 1918, spoke of the "increased confusion and great disturbance" that British air attacks had created among his troops.[33] Such successes sealed a general acceptance that close air support had a role to play in future wars, although that role remained ill-defined.

---

[29] Lee Kennett, "Developments to 1939," in Cooling, ed., *Case Studies in the Development of Close Air Support*, pp. 23–24.

[30] Jones, *War in the Air*, vol. 4, p. 166.

[31] Greenhous, "Close Ground-Support Role," p. 24. [32] Ibid., p. 25.

[33] Air Corps Tactical School, Attack Aviation Course, 1937, USAFHRA 168.7045–30, p. 5.

The bulk of the RAF's postwar analysis focused not on the moral and material success of the attacks, which was notoriously difficult to assess (much of the evidence being anecdotal in nature), but rather on their high cost. British official historians, speaking of the Cambrai battle, noted that the casualty rate ensured that:

> a squadron of highly skilled and experienced pilots, flying first-class fighting aircraft, would be, so long as it was employed on concentrated low-flying attacks on front-line troops in prepared defensive positions, required to be replaced about every four days . . . only in exceptional cases, or in extreme emergencies, would such a rate of wastage be justified.[34]

The perception that close air support was a costly luxury and an aberration would dominate the next quarter-century and contributed materially to the RAF's failure to advance its close-support practices during the interwar period.

Finally, much has been made of the success of British close air support in the secondary theaters of World War I, most notably against the Ottoman Empire in the Palestinian campaign.[35] Indeed, RAF support to General Sir Edmund H.H. Allenby's offensive, which culminated in shattering the Turkish Seventh Army by air assault at Wadi el Far'a on 21 September 1918, seemed to highlight the value and potential decisiveness of close air support. Yet the Palestine campaign took place under exceptional conditions. The RAF had already established air dominance over the area, the opponent was demoralized and already in retreat, and the terrain was unusually favorable for air attack, while affording Turkish forces limited cover. The experience had more relevance to the RAF's postwar role in policing the British Empire than to any hypothetical war against a continental opponent. It was prudent and understandable for the British air staff not to fixate on the results of the Palestine campaign in crafting postwar doctrine. What they chose to do – essentially ignore it entirely – was neither.

The United States

The United States was a latecomer to the European war. Its air service entered the conflict a virtual *tabula rasa*, while its allies and principal opponent had progressed towards sophisticated operational doctrines by 1917. Although the postwar development of American air power owed

---

[34] Jones, *War in the Air,* vol 4, p. 257.   [35] Hallion, *Strike from the Sky,* pp. 29–36.

much to Mitchell's personal contacts with Trenchard, during the war it was the French air force, rather than the British, that had the greatest influence on American training and practice.[36] American ground attack aviation accordingly did not reach even the RAF's level of sophistication during World War I, but there were strong indications as the war drew to a close that this mission would play a leading part in the future development of an American air arm.

The body of American experience in World War I consisted of a few operations in which air attacks on ground troops, usually carried out by pursuit aircraft, had noteworthy effect. This small sample of experience served as the basis for conclusions about the future of close support aviation after the war. These actions (most notably at the battle of St. Mihiel in September 1918) usually involved targets of opportunity, such as retreating German columns caught in the open. An Air Corps Tactical School lecturer described the St. Mihiel battle as follows:

> On September 12 American and French pursuit airplanes found the VIGNUELLES-ST BENOIT ROAD filled with the enemy's retreating troops, guns, and transport. The road was a forced point of passage for such of the enemy as were endeavoring to escape from the point of the salient. All day long our pursuit airplanes harassed these troops with their machine gun fire, throwing the enemy column into confusion. The result of these attacks was that all movement on the road was blocked.[37]

Commentators reported equally dramatic results during the Meuse-Argonne offensive; they emphasized the effect on German morale of the mere appearance of American planes.[38] Yet the use of pursuit aircraft in a ground attack role nonetheless remained a "fourth-priority mission."[39] Even during a period of successful American air intervention in ground battles, as at St. Mihiel, the air battle involved pursuit aircraft mostly in their main task: contesting with the German *Jagdstaffeln* for air superiority.

The methods of liaison between American aircraft and ground forces lagged behind German, and even British, practice. In place of standardized identification methods, American units employed "panels, Bengal flares, projectors, signal lights, and various makeshift methods, such as mirrors, pocket torch lights, handkerchiefs, overcoat linings, etc."[40]

---

[36] ACTS, Attack Aviation Course, p. 2. [37] Ibid., p. 5.

[38] Colonel Edgar S. Gorrell, Final Report, in Maurer Maurer, ed., *The United States Air Service in World War I*, vol. 1 (Washington, DC, 1980), p. 7.

[39] "Tactical History of the Air Service," in Maurer, *US Air Service*, vol. 1, p. 207.

[40] Maurer, *US Air Service*, vol. 2, p. 208.

There were a number of efforts to carry out air-ground training during the preparations for several major ground assaults.[41] In addition, air units attempted to work out specific methods for supporting infiltrating troops, and for separating attacking enemy infantry from their accompanying tanks.[42]

As the war drew to a close, American air commanders had begun to think through the problems associated with employing attack aircraft and to make recommendations for the next stage in the development of the weapon. Mitchell produced a "Provisional Manual of Operations," dated 23 December 1918, which reflected much of the contemporary thinking about ground support aviation.[43] He noted that pursuit aviation, in addition to its primary task of winning air superiority, had to perform four additional "special missions." These missions included close escort for observation planes, cooperation with day bombardment squadrons, attacks on enemy observation balloons, and "attacks on ground troops." Use of pursuit aircraft in the latter role promised "valuable results during a major operation."[44]

Yet Mitchell clearly regarded use of fighters for this purpose as an expedient, and an unsatisfactory one at best. Evidently much impressed by the German example, he recommended that the air service concentrate "attack" aircraft (and the term would be the accepted American nomenclature until the middle of World War II) into formations specifically trained and equipped for the role. In language strongly reminiscent of the German high command's directives for the *Schlachtflieger*, he maintained that

> the successful employment of attack squadrons depends upon their concentrated, continuous, uninterrupted engagement at the decisive time and place. This condition limits their use to that particular portion of the battle front upon which the entire operation depends, and prohibits their distribution over relatively unimportant portions of the front line.[45]

Significantly, they were not to be used for air-to-air combat if avoidable.

Mitchell's formulation had much to commend it. It advocated the creation of a specialized branch of aviation to deal with a particular operational problem, instead of relying upon units and personnel whose real

---

[41] Ibid., p. 208.   [42] Ibid., p. 154.

[43] Kennett, "Developments to 1939," pp. 42–43.

[44] General William Mitchell, "Provisional Manual of Operations," in Maurer, *US Air Service*, vol. 2, p. 286.

[45] Mitchell, "Provisional Manual," p. 291.

expertise lay elsewhere. It was, moreover, primarily a doctrinal, rather than a technical, solution to the problem of air support, and thus could be implemented fairly quickly.

Despite the late start, the United States emerged from World War I with perhaps the best opportunity to create an air force with a true close support component. It had the benefit of a growing aviation industry, eloquent champions of this form of air power, and a chance to study and absorb the wartime experiences of all the major belligerents. Yet the United States, like Britain, would enter World War II with only a vestigial capability for close air support.

INTERWAR GERMANY: FROM WEIMAR
TO BLITZKRIEG ON POLAND

The fact that Germany launched World War II with a *Blitzkrieg* against Poland, a campaign featuring close cooperation between the German Army and the Luftwaffe, demonstrated that the Germans had not neglected close air support to the same degree as the Western Allies. Yet it would be wrong to assume an uninterrupted evolutionary path from the *Schlachtflieger* of 1918 to the Stukas of 1939–1941. The story of German close support features discontinuities and neglect as well as successful extrapolation of World War I lessons into the era of mechanized warfare. Some of the factors constraining German advocates of army support aviation resulted from unavoidable external factors. Germany's status as a defeated power, with a minuscule professional military operating under the eye of the Inter-Allied Control Commission imposed by the Versailles Treaty, seriously restricted all German military technical and tactical progress. Significant institutional factors within the German air arm – both the clandestine and, after 1935, open varieties – also conspired to relegate close army support to a low priority.

A recent commentator on early close support aviation noted that "the vanquished . . . learned more than the victors."[46] Indeed, the German attempt to assess and digest the operational and tactical lessons of World War I far surpassed those of the Allied powers. While the British and the Americans conducted postwar analyses, including ambitious official history programs, the German effort was more diverse and more oriented to "lessons learned"; consequently, it would play a crucial role in the German victories of 1939–1941. Its greatest successes lay in the operational realm: the creation of a workable doctrine for the employment of the

[46] Greenhous, "Close Ground-Support Role," p. 28.

panzer force is perhaps the best-known example. Close air support, almost by definition an operational and tactical problem, was one seemingly tailor-made for solution by the German military's analytical methods.

The German effort to learn the aerial lessons of the last war took a number of forms, in the belief that the wider the investigation, the more valid would be any conclusions drawn from it. The camouflaged air service during the 1920s carried out an examination of wartime air operations with an eye towards the future employment of air power. A small group of ex-Imperial Air Service officers, known as the *Gruppe Luftstreitkräfte* (Air Power Group), received the task of producing a systematic analysis of air power in World War I. Its members, assisted by trained archivists, possessed access to relevant documentation.[47] Exploitation of available resources, augmented by the personal expertise and recent experiences of the officers, resulted in some useful conclusions. The group produced a number of studies, many concerned with ground support operations. One in particular dealt with the methods of adaptation employed by German squadrons operating on the Western Front under material and personnel shortages.

Individuals connected with the group also contributed articles to collected works dealing with the lessons of World War I, thus ensuring that the "air power advocates" had a discernable voice in the debates concerning the operational art that took place throughout the 1920s.[48] The prevailing atmosphere was one of intellectual freedom and interservice cooperation. The *Gruppe Luftstreitkräfte*, after a long and somewhat tortuous evolutionary path, became the Military Science Department of the Luftwaffe general staff, one of the most visible and prolific agencies charged with assisting the Luftwaffe command in analyzing operational lessons.[49] This development, however, lay in the future. In the post-World War I period, other agencies took the lead in German thinking about close air support and other aviation matters.

A second postwar analysis project produced a more tangible and immediate impact upon thinking about air support. In 1919–1920, Seeckt,

---

[47] For the early history of this venture, see Chef Genst. 8. Abt., Nr.3237/44 g-Chef, 31.21.1944, "Sonderstudie Heft 7: Überblick über die Entwicklung und Zielsetzen der Kriegswissenschaftlichen Arbeit in der Luftwaffe von 1935 bis 31.12.44," National Archives and Records Administration (NARA) Microcopy, T971/33/83ff.

[48] Arndt, "Fliegerwaffe," pp. 310ff.

[49] Air Ministry/USAFE Intelligence Report No. 109, "The GAF Historical Branch in the Service of the General Staff," 3 June 1946, USAFHRA 512.62512M-109.

the head of the *Reichswehr*'s general staff in disguise, the *Truppenamt*, set in motion a comprehensive examination of German air doctrine, combat effectiveness, and weaponry.[50] Nearly sixty teams of air officers participated. One study group recommended, among other things, that each army division receive organic close air support: a forty-two aircraft formation comprised of observation, light, and heavy ground attack aircraft. Another study challenged this contention and argued that a single air commander should control all air assets.[51] The latter view prevailed; indeed, all German practice in World War II adhered to the basic principle of centralized Luftwaffe control of the close support force.[52] Most importantly, the efforts of Seeckt's analysis group formed the basis for the relevant *Reichswehr* manuals of the period.

Finally, the *Reichsarchiv*'s official history program augmented the pragmatic evaluation effort initiated by Seeckt. It employed many of the same personalities and reached similar conclusions.[53] The German post-World War I analysis, in all its permutations, was one of the most successful in the twentieth century. Although inhibited by Allied controls, a shattered economy, and the limited size of the post-Versailles *Reichswehr*, the Germans succeeded in meeting the intellectual challenges that World War I experience and the advance of aviation technology posed. The ingredients of the successful close air support system that the Germans employed against the Poles, French, and Soviets, including the use of air liaison officers, recognition signals, and communications channels, all existed in the military manuals of the 1920s and early 1930s.[54] Army officers were keenly aware of the potential of air power and sought means of integrating it with ground forces, especially the nascent panzer arm.[55] But if the Germans succeeded in mastering the theoretical aspects of close air support, the application of the techniques and the provision of acceptable technology were another matter.

[50] For a complete discussion of this effort, see Corum, "Old Eagle as Phoenix," pp. 13–21.
[51] Ibid., p. 17.
[52] Karl-Heinrich Schulz, "The Collaboration Between the German Army and the Luftwaffe: Support of the Army by the Luftwaffe on the Battlefield," US Army MS #B-791a, p. 8.
[53] Reichsarchiv, *Der Weltkrieg, 1914 bis 1918* (Berlin, 1925–1944).
[54] James S. Corum, "The Luftwaffe's Army Support Doctrine, 1934–1941," paper presented at the Annual Meeting of the Society for Military History, Fredericksburg, VA, April 1992.
[55] Kommando der Panzertruppe Ia op Nr 3611/36 geh., 9.12.36, Betr.: Zusammenarbeit zwischen Panzertruppe und Luftwaffe, NARA T405/6/4834188–190. See also Heinz Guderian, *Achtung-Panzer! Die Entwicklung der Panzerwaffe, ihre Kampftaktik und ihre operativen Möglichkeiten* (Stuttgart, 1937), pp. 207–208.

While many authorities, particularly those seeking to deny the National Socialist regime credit for Germany's resurgent air power, argue that the foundations for the Luftwaffe's success in 1939–1941 were the product of Weimar, not Nazi, efforts, the fact remains that German aerial rearmament before 1933 was seriously circumscribed. Research and development on specific areas of aviation technology, such as high performance engine development and high-octane gasoline, proceeded slowly during the Weimar years.[56] In some cases the German aircraft industry never made good the resultant loss of time and effort. The failure of the Luftwaffe to field an acceptable heavy bomber was one of the more obvious consequences. Likewise, realistic training that involved air and ground operations was lacking. One German commentator recalled that at the annual fall maneuvers, "the army would rent a few obsolete private planes and have them circle above the sham battles, just to give an idea of what it would do if it had any military planes."[57] The *Reichswehr*, in a secret treaty laden with even more secret clauses, established a clandestine training school for aviators at Lipezk in the Soviet Union in 1923. Such ventures and the reliance on the civil airline Lufthansa as a "shadow air force" were expedients, stopgaps that could not ameliorate the lack of a genuine military air service which could put theories to the test. In fact, the Luftwaffe general staff, assessing the development of army support aviation in 1944, conceded that the Americans had dominated the field in the 1920s and 1930s.[58]

The Nazi accession to power in January 1933 fundamentally changed the nature and orientation of military aviation in Germany. Hitler had won the support of the professional German military with a pledge to dismantle the Versailles system and initiate open rearmament. German military planning would now operate in a context far different to that under the Weimar government, where planning had focused on a border war with Poland, a pyrrhic attempt against the French in the Ruhr region, and other limited or improbable contingencies.[59]

The effect on the air service, officially reborn as the Luftwaffe in March 1935, was no less dramatic. The National Socialist cast of the young service was undeniable. This influence, coupled with the contributions of a

---

[56] Edward L. Homze, *Arming the Luftwaffe: The Reich Air Ministry and the German Aircraft Industry, 1919–1939* (Lincoln, NE, 1976), chs. 1 and 2.

[57] "Hauptmann Hermann" (pseud.), *The Luftwaffe: Its Rise and Fall* (New York, 1943), p. 79.

[58] "German Ground Attack Arm," p. 3.

[59] See Robert M. Citino, *The Evolution of Blitzkrieg Tactics: Germany Defends Itself Against Poland, 1918–1933* (New York, 1987).

new generation of German air officers, would take the evolution of German close air support in a somewhat different direction. In a manner similar, if somewhat less pronounced, to the processes underway in Britain and the United States, a more far-reaching concept of air power subsumed close air support. In the German case, the concept was known as *operativer Luftkrieg* ("operative" or "operational air warfare").

The mid-1930s was a heady time for military aviation in Germany. The development of doctrine for a first-rate air power proceeded apace, and the Luftwaffe issued specifications for aircraft with which to implement this doctrine. In 1935, the Luftwaffe's primary doctrinal statement (revised in 1940 but never superseded), L Dv 16, *Luftkriegführung* (Conduct of Air Warfare), appeared. This manual fixed close air support in a subsidiary position within the overall mission of the air force. While it acknowledged that, in certain circumstances, all types of air units (including heavy bombers) might be profitably employed in direct support of the army, "use of aircraft against targets within the range of our own artillery" would be the exception, rather than the rule.[60] Moreover, L Dv 16 suggested that close air support would have little effect on well-entrenched, first class troops, and pointed out that attacking enemy columns and installations in the rear areas was a more worthwhile application of air power.[61] In fact, German doctrine of the time (and through the first years of World War II), did not recognize CAS as a distinct mission. Instead, German doctrine spoke of two forms of army support, "direct" and "indirect."[62] While the American definition of close air support is somewhat analogous to the German term "direct army support," the German category embraced some forms of air interdiction that were beyond the scope of the American concept.

The publication of *Die Luftkriegführung* initiated a wave of thinking and writing about air power. Many of the ideas were incorporated into the curriculum of the *Luftkriegsakademie*, or Air War Academy, and accordingly emerged as a kind of orthodoxy within the Luftwaffe.[63] Much of the theorizing revolved around the developing concept of *operativer Luftkrieg*. Although the concept was somewhat nebulous, "operational air warfare" envisioned the use of an independent air force as a component of a "total war" against an enemy nation's military and other sources of national pow-

---

[60] L Dv 16, *Luftkriegführung* (Berlin, 1936), paragraph 132.
[61] Ibid., paragraph 130.
[62] Von Rohden study Nr. 1b, "Die deutsche Luftwaffe im Kriege gegen Russland," NARA T971/26/231.
[63] Air Ministry, *Rise and Fall of the German Air Force*, pp. 42ff.

er. In practice, *operativer Luftkrieg* envisioned the use of bomber fleets against an enemy nation's air force, industry, communications, and supply systems. What distinguished *operativer Luftkrieg* from British or American concepts of "strategic bombardment" was the close integration of air action with the operations of the army. Germany was a continental power, and as such its air force could never leave surface operations out of its calculations.[64] But despite the fact that many of its senior officers transferred from the *Reichswehr* at a late date, the Luftwaffe was not encumbered by an "army support mentality," as some historians have suggested. In fact, many of the "retreads" from the *Reichswehr* turned out to be fanatical defenders of independent air power and the so-called "strategic airwar."[65] The perception that the Luftwaffe leadership, riddled with dive-bomber zealots such as stunt pilot Ernst Udet, willingly jettisoned "strategic bombing" in favor of tactical aviation sprang up only after the war.

In the context of German doctrinal development in the 1930s, close air support assumed a subordinate role. Luftwaffe planners acknowledged the importance of the *Schlachtflieger* units in World War I, but reasoned that the thirty-eight ground attack *Staffeln* in 1918 were a response to the peculiar conditions of trench warfare and the specific need to effect a breakthrough of British defensive line.[66] The theoretical Luftwaffe of the 1930s was a force that could play a part in an all-arms, operational-level victory by gaining air superiority, assisting the army in enveloping maneuvers, and striking targets far behind the front. The twin-engine medium bomber, with the potential to shatter Prague and Warsaw as well as rail centers, mobilization points, roadways, and other interdiction targets, was the true symbol of the rearmed Luftwaffe on the eve of war, rather than the single-engine Stuka dive bomber.

A powerful obstacle to the development of army support aviation was the lack of a single inspectorate, or *Waffengeneral*, responsible solely for army support aviation. Thus, the Junkers 87 dive bomber force, although associated in the public mind with close air support, was under the control of the bomber inspectorate (*Inspektion 2*), which planned to employ the Stuka in a wide range of roles, including airfield attacks and pinpoint bombing of industrial targets.[67] This arrangement led to the anomaly of the Stuka units in August 1939 appearing on the table of organization and

---

[64] Herhudt von Rohden, *Vom Luftkriege* (Berlin, 1938), chapter 1.

[65] Von Rohden study, "Die deutsche Luftrüstung, 1935–1945," NARA T971/26/316.

[66] Paul Deichmann, *German Air Force Operations in Support of the Army: USAF Historical Study No. 163* (New York, 1968), p. 32.

[67] Boog, *Luftwaffenführung*, p. 578.

equipment as part of the strategic bombardment force.[68] The Junkers 87, incidentally, was a far from ideal close support weapon: although its bombing accuracy was excellent, it was slow and vulnerable to ground fire. A handful of Henschel 123 ground attack biplanes, along with Messerschmitt Bf 110 twin-engine light bombers that might occasionally conduct ground attack, were meanwhile under the Inspectorate for Fighters (*Inspektion* 3), which understandably devoted little attention to the close support problem. While the Luftwaffe trained fighter units in dive bombing attacks,[69] in reality such units had little or no experience with this mission. The longstanding lack of an inspector general for close support aviation meant that there was no high-level advocacy of new technology and methods of training and employment for this specialized weapon. The Germans failed to correct this problem until fall 1943, when a major Luftwaffe reorganization took place.

How, then, did German CAS become so proficient? On the surface, many of the same factors that crippled close air support in the British and the American air arms were operating. The Luftwaffe, although its share of the rearmament budget was buttressed by Hermann Göring's strong political position, sought to cement its position in interservice matters as an independent branch of the Wehrmacht. The answer to this question lies in the same elements that made German army support aviation superior to its Allied counterparts in World War I: intellectual and doctrinal factors. An historian of German military performance in World War I has isolated the essential ingredients of the German success:

> First, innovation "bubbled up" from below, as German soldiers tried new techniques on an ad hoc basis in the field. Second, higher authority played a role not only in allowing this ad hoc activity . . . but in selecting out tactical innovators and giving them authority to apply their ideas further afield. Higher authority . . . also insisted on the best existing practice becoming the norm.[70]

The response to the combat experience gained in the Spanish Civil War demonstrated the success of the German process of institutionalizing innovation. The Luftwaffe's participation in that conflict, in the form of an autonomous air component known as the Condor Legion, is often cited as "creating" the concept of close support.[71] In fact, little took place in

---

[68] "German Ground Attack Arm," p. 4.   [69] Corum, "Army Support Doctrine," p. 11.

[70] Bradley J. Meyer, "Operational Art and the German Command System in World War I" Ohio State University PhD dissertation, 1988, p. 400.

[71] Raymond L. Proctor, *Hitler's Luftwaffe in the Spanish Civil War* (Westport, CT, 1983), pp. 257–258.

Spain that had not already been delineated in German manuals dating back to 1918.[72] The conflict did, however, demonstrate how poorly trained most German units were in direct battlefield cooperation. Under the Legion's chief of staff (and last commander) Wolfram von Richthofen, modifications in principles of employment and training corrected many of the deficiencies. Practical and valuable experience with air liaison, signals, dive bombing, and other innovations resulted.

Yet German sources indicate that success in Spain did not result in any major reorganization or expansion of the Luftwaffe's close support forces. Some German commentators viewed the Spanish war in the same fashion as RAF officers treated the air control experience – an interesting application of air power, but of little relevance to a major war against the great air powers.[73] The official attitude towards close air support remained somewhat ambivalent. In 1939 the Luftwaffe possessed only one flying unit, *Lehrgeschwader* 2, equipped with the Henschel 123 biplane, as a dedicated close support force.[74] The more numerous (and more capable) Ju 87 Stuka groups still had to carry out numerous tasks, and in any case represented only about 15 percent of the Luftwaffe's force structure. The emphasis placed upon close support in the combat units of the Luftwaffe of 1939 was far less than in the *Luftstreitkräfte* of 1918.[75]

The Luftwaffe's organizational structure for the early campaigns of World War II reflected the vision and preferences of its leadership. Its striking power grouped into a number of autonomous *Luftflotten* (air fleets, analogous to USAF numbered air forces). These units subdivided into *Fliegerkorps* (air corps), which possessed a mixture of fighters, bombers, dive bombers, and reconnaissance planes. The air corps commander would, in times of close cooperation with army formations, carry out direct liaison with his army counterparts. A system of air liaison officers (*Fliegerverbindungsoffiziere*, or "*Flivos*") ensured that "a constant exchange of ideas," situation reports, and requests occurred between Luftwaffe and army at all levels. Finally, Richthofen was ordered to set up a "*Fliegerkorps z.b.V*" (Special Duties Air Corps), a means of carrying on the innovation and experimentation with close air support that he had begun in Spain.[76] This unit later became *Flieger-*

---

[72] Williamson Murray, "The Luftwaffe Experience," in Cooling, ed., *Case Studies in the Development of Close Air Support*, p. 76.

[73] Herhudt von Rohden, "Luftwaffe und Gesamtkriegführung," in Major Dr. Eichelbaum, *Das Buch von der Luftwaffe* (Berlin, 1938), p. 20.

[74] "German Ground Attack Arm," p. 4.

[75] See Arndt, "Fliegerwaffe," p. 352; and Völker, *Die deutsche Luftwaffe 1933–1939*, p. 189.

[76] Schulz, "Collaboration Between German Army and Luftwaffe," p. 3.

korps VIII, by 1941 the Luftwaffe's premier (and only true) close support formation.

On the eve of war, the chief of the Luftwaffe general staff, Hans Jeschonnek, issued a set of instructions for Luftwaffe units engaged in direct support of ground elements. He noted that close air support represented a difficult technical proposition, fraught with hazard if the enemy possessed air superiority or ground troops that were well trained, equipped, and entrenched. He noted accordingly that "the requests of the army for direct support may consequently only be answered if the possibility of bringing about an unconditional, immediate, and sensible result exists."[77] At the same time, Jeschonnek realized that the mission could at times tip the balance. Therefore, he stressed the need for Luftwaffe and army commands to work closely together on solving the technical and tactical problems that would inevitably arise. This spirit of cooperation, coupled with the Luftwaffe's versatile and flexible force structure, brought dramatic results in the campaigns to come. The Luftwaffe's developing close support capability, limited though it may have been, was just sufficient to give the Germans a decisive edge in the campaigns of 1939–1941.

BRITAIN BETWEEN THE WARS

Britain's strategic situation in the interwar period was not conducive to rearmament or to significant military innovation or adaptation. This fact particularly applied to the newly created Royal Air Force which, although elevated to the status of a separate service in the wake of popular outcry over the 1917–1918 German air raids on Britain, had to fight for survival against arguments that the army and Royal Navy could better utilize air assets.

Compelling external factors led the RAF to neglect the mission of close air support. Even by the parlous standards of the 1920s and 1930s, the mission of supporting a modern army on the battlefield remained fixed at the bottom of the RAF's priority list. The effect of the Ten-Year Rule, which enjoined the services not to prepare for any major conflict within a ten-year period, certainly played a role, as it made little sense to invest resources in developing specialized aircraft types that would doubtless be obsolescent years before needed. Moreover, the debate over the "continental commitment" made it doubtful that a large British ground force

---

[77] Der Reichsminister der Luftfahrt und Oberbefehlshaber der Luftwaffe, Genst. 3. (takt.) Abteilung (II) Nr. 1240/39 geh., "Richtlinien für den Einsatz der Fliegertruppe zur unmittelbaren Unterstützung des Heeres," NARA T321/76/no frame #s.

would again fight on the continent.[78] Finally, the state of the economy seriously hampered the evolution of British air power. Although it is an oversimplification to blame interwar penury for all of Britain's military ills in 1939–1941, the RAF certainly felt the constraints of massive defense cuts.[79] Britain's military aircraft industry barely survived the lean years; only judiciously distributed orders for small series of aircraft kept Britain's research, development, and production capability viable.[80]

In this hostile fiscal climate and given the priorities of the RAF, air support received short shrift. When the war broke out in September 1939, the RAF possessed a modern fighter defense system, as well as the foundation of a "strategic bombardment" force. But it possessed virtually nothing in the way of an army support capability, despite the strides made in a somewhat related area of air power application, "air control."

Simply put, the *raison d'être* of the RAF through much of the interwar period was not to prepare for a future war against a real or imagined constellation of hostile powers. It was, rather, to insure its own survival as a separate service. Roles and missions such as close air support, which as a rule did little to promote the need for an independent RAF and in fact mitigated against it, would naturally go begging. To evaluate RAF development solely upon the criteria of its performance in France and Flanders in 1940 ignores the important point that the mission of supporting the army on the continent was one it had not sought or been led to expect.[81] Indeed, it had no reason to prepare for this type of war until a few years prior to the outbreak of the conflict. While the lackluster preparations in those final years are worthy of harsh criticism, the RAF in the earlier period had other problems with which to contend. Furthermore, few could fault RAF Fighter Command in its defense of the home island during summer 1940, or deny that the resources applied to developing the Spitfire, the Hurricane, or the radar chain during the last year of peace had been ill-spent.

The RAF's historical assessment of its experiences in World War I contributed to its neglect of close air support. Like the German and American air arms, the RAF lost no time in analyzing the lessons of World War I. In fact, the British official history effort, headed by Sir Walter Raleigh and R.A. Jones, was probably the most polished and impressive of them

---

[78] John Terraine, *A Time for Courage: The RAF in the European War, 1939–1945* (New York, 1985), pp. 62–63.

[79] Smith, *British Air Strategy*, p. 2.

[80] Mason, *Battle over Britain* (Bourne End, UK, 1990), pp. 34–40.

[81] Allan R. Millett and Williamson Murray, eds., *Military Effectiveness*, vol. 2, *The Interwar Period* (Boston, 1990), p. 99.

all. The six-volume series, *The War in the Air*, served both as a restrained and comprehensive analysis of operational experience and a cleverly constructed plea for an independent air force.[82] Accordingly, the series dedicated a disproportionate amount of space to analyzing the German air raids on Britain and the gestation of Trenchard's Independent Air Force and its strategic mission. The lessons for close support were simple and not encouraging: "Except in certain and rare circumstances, the results were disappointing when compared with the losses sustained."[83] This brief statement was essentially what passed for "close air support doctrine" at the highest levels of the air staff until well into World War II.

Despite the inhospitable climate of opinion, close air support was not completely neglected in Britain. Some people still thought about and experimented with close support in the interwar period, and in fact there were several sources of invention and innovation. Yet remarkably little change occurred in the RAF's position on air support. The 1932 edition of the War Office manual, "The Employment of Air Forces with the Army in the Field," pointed out that the RAF's main task would be to gain air superiority over the battlefield. Such close air support as could be expected fell under the rubric of "low-flying attacks." While concluding that fighter aircraft would be most suitable in that role (a direct result of the Western Front experience) the manual placed severe restrictions upon their use:

> The moral effect of this means of attack is great, apart from any material damage inflicted. If, however, low-flying attacks are made against well-trained and well-equipped troops of high morale, severe casualties among attacking aircraft are likely to result; it is very important therefore that the potentialities of aircraft in low-flying should not lead to a misuse of fighter aircraft, the primary function of which is to participate in acquiring and maintaining air superiority . . . [W]hen air superiority is still in doubt, fighter aircraft will seldom be diverted from their normal functions of securing it.[84]

Seven years later, in November 1939, an air staff memorandum echoed these sentiments, allowing that close air support was only justified in special circumstances. These occasions included pursuing a beaten enemy, in an emergency when stemming an enemy breakthrough or covering a re-

---

[82] Jones, *The War in the Air* (6 vols.); Smith, *British Air Strategy*, p. 59.

[83] Air Ministry, *Air Support*, p. 9.

[84] War Office, *The Employment of Air Forces with the Army in the Field*, (London, 1932), p. 26.

treat, or, on extremely rare occasions, a planned attack on a highly organized enemy defense. The air staff maintained that there was "no justifiable historical example" for the latter contingency, a curious observation
in light of the painfully effective German use of close support against the
British in March 1918.[85]

With the highest levels of the RAF providing little impetus for development of tactical aviation, any discourse on close support had to come
from other sources. In common with the German military, although considerably less successfully, advances in close air support took place at the
operational and tactical levels. Joint maneuvers, experiments with the armored force after its creation in 1927, and discussions in such forums as
the Royal United Service Institute and the Imperial Defence College provided at least a theoretical base for ground and air cooperation. In addition, such discussions ensured that a few officers had what might be
termed "joint warfare" experience, as even a recognized common vocabulary between army and air officers did not officially exist.[86]

However, almost complete stagnation, or even regression, characterized the 1920s. An officer of the Royal Tank Corps noted in 1926 that,
"We are to-day in some respects behind where we stood in 1918."[87] Realistic training was lacking, illustrated by the 1925 maneuvers, in which
the two fighter squadrons taking part, despite representing opposing
sides, attacked "in an impartial manner" the friendly troops engaged in a
mock assault.[88] The maneuvers involving the armored force were the ones
that offered some possibility for future innovation. In maneuvers in
1927–1928, the tank force gained some rudimentary practice in coordi-
nating with fighter aircraft which, although still urgently required for the
air superiority mission, could play a role in silencing enemy antitank guns,
thus paving the way for the advance of the armor.

Low-level initiatives kept such thinking alive. Wing Commander T.L.
Leigh-Mallory, later to rise to prominence as a fighter and tactical air commander in World War II, recommended the formation of a specialized
tank cooperation squadron. Even he acknowledged that "the number of
fighter squadrons available in the early stages of a war may make this pro-
cedure impractical."[89] Another RAF officer despaired of finding a true

---

[85] Air Ministry, *Air Support*, p. 10.
[86] Kennett, "Developments to 1939," p. 28.
[87] Lieutenant Colonel Frederick A. Pile, "The Army's Air Needs," *Journal of the Royal United Services Institute (JRUSI)*, 71, 1926, p. 726.
[88] Ibid., p. 727.
[89] Wing Commander T. L. Leigh-Mallory, "Air Co-operation with Mechanized Forces," *JRUSI*, 75, 1930, p. 571.

solution to the armor-air cooperation problem and suggested that "it really seems that we shall have to attach a single-seat fighter squadron to a tank company and let them work out the problem. Perchance then a perfect antidote may be discovered."[90] In microcosm, this offhanded suggestion illustrates how the solutions to most close air support problems eventually emerged. At the highest levels, the RAF frowned upon such initiatives: the air staff formally requested the army to cease its experiments with the RAF; it objected to the "request that the pilots violate RAF doctrine" in this manner.[91]

Despite signs of a willingness to experiment, most army-air force maneuvers in Britain during the interwar period displayed an appalling lack of practical cooperation. Indeed, the services possessed "three virtually independent philosophies of war."[92] Accordingly, an observer reported that the army maneuvers of 1930

> leave the impression that if either side had been exposed to the attack of even so small a force as four squadrons of modern aeroplanes, the issue would have been seriously affected. In the course of the exercises, most attractive targets for aeroplanes were noticeable; typical of such was a headquarters formation halted on a road, which did not make the slightest attempt to seek cover for at least half an hour; large convoys of mechanical transport climbing Amesbury Hill in close order; and a motor battery proceeding across the open plain in an order that offered an ideal target for single-seater fighter attack.[93]

Beyond indicating the existence of service parochialism, the above observation demonstrated that neither service thought in terms of the other's operating medium. The failure was mutual.

As late as March 1939, little had changed at joint training exercises. General Sir Archibald Wavell's "withering" verdict on the state of RAF close air support is justifiably well-known. The maneuvers

> showed conclusively that the RAF had given little or no thought to the problem of close support of ground operations, that their pilots had not been trained to this form of war, and that the results on the targets provided were extremely poor in consequence . . . I doubt whether the exercise occasioned even a ripple of thought about close support to pass over the minds of the air staff.[94]

[90] Air Commodore C. R. Samson, "Aeroplanes and Armies," *RUSI*, 75, 1930, p. 678.
[91] Jonathan M. House, *Toward Combined Arms Warfare: A Survey of 20th Century Tactics, Doctrine, and Organization* (Ft. Leavenworth, KS, 1984), p. 47.
[92] Smith, *British Air Strategy*, p. 76.   [93] Samson, "Aeroplanes and Armies," p. 676.
[94] John Connell, *Wavell: Scholar and Soldier* (New York, 1965), p. 204.

Lest Wavell's observation be dismissed as the product of service bias, a senior RAF officer, speaking of the same period, admitted that in the interwar era "the RAF forgot how to support the army."[95]

One reason for the neglect of close air support was the fact that the mission afforded the army a means of operational control over certain types of aircraft. The British practice of providing the army with supporting air power envisioned the commitment of an "air contingent" comprised of bombers, fighters, and "army co-operation" (equipped with observation and liaison aircraft) to the aid of the field army. The bulk of this force was intended to provide indirect support, including the acquisition of air superiority and the carrying out of deep interdiction strikes. The army cooperation squadrons (flying observation and liaison aircraft) were the only ones intended to work closely with the ground forces.[96] Although methods for employing these squadrons (which numbered only four in 1926) were deemed "excellent," no analogous methods of cooperation had been worked out for the fighter or bomber squadrons.[97]

This state of affairs inevitably led to suggestions from many quarters that the army have its own air component. The arguments in favor of this measure, which have recurred to the present day, were not without merit. Advocates pointed out that the average RAF officer was ill-informed about army operations and methods. Although not requiring actual transfer to the army, a close support pilot needed training in that specific and demanding mission, not simply detachment from more familiar tasks when the situation absolutely demanded it. Finally, the recommendation would achieve "unity of command" in a given operational area. One enthusiast for this scheme pointed out, "The claim that the military aeroplane is solely the business of the air force is as ridiculous as one that all 3-ton lorries should be controlled by the army."[98]

Opponents of this view (and there were many, given the desire of the RAF to maintain its independent status) decried the proposed solution as "cumbrous," inefficient, and wasteful, one that required needless duplication of training, procurement and production resources. The RAF argument anticipated the Luftwaffe's wartime solution to the issue: observation machines would operate under the control of the army formation they supported, but were part of the air force's organizational structure and crewed by RAF airmen. Offensive close air support units remained

---

[95] Terraine, *A Time for Courage*, p. 383.
[96] War Office, *Employment of Air Forces*, p. 14.
[97] Pile, "Army's Air Needs," p. 504.
[98] Captain I. O'B. Macgregor, "The Army and the Air," *JRUSI*, 80, 1935, p. 504.

under air force control; air leaders also invoked the argument for "unity of command." There was, in fact, at least one semi-serious proposal (by an army officer, incidentally) to "stop limping after Germany and take a bold jump ahead of her. Let us cut the Gordian knot by transferring to the Royal Air Force the tanks and armored cars."[99] As it happened, air support in the RAF (and the other air forces under examination) would be a problem solved through improved liaison, rather than by amalgamation.

No real mechanisms, therefore, existed in the interwar RAF for implementing or perfecting close air support. When considered at all, the task was only a second- or third-priority mission for units and aircraft types intended primarily for other roles. Accordingly, the British aircraft industry made no attempt to produce specialized close support aircraft; no further flirtation took place with armored aircraft on the German pattern of 1918. British policy emphasized developing sound basic designs, adaptable to multiple roles.[100] While this policy eventually paid dividends in the case of the de Havilland Mosquito fighter-bomber, it did not produce an acceptable army support aircraft for the campaign in France.

Evidence exists of belated proposals in 1938 to fill the gap in the RAF's inventory. Specifications were put forth for a dedicated close support aircraft, possessing speed, maneuverability, bomb carrying ability, and "a large battery of heavy machine guns, all firing forwards and slightly downwards."[101] The response to such proposals was overwhelmingly hostile. Commentators noted that heavy aircraft losses would ensue and maintained that the geometric dispersal of machine gun bullets from a fast-flying aircraft made it an ineffective weapons system. This last argument is not quite so ludicrous as it first appears, as even forces more attuned to close air support, such as the United States Marine Corps and the Luftwaffe's close support formations, found that the transition from slower biplanes to modern, all-metal, high-speed monoplanes posed great technical problems. At a November 1938 symposium dealing with "The Influence of Aircraft on Land Operations," a commentator referred to "the shocking misuse of aircraft as I believe exists in the United States – the building of special 'ground strafing' machines."[102] Another speaker concluded, "I am still inclined to think that [close air support] is an uneconomic use of aircraft as they are today. If the attack on land forces de-

<hr />

[99] Captain Francis Milton, "Army and Air Force Cooperation: A Suggested Solution," *JRUSI*, 85, 1940, p. 666.

[100] Kennett, "Developments to 1939," p. 34.

[101] Colonel M. Everett, "Fire Support from the Air," *JRUSI*, 83, 1938, p. 591.

[102] Major-General R. J. Collins, "The Influence of Aircraft on Land Operations," *JRUSI*, 84, 1939, p. 68.

velops to a very great extent we may have to follow suit, but I do not think that that time has come yet."[103]

This prevalent attitude led the RAF leadership to overlook the experience gained in their own "small wars" and in foreign conflicts such as Ethiopia and the Spanish Civil War, and even the German *blitzkrieg* victory over the Poles in September 1939. Some RAF officers took heed of the successes gained by air power in Spain, with the qualification that neither side possessed adequate anti-aircraft defenses, and the fighting value of the defeated forces involved – Italian General Bergonzoli's volunteers at Guadalajara in 1937, for example – was "far inferior to that of regular soldiers."[104] The air staff concluded, "All war experience seemed to prove that low-flying attack – which was the usual method of providing close support – on the battlefield against unbroken troops, deployed and in position, inevitably involved a very high rate of aircraft casualties."[105]

As the RAF Advanced Air Striking Force prepared to cross the Channel to face the Wehrmacht in the Battle of France, its close support doctrine had made only minuscule progress from summer and fall 1918. There was one notable exception in the otherwise dismal story of close air support in interwar Britain. This exception was the policy of "air control," the use of the RAF as a police force on the fringes of the British Empire. The relative success of this development is not only useful as a case study in military innovation, but it sheds some light on the question of why "mainstream" RAF close air support practice fared so poorly by comparison.

Air control grew out of the immediate postwar period, when the survival of the Royal Air Force was in doubt. The leadership of the RAF desperately searched for a mission that would justify preservation of the service. An opportunity appeared in 1919, with the uprising of the "Mad Mullah" of Somaliland. In the expedition that followed, a handful of RAF aircraft dispersed the dervishes at a minuscule cost to the British taxpayer of £77,000.[106] The concept of using aircraft, either independently or in conjunction with mobile columns of troops and armored cars, as an efficient and economical means of controlling the Empire saved the RAF. Malcolm Smith has noted, "In that period [1919–1934], indeed, the success of the RAF in imperial policing was a great deal more important to the standing of the force than any idea of a strategic air offensive in some unseen major war of the future."[107]

---

103 Ibid.
104 Captain Didier Poulain, "Aircraft and Mechanized Land Warfare: The Battle of Guadalajara, 1937," *JRUSI*, 83, 1939, p. 366.
105 Air Ministry, *Air Support*, p. 10.   106 Smith, *British Air Strategy*, p. 28.
107 Ibid., p. 14.

In a number of "small wars," defined by one commentator as "operations against wild men in wild places," at such locales as the Northwest Frontier, the Aden Protectorate, and the Sudan, fairly primitive aircraft (most often RE 8s, Bristol F2B Fighters, or DH 9s of World War I vintage) carried out the air control mission. The methods of cooperating with ground forces were far in advance of anything being practiced in Britain. While RAF squadrons were carrying out indiscriminate mock attacks on ground troops in army maneuvers on Salisbury Plain, the situation in the far-flung reaches of the Empire was significantly different. Air Commodore C.F.A. Portal in 1937 noted:

> The use of the air-bomb and machine gun in close support of troops on the ground has proved of the utmost value in police operations on the Indian frontier and elsewhere. It was brought to a very high state of perfection in the recent operations in Palestine where small bodies of troops were often held up by the fire of armed bands occupying strong positions. When this occurred, a W/T message was sent by the troops and so good was the organization that at almost any point in Palestine a formation of bombers would arrive within fifteen minutes of the origination of the message.[108]

Indirect benefits gained from these operations included a small cadre of air officers sensitized to the problems of supporting mobile operations. It is interesting to note that, in the pages of the *Journal of the Royal United Service Institution*, some of the more innovative writing on air support came from the pens of RAF officers with experience fighting in the "wild places."

The air control venture possessed a number of features that led to successful tactical support. The commands involved were usually small, and there was ample time and opportunity for establishing personal and command relationships and procedures for effective liaison and cooperation. Aircrews operated over the same area for long periods and gained an appreciation for the nature of the terrain and the problems associated with operating in and over it. Perhaps most importantly, their primary mission was air control – they were not being diverted from air combat or bombardment in order to support a beleaguered British column or attack recalcitrant tribesmen. And finally, the RAF embraced the mission because it furthered its chances for survival. In this instance, the operational needs of the situation coincided with the RAF's institutional imperatives.

---

[108] Air Commodore C.F.A. Portal, "Air Force Cooperation in Policing the Empire," *JRUSI*, 82, 1937, p. 346.

Yet the relative success of the venture should not lead one to overemphasize its importance. The experience was not directly applicable to warfare against "civilized" powers, although the methods of liaison and communication learned were certainly valuable. As the RAF in the late 1930s began to think earnestly about the coming war with Germany, the experience with air control seemed less and less relevant. As Wing Commander R. Saundby pointed out in a lecture to the RAF Staff College in 1936:

> In the past, we have been glad to take over the responsibility of control, sometimes in none too favorable circumstances, because we had no other chances of showing what we could do. The rearmament of Germany, which has emphasized with dramatic suddenness the problems involved in Home Defence and European war, has relegated the subject of small wars to its proper place in our microcosm.[109]

The strategic air offensive dominated that microcosm in 1939. That role also promised to guarantee an independent RAF. Yet the British were to pay dearly for their neglect of tactical aviation in the coming battles of France and Flanders. Unlike the Germans, or even the Americans, the British did not create much of an intellectual, much less practical, foundation for using their air force in support of the army. Their own process of innovation still lay ahead.

THE UNITED STATES: WHY ATTACK AVIATION?

Looking back on the pre-World War II doctrinal and institutional development of the army air corps, the official historians Wesley Frank Craven and James Lea Cate took a predictably teleological approach:

> It should be sufficient here to describe the three paramount trends in the period: the effort to establish an independent air force; the development of a doctrine of strategic bombardment; and the search for a heavy bomber by which that doctrine could be applied. An approach so circumscribed as this will inevitably omit much that is important and interesting ... Yet so important were those trends in the development of the air force which went to war in December 1941 that it is legitimate to confine the present discussion to them even at the risk of overemphasis and oversimplification.[110]

---

109 Wing Commander R.H.M.S. Saundby, "Lecture: Small Wars, with Particular Reference to Air Control in Semi-Civilized Countries," RAF Staff College, 19th Course, 1936 (on file at Air University Library, Maxwell AFB, AL), p. 12.
110 W.F. Craven and J.L Cate, The Army Air Forces in World War II, vol. 1 (Washington, DC, 1983), pp. 17–18.

While one might take Craven and Cate to task for viewing the World War II experience of the U.S. Army Air Force as merely the triumph of strategic air warfare (and indeed, the seven-volume series gives unjustifiably short shrift to tactical aviation), in one sense their portrayal of the intellectual currents in the interwar army air corps is accurate. Army support aviation, despite the early push given it by Mitchell and others, atrophied in the United States as other forms of air power employment took the lead.

The air service (which became the air corps in 1926) initially accepted the ground support mission.[111] Early doctrine suggested an extension of the successful system used at St. Mihiel. From the first postwar years, an emphasis on both concentrated and independent action by aircraft in army support was evident.[112] Attack aviation, in the theory of the late 1920s and early 1930s, was to be employed *en masse*, against a mixture of close support and interdiction targets. Methods for actually carrying out this mission lagged well behind the theory.

As suggested earlier, much of the doctrinal evolution of the air corps had less to do with planning for a future war (scenarios for which, in the 1920s at least, seemed farfetched) than with carving out an independent mission for the young branch. The struggles with the U.S. Navy over coastal defense were one obvious manifestation of the battle for survival. While it would be an oversimplification to state that attack aviation fell into disrepute simply because "strategic bombardment" did more for the independent existence of the air corps, the suggestion of the subordination of air force units to the ground forces, coupled with the significant technical difficulty of the mission, forced American attack doctrine away from a pure close support role into a mission more closely approximating the German concept of "indirect army support."

In a 1934 article somewhat plaintively entitled "Why Attack Aviation?" Colonel Horace Hickam, commander of the air corps' sole attack unit, indicated that thinking about attack aviation had retreated rather far from the original battlefield support concept:

> [I]t becomes evident that Attack is not a FLYING MACHINE GUN BATTALION or CAVALRY IN THE AIR any more than it is LIGHT ARTILLERY IN THE AIR . . . The function of attack in the air force team is, in brief: To increase its relative strength in bombardment by destroying enemy bombardment; to provide protection to bombardment by destroying enemy pursuit; to neutralize anti-aircraft artillery defens-

---

[111] Thomas H. Greer, *The Development of Doctrine in the Army Air Arm 1917–1941* (Washington, DC, 1983), p. 39.

[112] Ibid., p. 39.

es; and to reduce the general efficiency of the enemy air force by disruption and destruction of his base facilities.[113]

By the late 1930s, when the army air corps was expanding, its emphasis lay in creating a "strategic bombardment" force; as a result, medium or light bombardment almost completely subsumed attack aviation, with missions that included primarily attacks on enemy airfields and interdiction.[114]

The role of professional military education, in this case the Air Corps Tactical School, is a well-known facet of the development of the "strategic bombardment" doctrine. Historians have generally viewed the story of the doctrinal struggles of the 1930s as a clash between pursuit advocates, most notably Claire Chennault, and bombardment protagonists, epitomized by Harold George and Haywood S. Hansell.[115] No matter what the mission, much doctrinal thinking at the Air Corps Tactical School occurred in a vacuum; it took place almost entirely apart from national security or political imperatives. No less an authority than the school's commander "didn't know, or at least could not remember, what strategic assumptions underlay the development of air doctrine at that time. It was surely a question that was much evaded during the entire interval between world wars."[116]

As a result, the evolution of various doctrines depended on personalities, rather than on systematic method. In the case of attack aviation, its strongest proponent was Captain George C. Kenney; when he departed the Air Corps Tactical School in 1929, attack aviation suffered accordingly.[117] As a result, a lack of integration and harmony between the various branches prevailed at all levels of military training and education. A retired USAF general recently commented upon his experiences in the air corps PME system of the 1930s:

> [T]he class was divided into sections on bombardment, pursuit, attack and observation, and they were not related at all; neither was there any attempt to consider how any of them would work together . . . [at the Air Corps Tactical School] attack got a fairly low emphasis although they talked about strafing and dropping of fragmentation bombs on the troops but there was no indication of technique or anything like that.[118]

---

[113] Lt. Col. Horace M. Hickam, "Why Attack Aviation?," *U.S. Air Services,* February 1934, p. 16.

[114] Greer, *Development of Doctrine,* p. 121.

[115] For the bomber advocates' perspective, see Haywood S. Hansell, *The Air Plan that Defeated Hitler* (Atlanta, 1972).

[116] Greer, *Development of Doctrine,* p. 30.   [117] Ibid., p. 66.

[118] "Joint Royal Air Force/United States Air Force Seminar, held at the Royal Air Force Museum, Hendon, 29 October 1990," *Air Power History,* Winter 1991, p. 36.

The organizational and technological development of American attack aviation reflected the ambiguous and changing status of the mission within the air corps. The American reading of World War I seemed to indicate that "swing-role" aircraft and units would not be adequate for battlefield attack. The 1928 air corps "Advanced Flying School Attack Aviation Text" pointed out that "experience proved that success could not be obtained by assigning [observation] airplanes to this dual role, as pilots would become so engrossed in attacking the ground troops that they would neglect the primary mission of supplying information concerning enemy movements."[119] The obvious solution was to create specialized units that operated appropriate aircraft.

In one important respect the United States led the rest of the world during the interwar period in the field of close air support. The air corps operated what was long the sole dedicated army air support unit: the Third Attack Group. Established on September 13, 1921, out of "surveillance" units engaged in patrolling the border with Mexico, it was initially a reconnaissance unit in all but name. Originally composed of four attack squadrons, within two years two of the four squadrons were declared inactive due to budget cuts.[120] Further cuts shaved the personnel complement of the remaining squadrons from 130 to ninety. By the early 1930s, the Third Attack Group was undoubtedly a stepchild of a service that was itself a stepchild. In keeping with the shifts in doctrine discussed above, Third Attack adopted a conception of its mission that ultimately had little to do with close air support. Although the group managed to keep some semblance of army support aviation alive during the interwar period, its efforts swam against the tide. Light and medium bombardment eventually overshadowed its attack mission, and the success of U.S. Army Air Force's close support aviation in World War II would owe more to the pursuit tradition than to Third Attack Group.

The linear progression from interwar to World War II practice is easier to detect in the case of another service, the United States Marine Corps. Although marine aviation faced typical problems regarding close air support – most notably adapting its tactics to accommodate the use of fast, all-metal monoplanes – it was significantly more successful than the air corps in addressing the problems of supporting ground forces. Its small size, relatively homogenous officer corps, and above all its clear mission conception assisted in this process. Marine aviation existed primarily to support the corps' two primary tasks: prosecution of "small wars" and, later, amphibious landing operations upon hostile shores. This unclut-

---

[119] "Attack Aviation,"*1928*, USAFHRA 248.222–553, p. 4.
[120] Air Corps Tactical School, "Attack Aviation Course," p. 8.

tered intellectual and doctrinal focus permitted innovation to thrive in the interwar period. As one officer put it: "Marine aviation is not being developed as a separate branch of the service that considers itself too good to do anything else. Unlike the army air service, we do not aspire to be separate from the line or to be considered as anything but regular marines."[121]

The marine corps faced the same budgetary and related constraints as the air corps during the post-World War I era. The survival of its aviation component bore similarities to the case of the RAF in the same period. The marine corps argued persuasively that it could play a role in fighting guerrillas, a "small wars" mission that was analogous to the air control function of the RAF. While the implications of this experience for World War II may have been somewhat limited, they contributed to the general marine attitude that aviation was best employed in support of ground troops.[122] Operations in Haiti, and most notably in Nicaragua against General Augustino Sandino's forces, honed the skills of marine aviators. Although marine air units were flying DH 4s, or later Vought O2U two-seater biplanes, they developed tactics far in advance of what the air corps, or even the Third Attack Group, had thus far employed. The most significant of these innovations was dive bombing, first used in Haiti in 1919 and which by 1927 had become a "routine" mission of marine air.[123]

The doctrine that grew out of the operational experience reflected successful employment of close support aviation by the marines. In the 1940 edition of the USMC Small Wars Manual, no ambiguity appeared concerning aviation's proper role: "the primary mission of combat aviation in a small war is the direct support of the ground forces."[124] The manual recognized that the need for a sustained air superiority campaign would probably not be great, so all types of aircraft (fighter, observation, and dive bomber) might profitably participate in the attack mission. Little evidence suggests that the air corps took any interest in this specialized role for aviation: the attack aviation course at the Air Corps Tactical School at Maxwell Field in the 1930s contained only the briefest reference to the marine "small wars" experience, and that only as a result of personal con-

---

121 Lt. Col. Edward C. Johnson, Marine Corps Aviation: The Early Years, 1912–1940 (Washington, DC, 1977), p. 35.
122 Allan R. Millett, Semper Fidelis: The History of the United States Marine Corps (New York, 1991), p. 263.
123 Johnson, Early Years, pp. 55ff; Millett, Semper Fidelis, p. 247; Robert Sherrod, History of Marine Corps Aviation in World War II (Baltimore, 1987), pp. 22–27.
124 United States Marine Corps, Small Wars Manual (Washington, DC, 1940), Section VI, p. 17.

tact between the instructor, Major R.F. Stearly, and two marine aviators who flew in Nicaragua.[125]

With the formation of the Fleet Marine Force in December 1933, the marine corps accepted a more demanding mission – amphibious assault against hostile shores – and the problems for direct support aviation grew accordingly. The basic orientation of marine aviation (defined by the General Board in 1939 as "equipped, organized and trained primarily for the support of the Fleet Marine Force in landing operations in support of troop activities in the field")[126] remained essentially unchanged from the "small wars" period. Yet the more complex nature of amphibious landings on defended shores made it impossible for "pure" close air support to be applied as it had been in the jungles of Nicaragua. A substantial portion of marine air had to carry out the missions of air superiority and air defense. Its institutional relationship to the navy obligated it to provide squadrons for carrier operations and base defense.[127]

In the amphibious landing scenario, marine aviation began to face some of the same problems that had confounded the other air forces under examination. Exercises in the mid-1930s involving the 1st and 2nd Marine Aviation Groups revealed that the use of fighters and observation aircraft for close air support "interferes materially with the normal missions of these types, and is at best a makeshift expedient."[128] Moreover, the new generation of navy aircraft, exemplified by the Brewster F2B "Buffalo" fighter and the Curtiss SB2U "Vindicator" scout bomber, were too fast to allow employment of existing marine air-ground coordination procedures, developed for the DH-4.[129]

This combination of factors led to a retreat from the previous marine practice of providing extremely close fire support from the air to its ground forces. The Marine Corps Schools in 1940 articulated the new approach:

> When aviation is acting in close support of the ground troops its striking power should be used against only those targets which cannot be reached by the ground arms, or on targets for which ground weapons are not suitable or available. In almost all ground situations there are vital targets beyond the range of weapons of ground arms which can be powerfully dealt with by attack aviation. Therefore, the use of attack aviation to supplement the firepower of ground arms is generally discouraged, as it may result in the neglect of more distant and perhaps more

[125] ACTS, "Attack Aviation Course," p. 6. [126] Johnson, *Early Years*, p. 65.
[127] Millett, *Semper Fidelis*, p. 360. [128] Johnson, *Early Years*, p. 79.
[129] Millett, *Semper Fidelis*, p. 360.

vital objectives. As a ground rule, attack aviation should be used in lieu of artillery only when the time limit precludes the assembly of sufficient artillery units to provide the necessary preparation, and when such absence of artillery may involve failure of the campaign as a whole.[130]

While it may be an exaggeration to assert that marine aviation lost touch with the ground elements it was supporting, it certainly, on the eve of World War II, began to emphasize indirect air support over the direct variety. Once the Pacific War began, the doctrinal predilection of marine air power to support the ground forces would reassert itself, and marine CAS in the Pacific was as effective as its U.S. Army Air Forces counterpart in the European Theater of Operations in 1944–1945.

The marines made advances in close air support by emphasizing doctrinal improvements rather than new technology. The air corps pursued an additional technical solution to the problem. The selection of aircraft types for the mission followed the thinking of World War I on the subject. Mitchell had been most impressed by the success of the German *Schlachtflieger* units; he was equally enamored of the armored aircraft they received towards the end of the conflict. His concept of the ideal ground attack plane was a "flying tank" – heavily armored for low-altitude work.[131] Sherman, in his "Tentative Manual," seconded the sentiment: "the type of machine which is best adapted to ground straffing [sic] is necessarily an armored machine."[132] Sherman, however, recognized that such an aircraft would involve technological compromises, particularly concerning armor protection, performance, and reliability. The difficulty of reconciling this combination of factors dogged American attack aviation for the first decade or so of its existence.

The first attempt to produce such an aircraft, intended as an improvement upon the Junkers concept, was a complete failure. The Boeing GA-1 was described as

> a bi-motored, tri-plane, armored around motors and fuselage. It carried a pilot and mechanic, or two pilots; one gunner in the nose; one gunner in the rear; one gunner in the bottom of the fuselage firing below and to the rear; and one bomber [bombardier]. Due to its slowness, poor visibility and maneuverability, and general inefficiency as to useful load, it was never actively used, and was soon discarded.[133]

---

[130] Johnson, *Early Years*, p. 79.   [131] Greer, *Development of Doctrine*, p. 12.
[132] Maurer, *US Air Service*, Vol. 4, p. 375.   [133] ACTS, "Attack Aviation Course," p. 6.

The existence of this bizarre contraption showed that inventiveness, even if misdirected, was alive and well in the United States after World War I. The craft featured technical novelties such as a 37-mm cannon and 1/4" thick armor plating.[134] The failure of the GA-1 (and of its successor, the GA-2) led to a retrenchment, during which the Third Attack Group received aging DH-4s for the first five years of its existence. Throughout the 1920s, the group made do with slightly modified versions of whichever observation plane was currently in service. The air corps did not make another serious attempt to design an airplane specifically for ground attack until the end of the decade.

This new type aimed to reconcile the desires of the Third Attack Group, still the only air corps formation actively experimenting with the mission, with the technical resources available. In April 1929, a board of officers, ordered to produce appropriate specifications, met at Langley Field, VA, then home of the Air Corps Tactical School. The board, which included Third Attack's commander, Major John H. Jouett, and Captain George C. Kenney of the Air Corps Tactical School faculty, "strongly recommend[ed] against endeavoring to design an attack airplane capable of carrying out missions that should be properly assigned to other planes."[135] Yet the specifications ultimately agreed upon (which called for a two-seat, unarmored biplane with a top speed of 175 mph, four machine guns, and carrying up to 400 lbs. of bombs) were comparatively modest and similar to the capabilities of contemporary observation aircraft. In fact, the Air Corps Tactical School noted: "It is discouraging to note that the offensive efficiency of the airplane is only slightly greater than that of the wartime DH-4 airplane which was first adapted to attack use."[136] Limited to the existing powerplants, the board's recommendation could hardly be expected to herald a quantum leap in aircraft design.

The transition of American attack aviation to the metal monoplane era brought certain improvements, but also a number of complications, the sum of which was to retard evolution of the attack capability. Certainly, aircraft designs became more capable. The Curtiss A-12 Shrike (1933) and the Northrop A-17 (1936) were low-wing monoplanes, with greater speed and payload than the biplanes they replaced.[137] Other technical developments promised to enhance the impact that aircraft might have on

[134] Kennett, "Developments to 1939," p. 45.
[135] "Proceedings of a Board of Officers Convened at Langley Field, Virginia," April 1, 1929, USAFHRA 248.222–52, p. 4.
[136] Air Corps Tactical School, Maxwell Field, Alabama, "Present Status of Attack Aviation and its Trend of Development," 1931–1932, USAFHRA 248.222–55E, p. 5.
[137] Kennett, "Developments to 1939," p. 51.

the modern battlefield. Of these, the most promising seemed to be the use of "chemical spray," i.e., poison gas. "Because of its primary function of attacking personnel because of its low-altitude tactics, and because of the great superiority in efficiency of sprayed chemicals over chemicals carried in bombs, the task of using chemicals offensively by aviation has chiefly been delegated to attack aviation."[138]

But far from increasing the efficiency of ground attack aviation, the strides made in aircraft technology during the 1930s virtually expunged close air support from the air corps' roster of capabilities. The trend in attack aircraft development moved towards the twin-engine light and medium bomber designs. Two factors explain this shift: one was technological, the other was doctrinal. The same process of airframe and powerplant evolution that endowed bomber aircraft with performance superior to their 1930s pursuit counterparts favored the development of twin-engine attack aircraft.[139] These types, such as the Curtiss A-18 and the Douglas A-20 "Havoc," were less suited for close air support than for missions such as deep interdiction and attacks on enemy airfields.

While the development of the medium bomber was in many ways a successful and worthwhile endeavor, it did little to provide reliable air support to the ground forces. Proponents of attack aviation embraced the broader definition of their mission as a means of preserving attack's distinct identity.[140] This development coincided with the emphasis on the long-range bomber and the air superiority mission within the staff of GHQ Air Force. By the time of America's entry into World War II, attack aviation as originally conceived – designed to provide a "crushing moral and physical effect . . . in the battle between ground troops"[141] – was moribund. The result was "a near disaster on the battlefield, retrieved only by the common sense of tactical commanders on the spot."[142]

WORLD WAR II: INNOVATION REVITALIZED

All three of the air forces went to war with close support of ground forces as a secondary mission at best in their perceived scheme of air power employment. By the end of the conflict, all of the major powers had fielded

---

[138] ACTS, "Present Status of Attack Aviation," p. 5.
[139] Philip S. Meilinger, "The Impact of Technology on the Development of U.S. Pursuit Aircraft Between the Wars" (presented at the 23rd Annual Northern Great Plains History Conference, 1988).
[140] Hickam, "Why Attack Aviation?" p. 18.
[141] ACTS, "Present Status of Attack Aviation," p. 7.
[142] House, *Towards Combined Arms Warfare*, p. 130.

formidable tactical air formations, developed sophisticated means of controlling them, and achieved results from their employment which, if not always decisive, were certainly of great significance.

As might be expected, the German system of close air support initially proved the most successful on the battlefield, and the mechanisms for adaptation and innovation worked quickly to improve this already effective instrument. The German insistence upon air superiority as a prerequisite for any other type of air action was well-founded, and in Poland and France the Luftwaffe was free in short order to apply its combat power to the decisive point, to assist the army in effecting a breakthrough, or to shatter an enemy counterattack. Even so, the bulk of Luftwaffe missions in Poland and France went towards indirect, rather than direct, army support. Attacks on the enemy's reserves, rail traffic, and airfields were tasks more suited to the Luftwaffe's overall force structure. Air-ground liaison procedures were still primitive, and in the "fog of war," German aircraft routinely bombed and strafed friendly troops. Radio contact between army units and supporting aircraft was nonexistent, and the liaison system in place, although vastly superior to its vestigial Allied counterpart, was too slow and cumbersome. Several hours normally elapsed between an army request for air support and the actual attack.[143]

German victory in Poland brought about the same effort at analysis that had followed World War I, although the pressures of an expanding war accelerated it.[144] Stuka units and armored formations exercised relentlessly, attempting to perfect methods of identification, liaison, and communication.[145] By the time of the opening of the campaign against the Soviet Union, Richthofen's *Fliegerkorps* VIII at least could provide reliable, concentrated close air support on a sustained basis. The majority of German flying units in contrast were far less able to perform the difficult mission.[146] As a result, *Fliegerkorps* VIII found itself transferred, at a moment's notice, across the length and breadth of the Eastern Front as the tactical situation demanded. Its battle headquarters shifted location no fewer than eighteen times during summer and fall 1941.[147] Within the Luftwaffe, close air support remained a highly specialized task, and a few

---

[143] Williamson Murray, *German Military Effectiveness* (Baltimore, 1992), pp. 229–243.

[144] Ibid.

[145] Genkdo. XIX. Armeekorps, Ia, Anlagenheft z. KTB Nr. 2, "Zusammenarbeit mit der Luftwaffe," NARA T314/615/357.

[146] For the history of German close air support in the war against the Soviet Union, see Richard Muller, *The German Air War in Russia* (Baltimore, 1992).

[147] "Gefechtsquartiere des VIII. Fliegerkorps im Russland-Feldzug 1941," USAFHRA K. 113–309–3 v. 2.

overextended units bore the brunt of its execution. In sectors deemed less vital, German troops often went for months without receiving any air support at all.[148]

The later evolution of German close air support was bound up with the fate of the Luftwaffe as a whole. The combat effectiveness of the ground attack force grew dramatically as the war progressed. The Germans paid much attention to perfecting the methods of liaison and communication and in effect ironed out the problems that had plagued close support since 1917–1918. Army units received detailed questionnaires dealing with their experiences with Luftwaffe cooperation, and agencies such as the Luftwaffe operations staff, the Military Science Department, and the office of the *General der Schlachtflieger*, established in October 1943 as a separate inspectorate, analyzed the results.[149] By spring 1944 the result was a "sensible and generally intelligible" system of army-air cooperation, assisted by vastly improved aircraft types (most notably the Focke-Wulf 190F-8, a modified air superiority fighter) and radio equipment with frequencies in common with army units.[150] Manuals that appeared in late 1943 and early 1944 codified the system developed by Richthofen and his *Fliegerkorps* VIII staff and made the procedures standard for the entire Luftwaffe – a clear case of "innovation bubbling up from below." Such was the reputation of German close air support that Adolf Hitler, when lambasting Luftwaffe personnel for lack of fighting spirit in the final months of the war, specifically excepted the ground attack units from criticism.

Most of these later improvements, however technically impressive, had little impact upon the course of the war. They came about during a period when the Germans began to lose air superiority, particularly in the western theater. Fw 190 production was accordingly redirected towards the home defense fighter force, and the few ground attack units in the west equipped with the type could achieve nothing against the overwhelming Allied air superiority during the June 1944 Normandy invasion.[151] The close support units fighting in the east operated with greater effectiveness, but the same trends were operating there as well. Indeed, after spring 1944 the Luftwaffe in Russia, shorn of its bomber and most fighter units, was essentially a collection of regional close support commands, or *Flieger-*

---

[148] HGr. Mitte Ia, Anl. z KTB Erfahrungsberichte, Heft 3,65002/31, NARA T311/220/692.

[149] Oberkommando der Heeresgruppe Mitte, Ia Nr. 5924/43 geh., NARA T311/220/679.

[150] Oberbefehlsheber der Luftwaffe, Führungsstab Ia/Ausb. Nr. 1000/44 geheim, "Vereint schlagen: Zusammenarbeit Luftwaffe-Heer auf dem Gefechtsfeld," OKL/2085, NARA T321/243/no frame #s.

[151] Erfolgsberichte der Gruppe (III/SG4) vom 4.7.44–29.10.44, von Rohden 4376–477, NARA T971/39/430.

divisionen. All of the tactical brilliance that the Germans brought to the development of close air support availed them little after 1941.

Britain's experience with air support for the first years of the war was in grim contrast to Germany's. The hard lessons of the early campaigns led the British to adopt methods that were in many ways mirror images of their adversary's. Prewar air strategy had concentrated on the strategic air offensive and, in the final years before the war's outbreak, has also focused on the concept of "strategic interception" (forestalling an enemy invasion of the west) and home defense.[152] The Advanced Air Striking Force that accompanied the British Expeditionary Force to the Continent exhibited an inability to cooperate effectively with the ground forces. The Advanced Air Striking Force was ineffective on many levels. It lacked any kind of rational command and control system to enable it to assist the army. The short range of its aircraft types, the Fairey Battle and Bristol Blenheim, required a system of forward basing, and the German breakthrough and conquest of the Allied air bases in Northern France in May 1940 thoroughly disrupted the Advanced Air Striking Force's ability to intervene in the battle. The RAF ignored calls from the army to field a specialized army support force, since "an army air force would in effect be established and that would clearly result in a dissipation of effort."[153]

The cataclysmic defeat in May-June 1940, and the ejection of the BEF from the Continent, was the nadir of British military fortunes and of the RAF as a cooperative weapon. The experience of defeat proved a powerful catalyst for redressing the lack of army-air cooperation, although many false starts occurred before the British created a reliable system. Indeed, a "senior air force officer" announced in August 1940, that "the fact is that the RAF is now free to perform its proper function, that for which it has been organized and trained. It is no longer hampered by the necessity of acting as long-range artillery for the BEF."[154]

The initial push for reform came from the British misperception of the causes for the German success. British observers seized upon the Ju 87 dive bomber as the key to the *blitzkrieg* victories in France. As early as June 1940, events on the continent stampeded the British into ordering hundreds of obsolescent Brewster Bermuda and Vultee Vengeance dive bombers from the United States.[155] This clutching at straws resulted in

[152] Smith, *British Air Strategy Between the Wars*, pp. 110–111, 138.
[153] Air Ministry, *Air Support*, p. 15.
[154] Milton, "Army and Air Force Cooperation," p. 663.
[155] W. A. Jacobs, "Air Support for the British Army, 1939–1943," *Military Affairs*, 46, December 1982, p. 176.

an acquisitions disaster of the first order. After placing orders for 1,250 dive bombers, "by 1942 . . . the Air Ministry had no idea whether these orders had been canceled, increased, or decreased"; needless to say the British never operated dive bomber units against the Germans.[156] The early British fixation on the dive bomber overlooked, of course, the need for air superiority, a precondition for employment of dive bombers of which the Luftwaffe leadership was perfectly aware. In emphasizing aircraft design, the British initially failed to grasp the significance of the technological context – doctrine, command, and control – within which the Ju 87 operated.

Although this attempt to emulate German practice was a failure, valuable lessons emerged from a more detailed analysis of Luftwaffe close support methods during the summer and fall of 1940. A series of exercises, coordinated by Group Captain A.H. Wann and Brigadier General Sir John Woodall, conducted in Northern Ireland in fall 1940, put German methods to the test. The results of the studies, known as the "Wann-Woodall Report," set down four general principles for successful army-air cooperation. These principles included the need for a specially trained and equipped unit for close support; a specially trained army staff able to act as liaison officers between air and army headquarters; some type of joint command post; and a reliable communications system.[157] Subsequent exercises (most notably "Exercise Bumper" in fall 1941) indicated that "by far the greatest difficulty . . . was that caused by the very unsatisfactory performance of the communications system."[158] Operational experience would provide the means of solving individual problems, but the British had the intellectual foundation in 1941.

The necessary reforms were not implemented by a "top-down" approach. Army commanders recoiled at the idea of a centrally allocated, close support force under RAF control. Sir Alan Brooke, commander of Home Forces, argued that "his experience, during the retreat from Dunkirk, had illustrated very clearly the evils of the highly centralized system practiced by the RAF. He had never seen a single aircraft throughout the retreat."[159] Alan Brooke argued for a strengthening of Army Cooperation Command, which he envisioned as an organic air support force under the practical control of the army. The air staff, for its part, violent-

---

[156] Brereton Greenhous, "Aircraft Versus Armor: Cambrai to Yom Kippur," in Timothy Travers and Christian Archer, eds., *Men at War: Politics, Technology and Innovation in the Twentieth Century* (Chicago, 1982), p. 98.

[157] Shelford Bidwell and Dominick Graham, *Fire-Power: British Army Weapons and Theories of War, 1904–1945* (Boston, 1982), pp. 264–65.

[158] Air Ministry, *Air Support*, p. 28.   [159] Jacobs, "Air Support," p. 178.

ly objected to any suggestion that the army gain control over air power and dissipate it into "penny packets."[160]

The matter might have rested there, yet in an outlying theater of war a practical system emerged. It developed in a manner similar to the way in which the problems of "air control" had been solved. The focus of the reforms was the Royal Air Force contingent serving in the Middle East, which provided support for the Eighth Army. Many of the developments in North Africa paralleled almost exactly those suggested by the Wann-Woodall report – as historian John Terraine argues, "the fundamentals of the subject, on examination, are common sense."[161] The system of air support crafted by "Maori" Coningham and Arthur Tedder involved fighters such as the Hawker Hurricane – cannon-armed and carrying light bombs – controlled from a ground liaison network strikingly similar to the German *Flivo* system. When the Allies employed the system during the Normandy Campaign, in the expanded form of the Allied Tactical Air Forces, tactical aviation attained its highest level of wartime effectiveness.

Similarly, the United States entered World War II without a system of tactical air support. The promising start after World War I had disappeared, and the necessary organizational, theoretical, and institutional factors were lacking. The story of army air forces close support provides another example of the lack of success attending most "top-down" initiatives. The commander of army ground forces, General Lesley J. McNair, shared responsibility with General H.H. "Hap" Arnold for "the development . . . of ground-air support, tactical training, and doctrine."[162] McNair was openly hostile to the concept of an independent air force and struggled vainly to convince the airmen to take air-ground training seriously. In McNair's view, the success of air-ground exercises depended upon the army air forces providing sufficient numbers of modern aircraft and trained pilots. He found the army air force attitudes, which to him smacked of foot-dragging, a constant annoyance. An army commentator noted, "Reports from the field during maneuver exercises and after their conclusion were unanimous with regard to their failure." McNair despaired, "The trouble is that the air side of the setup has been too sketchy to permit effective training. I say this without criticism of the air force."[163] When the army air forces cobbled together a "flying circus" of

---

[160] Air Ministry, *Air Support*, p. 25.

[161] Terraine, *A Time for Courage*, p. 352.

[162] Kent Roberts Greenfield, *Army Ground Forces and the Air-Ground Battle Team*, Study No. 35 (Historical Section: Army Ground Forces, 1948), p. 7.

[163] Greenfield, *Air-Ground Battle Team*, 18; Kennett, "Developments to 1939," p. 56.

aircraft types intended to overfly army units as an aid to aircraft recognition, McNair denounced the measure as "largely eyewash."[164]

An "Air Support Board" convened at army air forces' request in December 1942 to redress the lack of close support capability produced equally limited results. While ostensibly concerned only with training matters, the board's proposal to rewrite air support doctrine caused McNair to fear yet another bid by the army air forces for independence. While army units in 1943 received additional air-ground training, it was widely known that the air staff in the Pentagon "openly scoffed" at the army ground force's own training scheme.[165]

The manuals that grew out of this doctrinal debate served less as a framework for training and operational practice than as elucidations of each faction's positions on the subject of air-ground coordination. FM 31–35, "Aviation in Support of Ground Forces," contained little on the specifics of air-ground coordination save to reinforce the principle of centralized army air force control of air assets.[166] Later, some army ground force leaders viewed even the landmark manual FM 100–20, "Command and Employment of Air Power," with the greatest suspicion; many in the army regarded it as little more than the army air force's "Declaration of Independence" and a cementing of a lackadaisical and neglectful attitude towards air support responsibilities.[167] For their part, air force commentators praised the document as containing the essential elements of the successful system that was eventually to emerge.[168] In fact, FM 100–20 codified lessons from the early part of the war and provided a basis for continued innovation. Greatly improved all-arms cooperation was the result.[169]

Two related factors, then, led to an alteration of the air force's position on close air support. These factors were the dismal performance of its units in Tunisia and Sicily, and the simultaneous observation of "British methods, wearing the authority of success."[170] McNair compared army

[164] Greenfield, *Air-Ground Battle Team*, p. 39.   [165] Ibid., pp. 32–40.
[166] Robert Frank Futrell, *Ideas, Concepts, Doctrine: Basic Thinking in the United States Air Force 1907–1960*, vol. 1 (Maxwell AFB, 1989), pp. 105–107.
[167] Greenfield, *Air-Ground Battle Team*, p. 47.
[168] Thomas J. Maycock, "Notes on the Development of AAF Tactical Air Doctrine," *Military Affairs*, 14, 1950, p. 186.
[169] Richard Greene Davis, *Tempering the Blade: General Carl Spaatz and American Tactical Air Power in North Africa, November 8, 1942 – May 14, 1943* (Washington, DC, 1989), pp. 109ff.
[170] Ibid., p. 29; David Syrett, "The Tunisian Campaign, 1942–43," in Cooling, ed., *Case Studies in the Development of Close Air Support*, pp. 161–165.

air force performance in Tunisia unfavorably with what had occurred in World War I while General Omar Bradley delivered the following verdict on air support in North Africa:

> We can't get the stuff when it's needed and we're catching hell for it. By the time our request for air support goes through channels the target's gone or the Stukas have come instead.[171]

Experience in Tunisia led to an appreciation for, and eventual adoption of, the system crafted by the commanders of the British Desert Air Force, and the U.S. Fifth Army employed the system, with few modifications. In expanded form, the system attained great success in Normandy and Northwest Europe. Significantly, the field manual codifying the new system did not appear until April 1945, demonstrating that the "trickle-up" approach to the problem of close air support was the one eventually adopted by all three belligerents. While an evaluation of the effectiveness of Allied tactical aviation in Normandy is beyond the scope of this paper (and recent studies have indicated that the effect on morale greatly exceeded the material damage inflicted), few doubt the role close air support played in the Allied victory.[172] Successful, although belated, innovation put this formidable weapon into the hands of Allied theater commanders when it was needed.

CONCLUSIONS

The preceding discussion suggests a number of conclusions about the process of military innovation. While the aircraft and the technology under examination seem almost quaint when contrasted with today's weapons systems, many of the factors that influenced early development have continued relevance to late twentieth-century air and surface forces. Many air power advocates sought to blame operational shortcomings of air forces in World War II upon inadequate funding, which peacetime demobilization, competition from the senior services, isolationism, antimilitarism, and other external factors all exacerbated. Mitchell's observation in 1919 that "normalcy was wrecking the nation's armed forces and particularly its air forces"[173] reflected a bitter truth that retains its edge in the post–Cold War world of the 1990s.

---

[171] Greenfield, *Air-Ground Battle Team*, p. 77.
[172] Ian Gooderson, "Allied Fighter-Bombers Versus German Armour in North-West Europe 1944–1945: Myths and Realities," *Journal of Strategic Studies*, 14, June 1991, pp. 210–231.
[173] Futrell, *Ideas, Concepts, Doctrine*, p. 32.

In a hostile climate, the development of a specialized weapon such as a close support force was bound to be adversely affected. Within the tiny – or in the case of Germany, nonexistent – air services of the 1920s and early 1930s, experimentation, training, aircraft development, and other types of technical and technological change were noticeably stunted. The checkered history of the United States Third Attack Group, which attempted to carry the concept of army support aviation forward with only a handful of aircrew and equipped with mostly inappropriate aircraft designs, illustrates the effect of these external constraints. Furthermore, the lack of real combat experience, as the lessons of trench warfare in World War I receded into the past, deprived the air services of the catalytic effect of sustained major operations that had vitalized reform and innovation in World War I.

Finally, all three air forces to a greater or lesser degree faced real threats to their institutional survival during the interwar period. A mission such as close air support seemingly promised to return air power to its World War I status as an auxiliary to the ground forces. The debate among army and air commanders regarding centralized or decentralized control of air support forces, one that raged on well past World War II, is the legacy of this understandable fear on the part of the young air leadership. To them, concentration upon missions, such as "strategic bombing," attaining "command of the air," deep interdiction, or *operativer Luftkrieg* not only maximized the theoretical contributions of air power to a future war, but furthered institutional imperatives as well. Close air support, so it seemed, did neither.

In the final analysis, close support aviation suffered because of a conscious neglect on the part of the respective air force leaderships. A reading of World War I led them to believe the mission was too costly, and of limited value against a first-rate adversary. The rapid advances in bomber design and technology opened up broader vistas and took the focus of air activity away from the forward battle line. The progression of American attack aviation provides a clear illustration of this point. It was concerned primarily with close support in the early 1920s, and shortly before World War II moved towards the medium attack bomber concept, which emphasized interdiction and attacks on airfields. These missions, though undoubtedly important, had only an indirect effect on the ground forces being supported.

The period does contain some significant exceptions to the general trend of neglect; these exceptions provide a number of valuable suggestions about the process of military innovation. The first concerns the value of official history or operations analysis programs. The Germans got by far the most out of their efforts to digest the lessons of World War I. While the German leadership remained in the dark about the political and grand strategic lessons, their reading of its operational and tactical lessons was on the mark. Because close air support was first and foremost an operational matter, it was the kind of problem for which they were able to devise effective solutions. Even if their capability initially lagged, the Germans, under pressure of circumstances, quickly reorganized existing units into effective close support formations. These formations played a significant part in the German victories of 1939–1941. Although the Allies also conducted a painstaking analysis of close support in World War I and, as in the case of Mitchell, sometimes came to much the same conclusions as the Germans, they failed to make effective use of the experience.

Likewise, the existence of agencies or channels of communication assigned to evaluate and disseminate combat experience materially advanced the innovation process. While such agencies were virtually nonexistent before the outbreak of hostilities, much of the rapid improvement in German close support (and somewhat later, in British and American) came about as a result of initiatives, often at corps level and below. More often than not, published "official close support doctrine" was merely the reflection of frontline practice arrived at months or even years earlier. Conversely, "top-down" initiatives, such as McNair's attempts to develop more effective close support procedures in 1942, frequently met with failure.

The lessons of the interwar period and World War II suggest that successful close air support in the first instance depended upon organization, doctrine, and training, rather than upon the existence of specialized weapons systems. "Strategic bombardment" in daylight required two essential technical ingredients: the long-range, four-engine bomber and the high-performance escort fighter to accompany it. Air forces that failed to procure the appropriate aircraft types, such as the Luftwaffe in the USSR in 1943–1944, found themselves in an unenviable predicament when they contemplated long-range bomber operations when their units operated only medium bomber types. Close air support did not depend upon the same essential "technical fix," with the lengthy production pipeline new aircraft types required. The army air corps, conversely, received ever more efficient attack types in the 1930s, but its ability to perform the mission hardly increased at all. The Luftwaffe, which dominated the field of close support for the first three years of the war, did so with aircraft types, such as the Junkers 87, that were approaching obsolescence and were not designed for close air support in the first place. In 1945 the United States

Army Air Forces got its best close support service out of the Republic P-47 "Thunderbolt," which had been conceived as an interceptor. In most cases, with the signal exception of the Il'yushin IL-2 "Stormovik," purpose-built ground attack aircraft – the GA-1, the A-36 (an early dive-bomber variant of the P-51 "Mustang"), and the Henschel 129 (a heavily armored "flying can opener"), to cite the most obvious examples – were bitter disappointments.

Finally, the creation of units specially organized and trained for the task of direct support was essential. During the "starving time" for army support in the interwar era, the most fertile sources for innovation and adaptation were the small, virtually self-contained flying units such as the USMC, the Condor Legion, and the RAF's "air control" detachments. During the course of World War II, formations such as Richthofen's *Fliegerkorps* VIII, the Desert Air Force, Kenney's Fifth Air Force, and the Allied Tactical Air Forces in Normandy took the lead in furthering the concept. Most of the technical and operational changes that heralded innovation were the result of personal contacts between ground and air commanders, often of relatively junior rank. The German military officially encouraged such contacts; in the Allied forces the process took time, although British and American commanders such as Tedder and Spaatz eventually proved just as creative and adaptable as their German counterparts. Had the outcome of World War II hinged upon the results of the first engagements, the Allied failure would have had disastrous consequences.

5

ADOPTING THE AIRCRAFT CARRIER

The British, American, and Japanese case studies

GEOFFREY TILL

In World War II, aviation had a decisive effect on the outcome of the naval campaigns against the Axis powers. Carrier and land-based maritime aircraft vastly extended reconnaissance at sea, offered the most effective means of protecting shipping – merchant as well as military – against air and submarine attack, and proved to be a potent means of attack on enemy warships and maritime commerce. Sea-based aircraft were essential for the conduct of amphibious operations because they were the principal means of attacking defensive aircraft and of securing command of the air. Carrier aircraft protected amphibious shipping and troops ashore and provided reconnaissance and fire-support facilities that were often decisive.

However, in the years between the wars, the British, American, and Japanese navies failed to realize fully the contribution that airpower could make to the conduct of war at sea. Policy makers, in all three navies, underestimated the extent to which naval airpower represented a significant departure in the way they would conduct their business.

Nonetheless, the following analysis will show that by the outbreak of World War II, the three navies differed only in the degree of their underestimation of the strategic significance of maritime air power. The British were the least perceptive, the Japanese, the most. But the differences among them, and indeed between them and other major navies not included in this review, were matters of degree, not a question of absolutes. An analysis of their approaches to the problem of maritime airpower offers interesting insights into the larger issue of effective military innovation. This study will focus on British experience and use Japanese and American developments as comparators.

THE ANALYTICAL PROBLEM

In 1912, a pioneering British naval aviator, Lieutenant Hugh Williamson, proposed construction of an aircraft carrier capable of operating wheeled aircraft. The Admiralty turned his suggestion down. Later, Williamson disgustedly recorded:

> No greater modification of any HM ships than that I proposed would have had the smallest chance of acceptance at that time. Prior to the First World War, the navy had had no war experience for a very long time; and a long peace breeds conservatism and hostility to change in senior officers. Consequently, revolutionary ideas which were readily accepted when war came, were unthinkable in the peacetime atmosphere of 1912. Circumscribed by the then existing limitations my proposal was the furthest one could hope to go.[1]

This portrayal of the prejudices of senior officers and existing bureaucracies as essentially conservative and inimical to the advance of challenging new ideas is common amongst innovators. It is also common amongst those who study technological development. The literature of naval aviation is full of accounts that represent the transition from the battleship to the carrier era as a duel between unimaginative, emotional, and often stupid battleship admirals against young, innovative, irreverent airmen.[2]

Such views are often informed by a conception of technological change as proceeding by a series of discrete jumps. The prescient recognize the jumps, while the conservatives do not. Often a particular event appears to work the conversion just as light and smoke changed St. Paul's beliefs on the road to Damascus. Thus, after an Armistice day fly-by, Admiral William F. Fullam, U. S. Navy, wrote in 1918:

> They came in waves, until they stretched almost from horizon to horizon, row upon row of these flying machines. What chance, I thought, would any ship, any fleet have against an aggregate such as this? You could shoot them from the skies like passenger pigeons, and still there would be more than enough to sink you. Now I loved the battleship, devoted my whole career to it, but at that moment I knew the battleship was through.[3]

---

[1] G. Robbins, "The Evolution of the Aircraft Carrier 1914–1916," PhD Dissertation, University of London, 1992. p. 46.

[2] Robert O'Connell, *Sacred Vessels: The Cult of the Battleship and the Role of the U.S. Navy* (Boulder, CO, 1991) is a recent example.

[3] Quoted ibid., p. 252.

From idea to actuality is rarely so simple. New technological departures are rarely discrete. In the interwar period, naval aviation was neither new nor separate. On the contrary, those involved considered it part and parcel of the general business of navies. In 1933, Admiral Chatfield, the British First Sea Lord commented:

> The air side is an integral part of our naval operation . . . not something which is added on like the submarine, but something which is an integral part of the navy itself, closely woven into the naval fabric. Whether our air weapon is present or not will make the whole difference to the nature of the fighting of the [f]leet and our strategical dispositions. That is a fact which will increase more and more, year by year.[4]

Naval decision makers therefore confronted a complex calculation as to whether or not new technology was a good investment. In the interwar period, they were in the business of striking balances between one aspect of fleet and ship design and another. They were not dealing with blacks and whites, but with shades of gray.

Thus, technological development is less a process of distinct jumps than a prolonged slither, to which decision makers react incrementally. One recent analysis of the U.S. Navy's attitude towards battleships summarized the bureaucratic response in this way: "Certainly, it did not reject new technology out of hand. Rather it attempted to do what the naval establishment has always done when faced with dramatic change – accommodate it within a familiar framework."[5] This remark was not intended as a compliment, but in fact it is. One of the most difficult issues involved in questions of technological adaptation is not whether but rather when the innovations will have the anticipated effect. Nearly always, decision makers prefer to base their policy on present likelihoods, not future possibilities.[6] They incline to look at the devil in the details and refuse to consider grand and vague visions of a misty technological future instead of the mundane practicalities of the present.

Innovators and their admirers so often represent this response as unthinking obscuranticism. Returning to Lieutenant Williamson's proposal, it is interesting to note that the Royal Navy rejected his far-sighted proposal for an aircraft carrier, not on the basis of the prejudice of diehard battleship admirals, but rather largely on the advice of the Royal Navy's

[4] Chatfield, 1st Meeting of Inskip Inquiry, 13 July 1936, CAB 16/151, Public Record Office (PRO).

[5] O'Connell, *Sacred Vessels*, p. 280.

[6] William M. McBride, "Challenging a Strategic Paradigm: Aviation and the U.S. Navy Special Policy Board of 1924," *Journal of Strategic Studies*, September 1991, p. 77.

other young pioneering airman, Lieutenant C.R. Samson. Samson argued that seaplanes were a better bet for the moment, because the notorious unreliability of aircraft engines made the flight of wheeled aircraft over sea too hazardous; he also suggested that Williamson's ideas about landing floated seaplanes on a deck were impractical.[7]

Finally, taking the technological determinism that lurks beneath presentations of the battleship versus aircraft issue too far is dangerous. Strategic, political, and economic considerations often determined naval responses to the prospects of maritime aviation. Navies had to adapt to more than technological possibilities.

THE EXPERIENCE OF WAR: 1914–1918

For British naval flyers, World War I was a story of success and expansion. The Royal Naval Air Service expanded from a force of 93 airplanes and seaplanes, 100 officers and 550 ratings to a large and experienced force of some 5,000 officers and 55,000 men, operating 2,949 airplanes and seaplanes.[8] Their ships and bases increased both in terms of quality and quantity. The British had entered World War I with a motley collection of improvised seaplane carriers and a few score fragile aircraft. During the war, naval pioneers solved many of the key problems involved in landing aircraft on ships and flying them off. By the end of the war, they were conducting carrier operations with aircraft that were operationally effective in shooting down enemy aircraft, conducting reconnaissance patrols, and dropping bombs and torpedoes.

The Royal Navy developed twelve aircraft carriers during the war, a process culminating in 1918 with HMS *Argus*, the world's first carrier capable of launching and landing naval aircraft. Its design identified and addressed all the problems associated with carrier operations for the remainder of the century. In many cases, early and workable solutions were in place or proposed, such as aircraft lifts and arrester gear. In others, the first round of long-running controversies such as arguments for and against island superstructures, the balance to be struck between wheeled aircraft and seaplanes, and so forth had been fought.

Moreover, the *Argus* was but one of a fleet of a dozen carriers of one sort or another at a time when no other naval power had even one. Events ashore followed a similar course. In 1914, the Royal Navy had six small seaplane stations and two airship stations. By 1918 this force had grown

[7] Robbins, "The Evolution of the Aircraft Carrier," p. 43.
[8] "The British Naval Effort, 4th August 1914 to 11th November 1918," Adm 167/57, PRO.

to fifty-two aeroplane and seaplane stations and thirty-four airship stations, with others under construction.

Operationally, the Royal Naval Air Service emerged from World War I with a wealth of experience in naval flying. The Royal Navy's flyers had pioneered virtually every aspect of naval aviation in direct support of the fleet. They had operated from carriers, from surface warships, and from large numbers of shore stations. They had provided air cover for fleet operations in the North Sea and elsewhere, spotted for naval artillery against sea and shore targets. They were adept at naval reconnaissance and in 1918 were planning a large-scale torpedo-bomber raid on the German High Sea Fleet in its bases, a strike that would have antedated the battles of Taranto and Pearl Harbor by more than twenty years.

British naval aircraft had also made an important contribution to the ultimately successful war against German U-boats. During 1918, in Home Waters alone, British naval aircraft on patrol and antisubmarine duty flew approximately 30,000 miles per day and attacked 118 German submarines. Consequently, the Royal Navy was reasonably successful in its adaptation to the new technology of naval aviation. A number of factors explain the level of its success:

- *Resources.* Resources were less of a problem during the war than before it. In the 1913–1914 estimates apparently the more urgent requirements of preparing the battle line for imminent war had squeezed naval aviation. Even after the onset of war, important resource constraints existed (especially in taking over vessels for conversion to carriers), but they became less important when it was clear the war would not finish before naval air affected the advances in question.
- *Required incentives.* The urgent operational incentives of war justified the costs of this policy. Specifically, the British were particularly anxious to produce weapon systems that could control German submarines, shoot down Zeppelins and other reconnaissance craft, and conduct reconnaissance missions. This priority reflected the Royal Navy's urgent requirement to force battle on the German High Sea Fleet. The Royal Navy did lack the stimulus that parallel efforts by other naval powers would have provided. The British were the leaders of the field and knew it. The Russians and the Germans were probably the next well-developed in naval aviation but concentrated their efforts on airships, seaplanes, and long-range land-based aircraft. The Americans and Japanese were interested but far behind. They, indeed, profited to a considerable extent from the pioneering efforts of the British. "So many of our ideas of naval (air) policy have been gained from the

British," concluded one U. S. naval officer in 1919, "that any discussion of the subject must consider their methods."[9] The Japanese also devoted much effort to learning as much as they could from British experience.[10] The British were later to discover that being a pioneer has drawbacks as well as benefits.

- *Enthusiastic open-mindedness.* The Royal Navy demonstrated a willingness to devote much effort to experimentation across all competing options. In effect, the Admiralty tried to back all the horses in the race, although it did have an early preference for seaplane-based options. In the period 1916–1918, the Royal Navy conducted scores of experiments, including the launch of wheeled airplanes from platforms on gun turrets of cruisers and battleships and from platforms towed behind destroyers, seaplane operations of all sorts, airship operations, and shore-based naval operations.

- *Sources of innovative thought.* Before and during World War I, a large but far from united group of air enthusiasts developed. Some, like Lieutenant H.A. Williamson, were influential and extraordinarily prescient. They often captured the attention of their seniors by conducting aerial "stunts," which made important points dramatically, even though, in many cases, their regular operational repetition was not feasible, for example, Lieutenant Longmore's first torpedo-dropping flight in 1912. He staggered into the air with a torpedo unlikely to damage an armored warship, in calm and enclosed waters with a bare allowance of fuel in order to save weight.[11] His achievement was symbolic not operational. Most early take-offs and landings-on were on a similar scale. In many cases, such stunts did have the desired effect of influencing senior officers. But many of the command level were enthusiasts, too. Friends in high places, like Winston Churchill, the First Lord, or Admiral Jellicoe, the Commander-in-Chief, were a major aid.

The British exhibited an early and commendable desire to profit from the war's lessons. They formed the Postwar Questions Committee, which focused, in particular, on naval aviation and the general submarine question. The American and Japanese navies followed suit.

Of course, dangers existed in relying on the apparent lessons of past

---

[9] Captain G. W. Steele, USN, quoted in S.W. Roskill, *Naval Policy Between the Wars,* vol. 1 (London, 1968) p. 244.

[10] Carl Boyd, "Japanese Military Effectiveness: The Interwar Period," in Allan R. Millett and Williamson Murray (eds), *Military Effectiveness,* vol. 2, *The Interwar Period* (London, 1988) p. 135.

[11] Robbins, "The Evolution of the Aircraft Carrier," p. 89.

experience, especially during a period of rapid technological change. Such lessons needed counter-balancing by analysis of other experience. As far as the British were concerned, there were three sources of new ideas. The first was a regular review of current tactical and technical experience gained through a combination of exercises and occasional theoretical enquiries. Actual war experience in the 1920s and 1930s was limited to the ambiguous lessons of the Abyssinian crisis and the Spanish civil war.[12] The Americans had an equally limited range of experience, but the Japanese derived unique benefit from the operational and technological lessons of their war in China.[13] The Japanese also gained enormously from the guidance extended by the British Sempill mission of the 1920s. Britain's hard-won experience thus provided a sound foundation for the development of Japanese naval aviation.

On their part, the Japanese conducted exercises with ruthless realism, which no doubt increased the predictive and analytical value, although the result sometimes came at the cost of many lives.[14] American exercises were bold and innovative too, although their results were sometimes a matter of dispute. According to air enthusiasts, the navy's senior leadership took Fleet Problem XII of 1931 unfairly as a demonstration of a carrier fleet's limitations in finding and stopping an enemy battle squadron.[15]

Both Japanese and Americans utilized the lessons of war-gaming at their naval staff colleges more than the British. War-gaming at Newport in 1925, for instance, heavily influenced the views of Captain Joseph Reeves, who went on to become a leading advocate of naval aviation over the next decade. In truth, potential lessons came from many sources and effective responses required decision makers to pick their way skillfully though a maze of often conflicting experience.

STRATEGY AND ADAPTATION

The British case

The strategic environment in which navies operate obviously heavily influences the manner in which they develop new technology. On the basis of their war experience, the British got off to a vigorous start in 1919. Ad-

---

[12] Alvin D. Coox, "Military Effectiveness of Armed Forces in the Interwar Period, 1919–1941: A Review" in Millett and Murray, *Military Effectiveness*, vol. 2, p. 261.

[13] Stephen Haworth, *To Shining Sea: A History of the United States Navy, 1775–1991* (London, 1991), p. 365.

[14] Yoichi Hirama, "Japanese Naval Preparations for World War II," *Naval War College Review*, Spring 1991, p. 67.

[15] O'Connell, *Sacred Vessels*, p. 286.

miral Sir Charles Madden, commander-in-chief of the Atlantic Fleet and later First Sea Lord, asked for construction and/or development of no less than twelve specialized carriers.[16] This quickly proved over-ambitious, and the plan settled down, after the Washington Treaty of 1922, to a twenty-year program designed to produce a squadron of seven fleet carriers by 1939, with other carriers, less capable, for an escort – trade – protection role. In the event, the British went into World War II with only four first-line carriers and three obsolescent ones.

This disappointing result prevented the Royal Navy from operating as many first-line aircraft as it had originally wanted, from concentrating its carriers as a squadron, and left it with no carriers for trade protection. Moreover, interwar circumstances compromised the standards of Fleet Air Arm in every way and significantly disadvantaged the British when compared to the Americans or Japanese.

So why and how had things turned out so badly? The argument that this state of affairs simply demonstrated that British admirals were insufficiently "air-minded" is, in itself, unfair. The size of the original carrier program clearly refutes such a charge. Economic constraints constantly forced program reductions on the Admiralty, but the navy accepted these cuts with the belief that naval aviation would have to take its share of the cuts, rather than on the basis that it did not matter. After all, the Royal Navy's battleship modernization program suffered, to an even greater extent.

The most severe criticism that one can legitimately make about the Admiralty's carrier construction program is that it adopted a policy of excessive gradualism. In the early 1920s many design issues were unresolved. These included the development and positioning of arrester wires, catapults, carrier "islands" and included questions of the optimum size of carriers, whether they should be specialized or general purpose, and so on.

On many of these matters, the British consciously and deliberately adopted a policy of "waiting to see" or of leaving it to the Americans and Japanese to set the pace in such incremental innovation. In contrast to British policy, the Japanese and Americans, being newer in the business, were willing to make bolder departures. Both, for example, followed their first carriers with a second generation that were three times as large. While gradualism had some beneficial consequences (perhaps most notably in the *Ark Royal* and the innovative armored carrier program of the late 1930s), the negative consequences outweighed the gains. The conscious decision to move slowly on the development of arrester wires and catapults, for example, was a se-

[16] G. Till, *Airpower and the Royal Navy 1914–1945: A Historical Survey* (London, 1979), p. 64. See also Plans Div. Memo, 13 May 1927, ADM 1/9271, PRO.

rious mistake that considerably reduced the carrying capacity of British aircraft carriers, the size of Fleet Air Arm, and the potential of both.

Significantly, the aviators of Fleet Air Arm and the Air Ministry supported both decisions, for what seemed the best of specific, technical, and aeronautical reasons;[17] once again, the problem was not simply a consequence of battleship admirals running the Fleet Air Arm. Nor was this policy of wait and see particularly new. The British had adopted this policy through much of the nineteenth century; the different outcomes lay in the fact that, while the British had possessed the industrial capacity to overhaul innovative competitors at that time, they no longer held such advantage after 1918. In the interwar period, the Royal Navy could not recapture a lead once lost.

Britain's relative industrial decline and severe financial difficulties were the real reasons for the deficiencies in Britain's carrier program. In the 1920s and early 1930s the economic situation, trade and budget deficits, and pre-Keynesian theory dictated savage economies in all fields of government expenditure, if the country were to regain financial health. Inevitably the navy and its carrier program took its share of the pain and grief. The Royal Navy kept its older carriers longer than anticipated and repeatedly postponed the start date for the four new carriers envisaged in the Admiralty's original plan. As the British had started with so many more carriers than either the Americans or Japanese, it was more difficult for the Admiralty to argue its case effectively. The Treasury, supported by the Air Ministry, was unimpressed by arguments based on quality and instead concentrated on quantity. Even when the government decided to proceed, Britain's relative industrial decline (which, of course, partly dictated the need for economy) meant that it took longer to refit, convert, and build carriers than originally envisaged.

The Air Ministry encouraged the Treasury's reluctance to authorize construction of new carriers, because it was far from enthusiastic about having to find the extra aircraft such ships would require. The Washington Treaty of 1922 did not in itself have a direct effect on the carrier construction since for all participants it was a question of building up rather than down to agreed limits. But the Treaty did create expectations about further naval arms control which made it inexpedient for the Admiralty to ask for more carriers or aircraft at a time when politicians were enthusiastically thinking about major reductions.

[17] Details of the Admiralty's approach to catapult development can be found in staff minutes of December 1925, ADM 138/637, NMM. See also Till, *Airpower and the Royal Navy*, p.72.

The Admiralty therefore remained under pressure from two directions. The government expected it to make significant contributions to a "peace dividend" at a time when defense expenditure only represented some 2 to 2.5 percent of Britain's GNP. There was also a widespread expectation of further progress in disarmament. The fact that Britain did not face a single and agreed threat in the interwar period further aggravated these problems. Instead, British strategists could discern at least four axes of threat, all of which materialized during the World War II. Each service, as well as the Treasury, evolved a set of working assumptions in line with the broad threat outlines, but each of which possessed quite distinctive elements as well. These four assumptions provided the parameters for British strategic policy making. Not surprisingly, these differing visions of the future led to interservice rivalry and often adversely affected the Admiralty's policy towards development of naval aviation.

First, the specter of economic collapse loomed large. This possibility required the Treasury to argue strongly for maximum economy in defense expenditure throughout the period. Even when rearmament began, the Treasury insisted that it should not occur at a rate that would imperil Britain's long-term economic health.

Second, the army tended to stress the two national requirements. Britain must police its Empire but it must also prepare an expeditionary force for service on the continent. Because the deterrent effect of the army appeared less than that of the other services, army estimates tended to fare less well.[18] But once again the critical test of war did much to substantiate the army's requests for more resources.

Third, the Royal Air Force focused on the air threat emanating from Europe, one aimed directly against London and the heart of the Empire. The air staff also argued that aircraft could perform peace and wartime roles of the other services more cost-effectively. In Iraq, the RAF demonstrated the potential of air policing; throughout the interwar period its leaders consistently claimed that land-based airpower represented the best defense of important imperial bases like Aden and Singapore, and indeed of the British Isles themselves, against sea-borne attack. The RAF demonstrated at least this point in summer 1940. The air staff's broadest claim was that "strategic bombing" campaigns would determine the outcome of future wars, and therefore it ought to receive top priority. Although such claims rested on an exaggerated view of the effectiveness of the

[18] Brian Bond and Williamson Murray, "The British Armed Forces, 1918–1939," in Millett and Murray, *Military Effectiveness*, vol. 2, p. 100.

"strategic bombing" in World War I,[19] the experience of World War II did much to justify such views.

Fourth, the Admiralty concentrated on the requirement to defend the Empire and its vulnerable sea communications against naval threats of the sort posed by the growing naval power of Japan. In home waters, the Admiralty saw the re-emergence of a Germany as a major naval adversary threatening Britain itself, British shipping routes, and sea-based support of European allies as the most serious threat of all. World War II entirely justified this view.

In fact, World War II proved that all four strategic visions were essentially correct. But in the interwar period they prompted a diffusion of effort. The notably poor interservice cooperation in an era of resource constraints only exacerbated the problem. The contentious atmosphere did not help the services develop either a unified strategic view or forms of war which, like amphibious operations or maritime airpower, cut across competing dimensions.

Even if the services had perfectly coordinated their efforts, the real difficulty lay in Britain's unfavorable strategic circumstances. On the one hand, Britain was an imperial power with worldwide possessions, requiring protection against growing threats. On the other hand, it was also a European power vulnerable to development on the continent of an adversary or a coalition of adversaries threatening the British Isles themselves, their shipping routes, and their strategic links with the continent. The sad fact was that the British simply did not have the resources to cope with this multiplicity of risk. Both in absolute and relative terms, Britain's resource base was declining.

Comparisons with Germany were particularly worrying. In the years 1935–1939, Germany produced an average of 20.4 million metric tons of steel a year against Britain's total of 11.8 million tons. In 1936 Germany was reported to be 80 percent self-sufficient in food and animal fodder and 40 percent self-sufficient in motor fuel. The British output of machine tools in 1939 was less than a fifth of the German, and in February 1940, the Treasury calculated that Britain's overseas assets and gold reserves would last another two years only if carefully husbanded. Six months later, with the continent lost and the Battle of Britain beginning, the Chancellor of the Exchequer forecast that Britain's gold and dollar reserves would be totally exhausted by Christmas.[20]

---

[19] Such is the argument of George K. Williams, "Statistics and Strategic Bombardment," unpublished D.Phil. dissertation, Oxford University, 1987.
[20] Till, *Airpower and the Royal Navy*, p. 196.

In such bleak and predictable circumstances, the first response of a country unable to balance the defense books is to seek security in the company of others. Britain was no exception. The trouble was the allies that Britain really needed, namely the United States and the Soviet Union, were both essentially unavailable, and the allies available, especially the French, had their own problems and liabilities.

These unfortunate circumstances had a serious impact on Britain's carrier and aircraft procurement programs throughout the 1920s and the 1930s, most especially when the Treasury and Air Ministry combined against Admiralty plans. Even though the Japanese and Americans embarked on much larger construction programs, the British delayed the building of Ark Royal for some six years on the basis of a succession of specious arguments. These included the view that Britain had more carriers than either of its rivals, that their relative antiquity was not particularly important, that an expansion of the navy's air arm would imperil the growth of the Royal Air Force, and that it would be better to wait and see how new foreign carriers turned out.[21]

Critics also raised such arguments against the Admiralty's aircraft procurement program. Specifically, the Air Ministry argued that the RAF and the Fleet Air Arm were inevitable competitors. Finite production facilities and common arms control ceilings meant that each machine allotted to the fleet "means one less immediately available for home defense against attack by the shore-based aircraft of European powers."[22] This was not selfish obstructionism; it was the air staff's view of their national duty, and the events of summer 1940 gave some justification to their arguments. The result, though, was clear. Compared to the U.S. Navy's visionary program of the Morrow Board of 1926, which aimed at the production of 1,000 naval aircraft by 1931, the Admiralty's plans remained at modest levels. As a result the Americans had moved ahead in numerical terms by 1926, and the Japanese by 1933. Despite an attempt in the 1930s to catch up, the British went to war in 1939 with 147 Torpedo/Spotter/Reconnaissance aircraft, forty-one amphibians, and thirty fighters. Another 192 aircraft with maritime missions were stationed ashore, some of which the RAF also regarded as "first-line aircraft." At this time, the Japanese and the Americans each had in excess of 600 front-line maritime aircraft.

[21] The extensive exchange of correspondence in Jan. – Feb. 1934 between Sir Bolton Eyres Monsell (First Lord of the Admiralty) and Sir Philip Sassoon (Under Secretary of the Air Ministry) over the construction of the Ark Royal nicely articulates both sides' arguments. Treasury 161/624/36130/34, PRO.

[22] Lord Thomson, Secretary of State for Air, 13 Jan 1930, ADM 116/3479, PRO.

Nonetheless, the Admiralty has to share some of the responsibility for falling so far behind. Its plans for expansion of its air wing were undeniably modest. No doubt this was partly the result of the Admiralty's conditioning by the fate of its more ambitious plans at the hands of the politicians. But its policies also reflected technical skepticism about the extent to which the Americans could really operate the numbers of aircraft they claimed to carry – a level of unwarranted skepticism resulting from the British failures to develop arrester wires, catapults, and deck-parks. The system of dual control also made it difficult for the navy to get the air crew it required. In retrospect the British remained more relaxed than the American and Japanese programs despite the threatening situation.

American and Japanese comparisons

The Americans and the Japanese devoted a higher proportion of their air effort to maritime matters than the British because neither nation possessed an independent air force. This was not a bureaucratic accident. Neither Japan nor the United States faced the prospect of aerial attack from nearby territory. Even at sea, Britain's strategic problems were more complex. The preoccupation of the Japanese and Americans was with each other. Their problems may have been great, but at least they were relatively simple. Few had much doubt over what the strategic priorities ought to be.

The American ability to point at the Japanese as a clear potential opponent was an asset in many ways; it provided a criterion against which they could judge their tactics and equipment.[23] Without friendly local air power in support, the U.S. Navy would clearly have to take its own. Either to relieve the Philippines from a Japanese siege or, from the mid-1930s, to recapture them, the Americans expected to move westward across the Pacific without relying on bases in the theater and against a determined adversary.

The same strategic problem confronted the Japanese. In Tokyo, there was little doubt about their likely opponent in a future war and about how that enemy would conduct himself.[24] The position adopted by the Americans at the Washington Conference, the redeployment of the U.S. fleet in the early 1920s, and the nature of its exercises reinforced the mes-

---

[23] Ronald Spector, "The Military Effectiveness of the U.S. Armed Forces, 1919–39," in Millett and Murray, *Military Effectiveness*, vol. 2, pp. 73, 82, 89.

[24] Kenneth J. Hagan, *This People's Navy: The Making of American Seapower* (New York, 1991), pp. 263, 269; Haworth, *To Shining Sea*, p. 359.

sage. The manifest naval and industrial strengths of the United States made it clear that the challenge confronting the Japanese Navy was "how to contend successfully against heavy odds."[25]

Not surprisingly, the Japanese were keen to explore novel avenues for reducing the odds, whether through the development of new super dreadnoughts like the *Yamato* and the *Musashi* (which they hoped would outrange American counterparts), long-range submarines like the Fleet Type 6, the "Long Lance" torpedo, highly capable destroyers and cruisers, and, of course, the maximum use of ship and shore-based aviation.

The American and Japanese navies also benefited in the political infighting required to capture the resources necessary to put their operational concepts into effect. For the Japanese, it was almost a mathematical issue. The Washington Treaty provided the Americans a 40 percent superiority, offset only by the 10 percent loss in efficiency for every 1,000 miles of Pacific Ocean they had to cross to reach Japanese waters. Such apparently easily quantified arguments were persuasive considerations in the battle for resources.

All the same, for Japanese as well as Americans, resources were tight, especially in the 1920s. The U.S. Navy, like everyone else, had to conduct itself on "sound business principles" and the economies taken were sometimes debilitating.[26] It is possible that resource constraints were neutral, in that they helped identify true priorities,[27] but in general they had the same limiting impact on naval aviation as they had on other branches of naval activity. In this respect, the Americans and Japanese suffered many similar constraints to the British. They had fewer aircraft and carriers than they wanted. But it was easier for Americans and Japanese to underline the consequences of their deficiencies, thus explaining their greater success in expanding the level of funding available to naval aviation from 1937. The situation was more complex for the British. They were unclear as to when they might reasonably expect to wage war, or who their allies might be. Time and time again, Britain's military leaders asked for and inevitably failed to receive guidance on these points. Their relative inability to predict the nature and scale of future conflicts, and to refine their strategic priorities, makes the British situation in the interwar period more analogous to the post–Cold-War world than either the American or Japanese experiences.

[25] Boyd, "Japanese Military Effectiveness," pp. 143–145; Hirama, "Japanese Naval Preparations for World War II," p. 71; Hagan, *This People's Navy,* p. 263, 269; and Haworth, *To Shining Sea,* p. 359.

[26] Spector, "The Military Effectiveness of the U.S. Armed Forces," pp. 71–72.

[27] V. Davis, *The Admiral's Lobby* (Chapel Hill, NC, 1967), p. 74.

ORGANIZATION AND ADAPTATION

The British case

The Royal Navy's efficiency in adapting to the new challenges of naval aviation was also compromised by the way in which circumstances combined to weaken Fleet Air Arm's ability to argue its case. World War I demonstrated that probably the biggest stimulus to development of new tactical or technical ideas in the conduct of naval warfare was and would be the formation of groups of committed enthusiasts. Strengthened by the company of the like-minded, such enthusiasts developed and explored their ideas, tested them against experience, and proselytized them in the wider body of the navy. Only occasionally did inspired individuals have an effect of this sort. Captain Henry Jackson's pioneering work in the development of wireless-telegraphy before the war is an example of this form of innovation. Normally, effective pioneers come in battalions rather than as single individuals.

In the interwar period bureaucratic circumstances virtually stopped Britain's naval aviators from developing their own collective sense of mission. Two factors limited their ability to act as an autonomous source of expertise and pressure. On one hand, the Royal Navy was institutionally determined to integrate its airmen into the mainstream of naval life. On the other, the existence of the Royal Air Force severely constricted the influence and freedom of maneuver of naval aviators.

Naval restrictions on naval flyers

The Royal Navy's bureaucratic determination to navalize its flyers is easy to explain. In the first place, the navy, as an institution, had always been averse to development of "private navies" within its ranks. *Esprit de corps* was valuable, but development of a service within a service was divisive. During its campaign to recover Fleet Air Arm from the Air Ministry, the Admiralty emphasized that it did not seek re-establishment of the Royal Naval Air Service, a "corps separate from the Royal Navy, possessing its own sources of supply, administration, recruitment . . . On the contrary, the Naval Air Arm will be one with the Royal Navy in all the above important points . . . an integral part of the navy as a whole."[28]

Accordingly the Admiralty made constant efforts to ensure that the Fleet Air Arm remained wholly integrated into naval life. For example,

---

[28] Rear Adm. Dreyer to Colwyn Committee, 12 Nov 1925, ADM 116/2374.

had the Admiralty recovered Fleet Air Arm in the early 1920s, responsibility for management of aircraft procurement and construction would have rested in a number of agencies in a manner similar to that existing in the United States before creation of the Bureau of Aeronautics; in that case the Bureau of Steam Engineering produced aircraft engines, the Bureau of Construction fuselages, and the other bureaus the rest of the components, a system that made the procurement program a "shambles, devoid of centralized co-ordination and control."[29] The Admiralty frowned on every semblance of autonomy; well into the World War II, when the operational value and the bureaucratic loyalty of the navy's flyers was beyond dispute, it remained morbidly sensitive even to the use of the phrase "Fleet Air Arm" because such usage was "apt to convey an impression that the Fleet Air Arm is not an integral part of the naval service."[30]

A number of reasons account for the extreme sensitivity to the appearance of autonomy. Experience in the World War I certainly reinforced such prejudices. In that conflict, British naval aviators had tended to an independence from the rest of the navy. Naval airmen had affected an insouciance of dress and manner that often irritated more conventional naval officers. Perhaps because the Royal Naval Air Service grew so much during the war (expanding to a force of some 60,000 by spring 1918) it developed interests in operational areas hardly connected with standard naval missions. Orthodox naval opinion believed that the business of defending Britain from air attack and of developing a "strategic bombing" capacity against Germany had so preoccupied the Royal Naval Air Service that it had neglected the real business of naval aviation, namely helping the battle fleet perform its duties and protecting British commerce.

There was also some divergence of view between the navy and many aviators as to the direction of future policy. Some of the more committed air enthusiasts publicly criticized the Admiralty's air policy and advocated establishment of an independent air force in which their ideas would receive a more sympathetic hearing. This only confirmed the Admiralty views that aviators were basically unreliable, and in the case of restive pioneers like Rear Admiral Sir Murray Sueter, almost traitors to their own service.[31] In the view of conservatives, flyers therefore needed to be properly integrated into the navy, for their own good and for that of naval flying.

Sound operational reasons argued also for integration. Orthodox opin-

---

[29] Till, *Airpower and the Royal Navy*, p. 124.
[30] AFO 2112/41 and staff discussions, ADM 116/5057, PRO.
[31] Till, *Airpower and the Royal Navy*, p. 114.

ion held that naval flyers ought to be seamen first and aviators second. They could respond to the strategic and tactical environment in which they operated only by this means. Unless aviators understood the principles and practices of naval warfare, they would not make a material contribution to its success. Aviators were to be thoroughly immersed in the training and procedures of the surface navy.

Therefore throughout the interwar period the Admiralty insisted that naval personnel alternated in service between the Fleet Air Arm and regular naval billets. General service officers, not aviators, commanded British carriers of the period. This was partly because of the loss of most naval air officers with the necessary seniority in 1918 with the creation of the RAF and partly because of a view that commanding difficult ships like carriers required, above all, good seamanship. The result was development of a Fleet Air Arm in which the leading lights were thoroughly naval in approach and unable to act as an internal lobby even had they wished to do so.

The Admiralty was always acutely conscious of the *political* dangers inherent in the appearance of autonomy. The Air Ministry's line throughout the period of Dual Control was that it was the sole source of aeronautical expertise for the navy and that naval airpower related more to air power in general than it did to naval concerns. Any bureaucratic departure that made naval aviation look different from the rest of the naval service therefore played into the air staff's hands. As a result, considerable anxiety gripped the Admiralty with the proposal to establish a Rear-Admiral, Aircraft Carriers [RAA] lest the Air Ministry use this tacit admission that the naval air service was different from the rest of the navy "as a lever to obtain further control of the Fleet Air Arm."[32] All this reduced the aviators' capacity to argue their case in the navy.

The effects of dual control

The lobbying power of British naval aviators was undermined from a second direction as well. Establishment of the Royal Air Force in April 1918 deprived the navy of most of its aviators, those who were by definition most expert in the business and the potential of naval aviation. As a direct result, the navy's senior and middle-ranking officer corps expressed significantly less interest and wielded less expertise in this new form of naval warfare.

The re-creation in 1921 of the Fleet Air Arm meant that the navy would

[32] Captain H. Fitzherbert, 25 May 1930, ADM 116/2771, PRO.

eventually produce a second generation of air-minded officers, but, in the meantime, a gap was bound to occur. Senior air-minded officers served in the navy in the 1920s and early 1930s, but fewer than would otherwise have been the case. While the U.S. Navy had one vice admiral, three rear admirals, two captains, and sixty-three commanders on flying pay in 1926, the British could manage only one rear admiral, and a handful of junior captains and commanders by 1939.[33] As a result, for some ten years, Fleet Air Arm lacked inspiring leadership that might have provided it with the vision of future prospects and mitigated the hostile economic conditions of the 1920s and early 1930s. This loss of expert leadership materially slowed progress in naval aviation.

Throughout the interwar period, responsibility for British naval aviation remained uneasily divided, rather than shared, between Admiralty and Air Ministry in the system of "dual control." Both sides had to agree to every advance in naval aviation; the fact that the two departments were, for much of the interwar period, in a state of acute and potentially mortal competition only exacerbated the difficulties. The system, the Admiralty argued with some justice, was essentially unworkable and only created tensions and disunity of purpose. How would the air force have liked it, they asked, if the government had decided that the army would provide all the RAF's air gunners? The result was a vicious and prolonged twenty-year bureaucratic dispute between navy and RAF over control of Fleet Air Arm.

In the personnel area, dual control reduced the flow of recruits into Fleet Air Arm, slowed their training, and impeded their promotion. As a result, it took the navy the better part of twenty years to replace the generation of flyers it lost in 1918 with the establishment of the Royal Air Force. The design and procurement of British naval aircraft suffered equally. The original agreement establishing the Air Ministry assigned responsibility for *all* aircraft to the RAF, partly because the navy no longer would design the aircraft it flew, partly because the navy lost the strong body of technically-minded officers it had developed in 1918. Once the Admiralty lost its links with the British aviation industry, it found it almost impossible to reestablish those connections. The Air Ministry's priorities were naturally with the production of fighters and bombers. Torpedo bombers, maritime patrol aircraft, carrier operation, even dive-bombers, therefore took second place. In terms of technical standards, British naval aircraft were soon to be significantly inferior to the carriers from which they operated. This was a sure sign that the procurement sys-

---

[33] Till, *Airpower and the Royal Navy*, p. 45.

tem imposed by Dual Control was to blame for the problem. By the late 1930s the results were naval aircraft distinctly inferior to American and Japanese standards, and which paradoxically justified the low expectations that the skeptical had of naval aviation in the first place.

Dual Control also emasculated the leadership of Fleet Air Arm, and deprived the service of the vision required to guide it through the difficult economic and political conditions of the interwar period. In the 1920s, the navy undertook no serious study of the future role of naval aviation because no specific staff institution or officer was directly responsible for producing one. A small naval air section supervised the naval air service. Moreover, the Air Ministry remained adamantly opposed to creation of a naval air bureaucracy within the Admiralty; Fleet Air Arm officers within the policy-making portions of the Admiralty bureaucracy became preoccupied by humdrum issues of personnel and material management (because there was no one else concerned) and, above all, with the long crusade to regain the air service from the Air Ministry. They had little time and energy available for strategic reflection about the future role development of naval aviation. Only at the end of the interwar period were the institutions of the Fleet Air Arm capable of conducting effective reviews of aerial progress and engaging in real analysis of the place of naval aviation in warfare at sea, but these crucial studies were not yet available by the time World War II began.[34]

American and Japanese comparisons

The main difference between the organizational environments of British naval aviation and its counterparts in Japan and the United States was that neither Americans nor Japanese suffered the consequences of an independent national air force. American sailors were well aware of the disadvantages faced by the British in this respect and helped fend off creation of an independent air force in the United States. Indeed, establishment of the navy's influential Bureau of Aeronautics represented a conscious bid to inoculate its airmen and their supporters against the idea of an independent air force. Thus, Rear Admiral Bradley A. Fiske commented in 1920:

> For the sake of the U.S.N. and the U.S. (of) America – let's get a Bureau of Aeronautics p[retty] d[amned] q[uick] – as p.d.q. as possible . . . If we don't get that Bureau next session, Gen'l Mitchell and a whole horde

[34] Ibid., p. 129.

of politicians will get an "Air Ministry" established, and the U. S. Navy will find itself lying in the street, and the procession marching over it. . . . If the Air Ministry is established, the navy will rank with medieval institutions in about ten years.[35]

Relations between the U.S. Navy and Army were never particularly good,[36] but this condition had little effect on development of naval aviation.

American naval air institutions were also stronger than their British counterparts. Here an effective and vociferous naval air lobby prepared the way for large expansion plans in 1926 by creating a strong bureaucratic machine to look after the interests of naval flyers. American naval aviation's highest representative was a political appointee, the assistant secretary for naval aeronautics, a post held at the beginning of the interwar period by David Ingalls, a naval ace of World War I. Moreover, the vigorous and popular Admiral William A. Moffett, chief of the new Bureau of Aeronautics (BuAer), provided effective leadership. As one commentator noted: "Moffett has tackled the subject with almost fanatical zeal, supported by the whole nation from the president downwards."[37]

Moffett was in charge of the bureau twelve years and would have lasted longer had he not died, paradoxically, in an air crash in 1933. When the chief of naval operations tried to block Moffett's third reappointment in 1929, President Hoover specifically overrode the attempt. This contrasts strongly with the case in Britain, where naval aviators ran the Naval Air Division only at the end of the interwar period. Most of its directors were in career terms high-flyers, but they were nonaviators who came and went with the usual turnover of less than two years. The bedrock of continuous experience in the Fleet Air Arm remained therefore in lower echelons, with the technicians, passed-over lieutenant-commanders, and civil servants who came and stayed. Such individuals could not make strategic policy.

The Bureau of Aeronautics had more responsibility than the Naval Air Division for such matters as design and construction of aircraft, management of air personnel, development of aerial warfare at sea, and even for the provision of funding for naval aviation. It negotiated informally with Congress and the White House and kept in close touch with the aircraft industry. With Moffett at BuAer, "flying stock went up in the Navy De-

---

[35] Quoted in McBride, "Challenging a Strategic Paradigm: Aviation and the U.S. Navy Special Policy Board of 1924," p. 75.

[36] Spector, "The Military Effectiveness of the U.S. Armed Forces," p. 86.

[37] Adm. B. Domvile, DNI, 19 Oct 1928, Adm. 116/3117, PRO.

partment[;] with an Admiral to fight our battles we began to get things done. . . . Best of all, we had a well-informed group of properly accredited officers to present our case to Congress when aviation matters came up."[38] BuAer was not always right (in fact with its perverse preferences for airships and small carriers, it was not), but at least it possessed a coherent policy for naval aviation, and there was far less fatal indecision and drift that so blighted British carrier and aircraft procurement programs.

BuAer's informal role in encouraging Congress to support development of naval aviation, a role that its British counterparts certainly did not play, was particularly important. The relationship between Moffett and Representative Carl Vinson, chairman for twenty-five years of the House Naval Affairs Committee, was critical, a debt acknowledged by the naming of the U.S. Navy's fourth nuclear carrier in his honor in 1982.[39]

American naval aviation also benefited from a pluralistic decentralization that characterized the bureau-dominated naval administration of the 1920s and 1930s.[40] Naval airmen got on with things to a much greater extent than did their British counterparts. Moreover, there was a significantly higher proportion of naval aviators from which administrative agencies of naval aviation could draw. The Morrow Board of 1925 issued the edict that only aviators could command carriers and air stations, presumably on the grounds that air experience was essential for administration and for commanding operations. This sentiment was not echoed in Britain.

Some historians, and indeed some naval flyers of the time, have lamented "the ingrained technological conservatism of the capital ship navy and . . . the inhibiting effect of a rigid seniority system on the evolution of naval weapons systems."[41] Admittedly, less imaginative members of the "Gun Club" in the U.S. Navy remained disinclined to give naval aviation its proper due. But the flyers themselves hedged their bets with qualifications.[42] The overwhelming majority held that for the time being the chief function of naval aviation was to support the battle-line. Nonetheless, the bureaucratic and administrative environment in the United States was more conducive for flyers to argue their case than it was in Britain.

The Japanese were less well placed than the Americans, but more fortunate than the British. For most of the 1920s, administration and devel-

---

[38] Quoted Till, *Airpower and the Royal Navy*, p. 127.
[39] Hagan, *This People's Navy*, p. 282. [40] Davis, *The Admiral's Lobby*, pp. 63, 76–77.
[41] Hagan, *This People's Navy*, p. 272.
[42] Ibid., p. 273; McBride, "Challenging a Strategic Paradigm: Aviation and the U.S. Navy Special Policy Board of 1924," p. 75.

opment of naval aviation was as decentralized as in Britain, although Japanese naval aviators were at least spared the trials that resulted from creation of an independent air force. The overall administration of naval aviation was in the hands of the navy ministry's general affairs bureau, its training and technical development administered by departments responsible for such matters in the navy as a whole. Technological innovation tended to be mainly encouraged at Japan's naval air stations, although these were under the authority of the chiefs of the major naval districts. Furthermore, the factionalized division of opinion between the navy ministry and the naval general staff complicated development. The latter planned strategy and operations, tended to fight against observance of the limitations of the Washington Naval Treaty, and remained skeptical about the potential of naval aviation. The former was responsible for administration and was much more supportive, but had little operational role. Not surprisingly, the Japanese Navy failed to build coherent approaches to fostering naval aviation, although it made progress in some aero-technical fields.

However, things changed in 1927 when an independent naval air headquarters came into existence in Tokyo. It was directly responsible to the naval minister, but outside the navy ministry's chain of command. It assumed overall control of all naval air training, development of airframes, engines, ordnance, and other air equipment and soon took over all technical research. The establishment of a naval air arsenal in 1932 reinforced this trend towards centralized direction. In 1938, the Japanese concentrated all maritime and shore-based air power into a combined naval air wing in a manner that would have been impossible in Britain.[43]

A few key figures were influential in advancing the cause of naval aviation. Admiral Yamamoto was particularly important in changing the Japanese navy's overall attitude towards naval aviation in his role as vice navy minister from December 1936 to August 1939. Successive chiefs of the naval air headquarters also made influential contributions to the evolution of policy. Moreover, the Japanese Navy was assiduous in maintaining good relations with the press and with Japanese industry through the *zaibatsu* system, all of which helped develop an atmosphere conducive to innovation.[44]

---

[43] Hirama, "Japanese Naval Preparations for World War II," p. 70.
[44] See Mark Peattie and David Evans "Wings Across the Sea: The Rise of Japan's Naval Air Power, 1920–1937," a 1992 draft of a chapter in their forthcoming book. See also Arthur J. Marder *Old Friends, New Enemies: The Royal Navy and the Imperial Japanese Navy* (Oxford, 1974) p. 44; and Boyd, "Japanese Military Effectiveness," p. 138.

The revolution in naval administration which began in 1927 explains the impressive surge in the development of Japanese naval aviation from the mid-1930s. It does much to justify the view that the bureaucratic/administrative environment is of decisive importance in the evolution of military capability and the process of innovation.

ADAPTATION AND DOCTRINE

The British concept of operations and doctrinal assumptions also affected the navy's capability to make the most of its air power. These concepts and doctrines were simple and most usefully summarized in a document of December 1918, which reviewed the experiences of the war.[45] The Allies had fought and won the war on land. But the passage and sustainment of those victorious armies had depended on the security of sea communications:

> All . . . depended mainly on the supremacy of the Allies at sea – guaranteed by the Grand Fleet – and on the carrying-power of the British mercantile marine. The navy and the mercantile marine of Great Britain have, in fact, been the spearshaft of which the Allied armies have been the point. Security of sea communications has been the cornerstone of Allied effort, not only in the military campaign, but in the supply and maintenance of Allied industries and peoples. . . . Conversely, the closure of our enemies' maritime avenues of supply seriously hampered their military effort and led to a slow disintegration of their resisting power. Their navies have been neutralized, their supplies cut off, their fleets of war and merchantmen held immobilized in their harbors. Their cry for freedom of the seas is an index of the effect of this noiseless pressure.

The views of Alfred Thayer Mahan appear clearly here, and the remainder of the document made it clear the navy, like him, thought that the necessary supremacy at sea depended on the superiority of the battle fleet: "The command of the sea . . . was secured before the actual outbreak of hostilities by the presence in the North Sea in full readiness for action of nearly all the modern ships of the British Fleet." It followed logically that the Admiralty's most important peacetime task would be the ensurance that the battle fleet was fully prepared. Both Arthur Marder and Steven Roskill, noted historians of the Royal Navy, have criticized the navy for its preoccupation with a major fleet action, not least because it led the

[45] ADM 167/57, PRO.

British to neglect the requirements of trade protection and amphibious operations. Even at the time, some were ready to take up the cudgels in the debate.

In 1938, for example, Admiral L.E. Holland (then assistant chief of the naval staff and eventually lost in the *Hood*) acknowledged the existence of such concerns by his argument that this concentration on the fleet action was the right policy to pursue for a war against the Japanese, for they had exactly the same conceptions of war. "It is generally accepted," he said, "that in the case of a war against Japan the same processes which brought about main battle in the past will be again at work and the final round will be fought between the main fleets." The Germans might be more of a problem because, for the foreseeable future, they would not have sufficient naval forces to make a sustained challenge. But eventually the pressure of British naval superiority would force them back to "main battle between concentrations of capital ships." Paradoxically, Holland's own fate demonstrated the ultimate correctness of his view.

The Admiralty generally viewed artillery action by line of battleships as the decisive factor in the main battle between fleets. Such a concept of operations molded Admiralty policy towards naval aviation and provided the criteria against which the navy could assess the utility and the relative priorities within the Fleet. Significantly, there was some anxiety that the navy's preoccupation with the requirements of decisive action might skew the development of naval aviation as a whole. Commenting in 1938 on what was the most substantial interwar analysis by the British of the future role of naval aviation in war, Admiral Backhouse, then commander-in-chief of the Home Fleet and later First Sea Lord remarked:

> The memorandum may be said to visualize, almost exclusively, action between two battlefleets both accompanied by carriers. This condition is only applicable at present in a Far Eastern war. It assumes also, that both fleets intend to fight – so that what might be described as an unorthodox form of battle results. While it is appreciated fully that these conditions must be given due weight, functions of the Fleet Air Arm in a war much nearer home must be considered also.[46]

The onset of war in 1939 overtook the preparation of a second volume on trade protection and related matters. This failure through lack of time and resources aptly symbolized the Admiralty's order of operational priorities in the interwar period. Both the United States and Japan also ne-

---

[46] This issue is discussed at greater length in Till, *Airpower and the Royal Navy*, pp. 169–171.

glected the requirements of trade attack and defense.[47] This preoccupation with fleet operations had far-reaching effects on development of British naval aviation, firstly in deciding the roles of naval aviation and secondly in determining provision of its equipment.

Doctrine and the role of British naval aviation

One can best assess the anticipated roles of naval aviation in the Royal Navy in the 1930s by following the patterns of imaginary battles culled from the paper and fleet exercise conducted at the time. The British expected that several aircraft carriers would form a portion of the battle fleet as it steamed towards its adversary. The prior war's experience suggested that the most difficult task would be to find the adversary, especially if he aimed to avoid battle. Naval commanders believed that aerial reconnaissance would be the most likely solution to this problem, and that mission was one of the Fleet Air Arm's main priorities in the interwar period.

If the adversary had any sense, he would try to avoid the Royal Navy, possibly to the extent of not even leaving harbor. For that reason, the British devoted much thought to how they might use the new naval air power to lever a reluctant foe out of the security of his bases. In 1918, at the suggestion of Admiral Beatty, the British were preparing a massed strike of torpedo-bombers on the High Sea Fleet at Kiel. While the war ended too early for the attack to take place, the British launched the first simulated large-scale torpedo attack on a fleet in its base against the Second Battle Squadron at Portland in September 1919. In the early to mid-1930s, the Mediterranean Fleet prepared to do likewise against the Italians; its planning laid the foundation for Fleet Air Arm's attack on Taranto in November 1940.[48]

The Royal Navy did not enter upon such operations lightly. One could not expect the enemy to be as supine as the Bolshevik fleet at Kronstadt in 1919 when the British attacked it during the war of intervention. In 1934, when the British analyzed the prospects for carrier strikes against the Japanese fleet, they felt carriers would run a considerable risk of damage, and doubted if the results of carrier-borne attacks against Japanese air bases would justify risking irreplaceable carriers before the main fleets had reached a decision.[49]

If the adversary were at sea, the next task was to "fix" him. Torpedo

---

[47] Boyd, "Japanese Military Effectiveness," p. 248.
[48] Till, *Airpower and the Royal Navy*, p. 166.
[49] COS 344 JP, 30 July 1934, CAB 53/24, PRO.

aircraft might attack key units of the enemy fleet and slow them down sufficiently for the main British fleet to catch up. "The torpedo machine is the only weapon we have," wrote Admiral Sir Roger Keyes in 1927, "which holds out the hope of fixing a retreating enemy, and as such may be all-important in the British Fleet." An attack on the enemy's flagship might impose "a sufficient reduction of speed . . . to make it certain that the enemy will be brought to action."[50] The battle of Cape Matapan in March 1941 illustrated the point exactly. As the British closed on their prey, additional torpedo, bomber, and fighter attacks would delay the enemy still more, and, most importantly, would make it difficult for the enemy to coordinate his line of battle sufficiently to make the most of his firepower. The Battle of Jutland and countless interwar exercises conclusively showed that disrupting the battleline could decide the outcome of the action even if the fleet sank or disabled no enemy battleships.

It was rare, in fact, for British sailors to predict that such attacks would actually sink many battleships. The vulnerability of surface warships to air attack was the subject of an unending series of physical and practical experiments as well as theoretical enquiries such as the Postwar Questions Committee of 1920–1921, the Bonar Law Enquiry of 1921, the ABE Committee of 1936–1939, and the VCS Committee of 1936. Whatever else may be said about its attitude to the air threat, the Royal Navy certainly cannot be accused of ignoring it.[51]

Uncertainty about the course of *future* development bedeviled these deliberations and experiments, but for the present it appeared difficult for aircraft, in the numbers likely to be available, to sink properly supported battleships. As the issue was basically a matter of conjecture it was natural that navies should give the benefit of the doubt to the battleship – a weapon system of established value, one at the center of their service.

The Air Ministry agreed. The air staff never pretended to be able to dispose of battleships with the Olympian ease assumed by the more violent of its partisans. "Let me again restate," said Air Marshal Hugh Trenchard, chief of air staff in the 1920s, "I do not claim to be able to sink a battleship." His successors in the 1930s said much the same to the ABE and VCS enquiries.[52] Their point was the more subtle and realistic one, that they could impede the battle fleet as it proceeded by attacking its supporting vessels, by making its life difficult in harbors, and by constantly harassing it at sea.

While Fleet Air Arm was attempting to locate and fix its victim, the en-

---

[50] Keyes, 10 Aug 1927, Air 9/26, PRO.
[51] Till, *Airpower and the Royal Navy*, pp. 158–159.   [52] Ibid.

emy would doubtlessly be attempting to do the same in return. The British fleet would therefore need to defend itself against enemy aircraft. In the early 1920s, most British naval opinion considered that the main defense against enemy air activity would be, as in World War I, defensive fighters operating near the fleet. Anti-aircraft gunnery would simply fill in the holes. By the mid-1930s, however, the positions had reversed. Most naval officers thought anti-aircraft gunnery was the main defense, while fighters would sweep up the survivors. Accordingly, when German aircraft attacked the Royal Navy's best carrier *Ark Royal*, three weeks after the outbreak of World War II, its commander stuck the Skua fighters below deck with their tanks drained of petrol and placed his reliance entirely on anti-aircraft gunnery.[53]

Two assumptions supported his approach. First, he possessed an absolute faith in the lethality of the anti-aircraft gunnery developed to meet the fears of the 1920s, a faith much exaggerated in the event. Second, the deficiencies of the fighters themselves made anti-aircraft gunnery appear the safer bet, although this rationale represented a self-fulfilling prophecy. Nonetheless, naval fighters had important functions within the fleet. They were expected to shoot down enemy spotter and reconnaissance aircraft and protect British aircraft embarked on similar missions. In exercises, they were often able to harass attacking enemy aircraft and warships. Assuming events unfolded according to plan, the final act of this grand drama would be the majestic sight of a line of battleships firing at an unseen adversary at ranges of approximately 25,000 yards. Here again naval aviators would have a role. They would maximize the impact of the weight of the fleet's firepower by providing direction to the battleline's fire: the contribution of such aerial spotters to the eventual outcome of the battle could well be decisive. Thus, the protection of spotters against enemy aircraft received top priority during the battle.

Doctrine: A preliminary commentary

Naval air power was obviously an integral and important element of every stage in the battle. The Royal Navy did not consider it a separate, independent activity. This assessment does much to explain the navy's insistence on Fleet Air Arm's integration into the general corpus of the naval service. Some historians have criticized the Royal Navy's preoccupation with the form of the decisive action because "conservative battleship admirals thus returned to the old prewar anti-intellectualism, their naval col-

[53] Ibid., p. 149.

leges and fleet maneuvers looking to another Jutland and virtually ignoring the promise of the aeroplane and submarine."[54] Such sentiments miss the point because, on the contrary, the Royal Navy was determined that Jutland must *not* happen again and its conclusion was that supportive airpower was the best hope to ensure that it would not.

Other ways existed for avoiding another Jutland. Historians often represent the battleship of 1916 as a finished weapon system incapable of further development. In fact, the reverse was true. In the interwar period the battleship developed greatly, with, for example, significantly improved fire control systems, better protection against bombs, shells, and torpedoes, and so on. At Jutland, the *Yamato* would have made mincemeat of the opposition. Battlefleet operations developed also, particularly with improvements in decentralized control and divisional tactics, and in night-fighting capabilities. It is in fact debatable which, in its own terms, made more progress in the interwar period, the battleship or naval aviation.[55]

Allied to the criticism of excessive devotion to obsolescent battleships is the canard, often voiced by disgruntled airmen themselves, that the gun club was against them and used its dominance of the senior levels of the Royal Navy to undermine air development at every step. Admittedly, military bureaucracies tend to conservatism in that they reflect the status of established groups whose pre-eminence rests on the proven effectiveness of existing rather than future weaponry. It is also easy to find individual gunnery officers who conform to this stereotype, but in general gunnery officers did not. Those like Chatfield, Dreyer, and Keyes were, on the contrary, at the forefront of the Royal Navy's struggle to recover Fleet Air Arm, precisely because they recognized its importance to achieving their aim in a literal as well as figurative sense.[56] The airmen's view that gunnery officers constantly under-rated air power in fact was more probably a contributing factor to formation of their own *esprit de corps* than an essay in reasoned analysis.

Doctrine and the carrier program

These conceptions of battle had a considerable effect on the development of naval aviation. The priority given to seeking command of the sea through decisive battle, for example, inevitably lowered the priority on

[54] C. G. Reynolds, *Command of the Sea* (New York, 1974), p. 484. See also Corelli Barnett, *Engage the Enemy More Closely: The Royal Navy in the Second World War* (New York, 1991), p. 44.

[55] Jon Tetsuro Sumida, "The Best Laid Plans. . . .The Development of British Battle Fleet Tactics," unpublished paper, Naval History Symposium, Annapolis, Sept. 1991.

[56] Till, *Airpower and the Royal Navy*, pp. 130, 161.

trade protection or escort carriers. Despite imaginative ideas about "mercantile aircraft carriers," this scale of priorities combined with constrained economic and industrial resources meant that Britain, a great trading nation, went to war in 1939 without a single escort carrier. In an era when battle fleet needs were far from met, the Royal Navy suffered under an even greater neglect of the requirements of trade protection.

The British expectations that carriers would probably operate with the battle fleet, very possibly in Europe's narrow, shore-bound waters, heavily influenced their approach to carrier design. This does much to explain Britain's pioneering development of armored aircraft carriers. The inclination to build carriers capable of *absorbing* heavy damage underlined the navy's lack of faith in the defensive capabilities of fleet fighters. Admiral G.C.C. Royle, the Fifth Sea Lord and Chief of the Naval Air Service a bare two months before the Norwegian campaign of 1940, usefully summarized the Admiralty's reliance on anti-aircraft fire. "The Fleet when at sea with its destroyer screen in place presents quite the most formidable target a formation of aircraft could attack and the presence of fighter aircraft to protect the fleet is by no means a necessity. They should be looked upon as an added precaution, if available."[57]

This policy was disastrously mistaken. But it was the product, not so much of naval conservatism nor even of *a priori* doctrinal assumptions, but of a whole set of quite minor technical considerations and limitations, advanced more often than not by flyers themselves. These included a general preference for what Churchill called "knock-about aircraft for general purposes," largely because the few aircraft the fleet could carry should be as versatile as possible. Delays in production of effective catapults and the general difficulties of carrier operations also tended to make the Air Ministry skeptical about the fighting potential of naval aircraft. Until radar's advent in 1938–1939, countless exercises suggested that the few fighters carried with the fleet stood little chance of intercepting inbound enemy aircraft. Finally, long delays in acquiring effective, remotely controlled target aircraft (brought about by the Air Ministry's desire not to endanger its pioneering work in missile technology) delayed the Admiralty's discovery of its anti-aircraft weaknesses, although the truth was certainly beginning to dawn by 1939. Interestingly, around 1934, the Japanese also came to the conclusion that naval fighters were of limited utility, and for much the same reasons as the British. All of this would suggest that doctrine is largely a function of technical reality and that the existence or absence of quite small-scale technical developments often has disproportionately large-scale effect on the potentialities of innovation.

---

[57] Royle, letter to First Lord, 25 Feb 1940, ADM 116/3722.

American and Japanese comparisons

The Royal Navy's conception of the role of naval aviation in fleet action differed little from those of the Americans and Japanese. "The strategic concept of the Japanese Navy," wrote one of its first aviators after World War II, "was undoubtedly based upon the doctrine of annihilating the enemy in decisive battle, with battleships the backbone of the Japanese Fleet." At least until the "Second Shanghai Incident" of 1937, the main functions of Japanese naval air were orthodox operations in support of the battleline. Even after the "Third Replenishment" program of that year, this mission remained the dominant conception, as construction of the super-battleships *Yamato* and *Musashi* underlines. The large-scale expansion of Japanese naval aviation came only in the "Third and Fourth Replenishment" programs of 1937 and 1939. According to one estimate, the Japanese spent three times as much as the British on battleship modernization.[58]

The U.S. Navy conducted its naval operations along similar lines, and like the British, its fleet exercises concluded with traditional set-piece engagements between two hostile forces of battleships. What historians have described as the battleship paradigm was widely accepted in the U.S. Navy, and the Americans spent five times as much on battleship modernization as the British.[59]

The critical differences in naval aviation were those of scale. The Japanese and Americans possessed far more aircraft than the British and apparently expected the preliminary air phase to take longer. As they had more aircraft, the scale of the threat they posed to surface warships was higher. American and Japanese aviators therefore had higher expectations of the results of their attacks on the enemy's battleships and carriers.

Strategic geography also affected conceptions of battle. As far as the Japanese were concerned the strategic challenge throughout the interwar period was the advance westward across the Pacific of a superior American battle fleet. Their first response was the so-called attrition-interception strategy by which they hoped to whittle away American superiority by submarine and air action. Eventually, a reduced American fleet would arrive in Japanese waters in such weakened condition that the Japanese battle fleet could dispatch it. Carrier-based and shore-based maritime aviation would be important at every stage of this process.[60]

---

[58] Sumida, "Best Laid Plans," p. 8.
[59] Ibid.; and O'Connell, *Sacred Vessels*, pp. 277–316.
[60] Hirama, "Japanese Naval Preparations for World War II," pp. 66, 73–74.

Naval aviation was even more important in the preemptive strategy that replaced the attrition strategy. In drafting this strategy, Admiral Yamamoto drew inspiration from the performance of Japanese carriers during the war with China. In that conflict, Japanese carriers involved themselves heavily in military operations ashore, and the navy discovered the potential of carrier aviation in projecting power over great distance as well as the value of long-range fighter escorts and high-performance aircraft. The origins of the famous Zero fighter lay in this discovery. The next advance, partly inspired by the British attack on the Italian fleet in its base at Taranto in November 1940, was to form carrier task forces to cross the Pacific and strike the American battle fleet at Pearl Harbor.

Geographic circumstances also inspired American strategic thinking about a Pacific war. The U.S. Navy's principal concern was its safety when advancing westward across the Pacific, beyond range of friendly land-based aircraft. The fleet would clearly need to bring its own air support, and this made the rapid development of naval aviation essential. As one recent authority concludes: "The U.S. Fleet would have to mount sufficient strength in the air to overwhelm the enemy's combined fleet and shore-based air force. Hence the necessity for large numbers of U.S. naval aircraft and the carriers from which to fly them."[61] In the Pacific, the U.S. fleet would have to take on sea- and land-based airpower simultaneously, rather than in sequence as was the more orthodox British expectation. The British always assumed the decisive fleet action would come first.

The strategic situation in the Pacific encouraged the formation of semi-independent carrier task forces. The U.S. Navy had to explore every aspect of the defense of vulnerable points like the Panama Canal and Pearl Harbor against sudden naval attack. In this context carrier groups launched their celebrated mass air strikes on those two locations in 1929 and 1933. Interestingly, the *Saratoga*'s success against the Panama Canal in 1929 led to its being "sunk" on three occasions. These exercises demonstrated the vulnerability of carriers operating independently of the battle fleet and played a part in persuading the U.S. Navy not to develop carrier task forces in the 1930s. The Japanese, and perhaps even the British, were more prepared to operate their carriers in company. In fact, the Japanese created the First Carrier Division in 1928, but arguments about the extent to which Japanese carriers should operate independently continued through the 1930s into World War II.

This brief comparison of the three navies in the connections they es-

---

[61] C. M. Melhorn, *Two Block Fox: The Rise of the Aircraft Carrier 1911–1929* (Annapolis, MD, 1974), p. 88.

tablished between doctrines of battle and the development of their air
wing, also demonstrates that the differences were matters of degree. They
confronted the same technical and tactical issues and often came to similar conclusions. By the end of the 1930s, though, the Japanese were most
advanced in naval aviation, and the British least. The critical difference
lay in numbers. The sheer size of the American and Japanese naval air
wings established combat potential that way outstripped the British. But
these size differentials were not the result of differing *a priori* assumptions
about the importance of naval aviation. They were the consequence of a
host of small-scale, incremental, and usually technical, issues, decided
more often than not by airmen themselves. The Americans gained enormously from their development of catapults, early arrester systems, and
the tactical creation of the deck park. Like the British, the Japanese kept
their aircraft in hangars and not on deck, but the numerical disadvantage
was more than outweighed by possession of a first-rate, shore-based naval
air wing. The British also suffered the disadvantage of having too many
first-generation experimental carriers in their fleet. From such technical-
tactical considerations did the three navies develop their strategic ideas
about the role and potential of naval aviation.

THE LESSONS OF WAR

The last question considered here judges how wartime experiences validated interwar anticipations and the extent to which the three navies ad-
justed to the harsh lessons of war. At first glance, much increased reliance
on maritime air power, both land-and sea-based, during the course of the
World War II and indeed the wide-spread recognition that "the aircraft carrier force will form the core of the Navy,"[62] suggest that a transformation
of attitude took place in the Royal Navy over the course of the war. This
conclusion seemingly suggests that the pressure of war finally forced naval
officers into a belated recognition of the true potential of naval air power.

There were, however, more continuities between the attitudes of 1939
and 1945 than at first meet the eye. This becomes clear when one analyzes attitudes toward the battleship. At the beginning of the war, these
ships were the centerpiece of the fleet. By the end of the war, however, we
still find the naval staff producing papers arguing that the battleship remained an essential component of the fleet, because they alone could deal
with their opposite numbers.[63] In 1944, the First Sea Lord told Churchill

[62] Admiral C. E. Kennedy-Purvis, Deputy First Sea Lord, 7 Sept. 1942, ADM 205/18, PRO.
[63] Admiralty Memorandum to the Cabinet, 2 July 1945, ADM 205/53, PRO.

that the battleship remained "the basis of the strength of the fleet . . . (and) . . . a heavier broadside than the enemy is still a very telling weapon in a naval action."[64] Cunningham's view was that sea power rested on a proper proportion of surface forces, carrier-borne aircraft, and shore-based air forces. The battleship retained a place in a balanced fleet, obviously less decisive than earlier, but still essential. This explains the Admiralty's continued pressure throughout the war for eventual resumption of battleship construction.

These notions were not as perverse as they might at first glance appear. In the European war, battleships did indeed dominate the outcome of naval battles (Narvik II, Mers-el-Kebir, Calabria, Cape Spartivento, *Bismarck,* Matapan, and *Scharnhorst*). In other battles, their presence in the wings dominated strategic formulation (Crete, the Mediterranean campaign generally, and *Tirpitz*). Battleships underway also proved just as difficult to sink by air attack as most prewar estimates had concluded.

Battleships also continued to be useful in the Pacific war, although here the existence of two carrier fleets, both seeking command of the sea, gave the war a very different character. Even if a grand set-piece encounter had occurred between the Japanese and American Pacific fleets, it is reasonable to argue that after the rapid destruction of each other's carriers and air forces, the matter could well have finally rested on the battlelines slugging it out in the manner most interwar sailors had expected.[65] The large numbers of Pacific battles which were *not* decided by maritime air power suggests limitations to carrier dominance in naval affairs, even here. Clearly in 1945 the carrier was more important than the battleship, but the real lesson of most of World War II was that both were essential components of a balanced fleet.

War experience also validated many of the assumptions and the policy decisions of the Admiralty during the interwar period in carrier design and procurement. Despite their relatively low aircraft complement, British armored carriers performed well in hazardous conditions in the Mediterranean and Pacific. This reinforced the Royal Navy's preference for quality rather than quantity in carrier design, a trait already developed in the interwar period. The British pursuit of excellence manifested itself in the much expanded wartime fleet carrier construction program, most notably in the *Ark Royal* and *Malta* classes of 1942. The British also insisted on

[64] Cunningham, 4 May 1944, Prem 3, 322/6, PRO.

[65] Jon Tetsuro Sumida, "Historical Presentations of 20th Century Naval Invention" (a paper delivered to the Conference on Critical Problems and Research Frontiers, Madison, Wisconsin, November 1991), p. 11.

significantly higher "Admiralty standards" of protection for escort carriers serving in the Royal Navy. British-built escort carriers had to meet more demanding standards than their American counterparts, and American-built escort carriers had to be upgraded, a process that took on average some three months before the Royal Navy would accept them into the fleet.

Resource constraints had a bearing on this preference for excellence for two reasons. First, the British could afford fewer carriers. Second, they wanted carriers to retain operational value as long as possible in an era in which advancing technology seemed always to require larger and more expensive carriers. This thinking meant, for example, designing carriers to cope with the requirements posed by the next generation of aircraft as well as the present one. There is little that is unexpected about this desire, for the resource constraints often paradoxically led to the pursuit of perfection and, because the best is often the enemy of the good, the acceptance of lower standards in the meantime.[66]

The essential problem for the British was that, just as in the interwar period, resource considerations resulted in considerable scaling down of carrier construction programs. In an era when it took on average thirty-eight months to build a fleet carrier, compared to the Americans' eighteen,[67] it was simply impossible for the British to build what they wanted. There were, moreover, other pressing requirements in the shipyards. As a result, the United States provided most of Britain's forty-four escort carriers.

The story was the same for aircraft. By March 1943, the United States was providing 22 percent of the Fleet Air Arm's torpedo bombers and 58 percent of its single-seat fighters. By the end of World War II over half the Royal Navy's aircraft were American-built. American aircraft were generally of better quality. The F4U Corsair and F6F Hellcat were much better than the Seafire or Firebrand and the Grumman TBM Avenger considerably outperformed the Fairey Barracuda.[68]

There were other continuities between the interwar and World War II experiences, not least keen strategic competition with the Royal Air Force. In 1942, after Midway, the Admiralty attempted to set in motion a major carrier construction program, but encountered strong opposition from the Air Ministry, which raised arguments exactly analogous to those of the 1920s. Sir Archibald Sinclair, the Secretary of State for Air, was skep-

---

[66] Spector, "The Military Effectiveness of U.S. Armed Forces," p. 75.
[67] Till, *Airpower and the Royal Navy*, p. 84.   [68] Ibid., p. 109.

tical of the need. "Is it certain," he asked, "that aircraft carriers have proved successful, particularly when operating within range of enemy-based aircraft, as at Midway?"[69] Another focus of rivalry lay in the allocation of long-range aircraft between the "strategic bombing" mission against Germany and antisubmarine patrol requirements of the Battle of the Atlantic. But it was in the Pacific War that naval aviation reached its greatest utility. In most Pacific battles, the aircraft carrier proved itself as the new capital ship. The U.S. Navy's carrier force became the principal means of naval strike and by the end of the war was capable of operating hundreds of aircraft for long periods of time. This was the chief means by which the Japanese Navy was destroyed and Japan's island territories attacked. In large measure, the impressive development of naval aviation was a function of the distinctive characteristics of the Pacific War – a war fought between two maritime nations directly struggling for command of the sea in an oceanic theater of operations. Not surprisingly, therefore, naval airpower developed to a much greater extent in the Pacific than it did in the European theater. Perhaps more surprisingly, this likely difference was predicted in the interwar period by the preparations of all three navies.

CONCLUSION

The remarkable extent to which prewar attitudes survived the trauma of war is evidence less of the innate conservatism of naval officers, than of the basic soundness of their views in the interwar period. It reinforces the impression that in the interwar period, and even in World War II, development of naval aviation was a process of innovation that was, and should have been, essentially incremental. It was a question of constantly tinkering with the way in which things were done rather than of wholesale replacement of one set of weapons systems by another.

This being the case, the success of innovation needs to be assessed against a rather lower standard than would apply in a situation where revolutionary change was a realistic possibility. Against such a criterion all three navies were reasonably successful, and one can trace specific failures to apparently minor technical errors and problems, such as the failure to develop arrester wires or a preference for multipurpose aircraft. Moreover, one can also attribute many of these decisions to the innovators themselves.

But in the three processes of adaptation under consideration here, what

---

[69] Sinclair, War Cabinet Meeting 6 Aug 1942, Prem 3, 322/10, PRO.

issues stand out as having been particularly important in the development of carriers and naval aviation? Certainly, the question of resource constraints was critical. The real limits to the effectiveness of the response to changing technical and tactical circumstances in all three navies were resource-based, rather than attitudinal or bureaucratic. A shortage of resources encouraged tendencies to play for safety, both in emphasizing the value of relying on tried and tested weapons systems and in the conscious pursuit of perfection in developing new ones. Resource constraints reduced the ability and willingness to test what they did and made it more difficult for the navies to cooperate with the other services.

For the British, fundamental uncertainty about what they needed to produce exacerbated this problem. The ability to predict who they would be fighting and when was especially uncertain, and this shortcoming had a direct bearing on the capabilities the Royal Navy needed to produce. Having more specific incentives, the Americans and Japanese were better placed, at least to know what they wanted. Knowing what they needed to do helped create a climate more conducive to innovation.

The British also suffered grievously from having divided the responsibility for the development of naval aviation between two services. The Americans and Japanese, on the other hand, benefited from an administrative system which allowed enthusiastic innovators a high degree of autonomy. This proved effective in keeping the innovators on board, provided decision makers with a wide range of options from which to choose, allowed the thinkers scope for independent reflection, and eventually proved capable of producing those friends in high places, without whose constant support innovation would have been more difficult.

6

INNOVATION IGNORED:
The submarine problem

GERMANY, BRITAIN, AND THE UNITED STATES, 1919–1939

HOLGER H. HERWIG

The submarine is not an honest weapon. It suggests the foot pad, the garotte, and the treacherous knife dug in an opponent's back when he is off guard.

Sir Archibald Hurd, 1902

The submarine defies neutral sentiment. To its advocates, it constitutes a cheap but effective counterthreat to superior surface forces, the weapon of the poorer against the richer power. The brotherhood of submariners often views itself as a "private navy," hides its arcane mysteries from outsiders, and defends in every sense its "silent service." To adversaries, the submarine is an unfair and unsportsmanlike weapon, a weapon of piracy and barbarity. Both positions appeared as early as the Hague Conference in 1899; Germany, Britain, and Russia were willing to ban underwater warships – albeit, on condition of universal prohibition, but France and the United States refused to abolish what they considered the cheapest means available to reduce British naval superiority.[1]

Karl Lautenschläger has identified five general principles that govern the use of submarines in naval warfare.[2] First, underwater warships possess no inherent immunity against countermeasures. Although they are difficult to find and largely invulnerable while submerged, submarines are

---

[1] For a brief overview of the legal parameters of the submarine debate, see Richard Dean Burns, "Relegating Submarine Warfare, 1921–1941: A Case Study in Arms Control and Limited War," *Military Affairs*, vol. 35, April 1971, pp. 56–62.

[2] Karl Lautenschläger, "The Submarine in Naval Warfare, 1901–2001," *International Security*, vol. 11, Winter 1986–1987, pp. 94–140; see also Robert E. Kuenne, *The Attack Submarine: A Study in Strategy* (New Haven, CT, 1965).

open to attack once they disclose their presence by attacking surface vessels. Second, navies have trouble integrating submarines into existing force structures and operational concepts. In most cases, navies have found it difficult to define a specific mission for the submersible. Conversely, submarines often receive combat roles before they can fulfill them.

Third, competing wartime demands on submarines often preclude their achieving full potential. Are they designed primarily for coastal defense, support of the main battle fleet, or patrol as independent raiders? Their ability to perform several missions, then, brings with it a tendency to divide the force among a number of possibilities. The exception to this principle was the American submarine offensive against Japanese merchant shipping after 1941.[3] Fourth, submarine campaigns provide neither quick nor simple routes to victory. In most cases, underwater warfare comes close to classic wars of attrition. It extends over entire oceans, encompasses fleets of submarines and antisubmarine craft, and requires prolonged effort to be effective. Lautenschläger simply terms it "major undertakings of profound complexity."

Fifth, submarine and surface fleets are not alternatives; navies almost never possess a clear either-or choice. Instead, surface and subsurface fleets possess parallel and complementary functions. Most naval planners require both. The true task is to balance forces in order to enhance a navy's ability to achieve its operational and strategic goals. One may add one more general point to Lautenschläger's principles: the basic function of submarines is naval attrition. They can accomplish that goal by engaging the enemy's main battle fleet, by mine warfare, by pitting submarine against submarine (usually with little effect prior to recent times), or by attacking enemy commerce.

Three major submarine wars have taken place in this century. Two featured German forces, in 1914–1918 and, again, in 1939–1945. The third was the United States Navy's campaign in the Pacific after December 1941. In none of these wars was commerce raiding either the planned or preferred strategy, but rather appeared as an *ad hoc* child born of necessity. And in none of these conflicts did navies prepare for antisubmarine warfare; in all cases they preferred offensive strategies based on large surface units.

WORLD WAR I AND ITS LESSONS

The World War provided the first test of the conduct of submarine war by the Germans and the Allied response. As with air war, many of the

---

[3] See in particular Clay Blair, Jr., *Silent Victory, The U.S. Submarine War against Japan* (New York, 1975), p. 61.

larger issues and operational concepts that determined the course of the next conflict occurred in World War I. But as with air forces, navies found it difficult to discern the patterns of the emerging war over sea lines of communications in the U-boat war of 1914–1918.

The Germans had begun unrestricted submarine war against Allied commerce in spring 1915 in response to the British blockade. However, they had only twenty-nine boats in service and could keep only one boat on station for every four in maintenance or other status. With this inadequate force the Germans sank an insignificant number of enemy merchant vessels, including, however, the *Lusitania*, which almost brought the United States into the war.[4]

Chastened by American outrage, the German government backed off its campaign of unrestricted attacks, but for the next year and a half its navy waged an increasingly effective political campaign at home to resume unrestricted U-boat warfare. Despite such efforts, the Germans lavished resources on construction of battleships, while starving submarine production. In January 1917 with the support of the army the navy forced the German government to resume unrestricted submarine attacks despite warnings that such an effort would bring the United States into the war.[5] In fact at the start of its campaign the navy had only 36 of its 100 U-boats on station on 1 February 1917.

Nevertheless, the Germans inflicted heavy losses on Allied shipping, but only because the Admiralty obdurately refused to introduce convoys. The British response to submarine war was thoroughly aggressive; it aimed to seek out and destroy the U-boats. Such an approach was akin to looking for a needle in a haystack; even in areas where U-boats concentrated, antisubmarine forces had little success. Meanwhile, Allied and neutral merchantmen sailed independently and this increased German opportunities for sightings and thus the number of attacks. Such attacks inevitably took place under conditions where there would be no interference from escorts, absent chasing false alarms. This led to terrible losses among allied merchantmen.

Only the threat of defeat forced the British to adopt the same convoy system that had thwarted French raiders in the Napoleonic wars. The use of convoys caused an immediate, drastic reduction in shipping losses. In

---

[4] For further discussion of the U-boat campaign in the Great War, see Holger Herwig, *"Luxury" Fleet, The Imperial German Navy, 1888–1918* (London, 1980), pp. 163–164.

[5] Ludendorff commented to a German industrialist at the end of 1916: "The United States does not bother me . . . in the least; I look upon a declaration of war by the United States with indifference." BA/MA, Nachlass Vanselow, F 7612, Ludendorff to Stinnes, 15 September 1916.

April 1917, U-boats sank 458 ships of 841,118 tons. By July the total had dropped to 365,000 tons; by September merchant tonnage lost had fallen to barely 200,000. Over the course of that month, 1,306 merchant ships sailed to Britain in eighty-three convoys; U-boats sank only ten of that total.[6]

The introduction of convoys raised two difficulties for the U-boats. First, it concentrated escorts in the immediate area where U-boats had to attack. Even more importantly, it reduced the submarines' ability to sight merchant shipping at the same time that it reduced the time available to attack. As Karl Dönitz noted in his memoirs:

> The oceans at once became bare and empty; for long periods at a time the U-boats, operating individually, would see nothing at all; and then, suddenly, up would loom a huge concourse of ships, thirty or fifty or more of them, surrounded by an escort of warships of all types . . . The lone U-boat might well sink one or two of the ships, or several, but that was a poor percentage of the whole. The convoy would steam on . . . bringing a rich cargo of foodstuffs and raw materials to port.[7]

One should not, however, attribute defeat of unrestricted submarine war entirely to the introduction of convoys. By 1918 the British had introduced aircraft to support convoys within range of the British Isles, while more effective escort attacks were taking some toll of U-boats. But the problem remained of how to attack submarines once they had submerged; not until after the war did research efforts produce asdic, the ancestor of sonar, but too late to influence the battle against submarines.[8]

By 1918 some German submariners were urging concentrated submarine attacks against convoys, what would be called "wolf packs" in World War II. But the German naval high command set itself firmly against such an approach and in the face of increasingly successful Allied countermeasures planned for unrealistic increases in submarine production.[9] By the war's end submarine and antisubmarine warfare were emerging into a full-blown war over convoy routes in the eastern Atlantic. Aircraft, intelligence, technology, operational concepts were all appearing as essential pieces of the puzzle. But the full complexity of the equation, as with that of air war, still remained opaque. In the postwar world, the problem for

---

6 Williamson Murray, "Naval Power in World War I," in *Seapower and Strategy*, ed. by Colin S. Gray and Roger W. Barnett (Annapolis, MD, 1989), p. 206.
7 Karl Dönitz, *Memoirs, Ten Years and Twenty Days* (Annapolis, MD, 1959), p. 4.
8 Although asdic grew increasingly sophisticated during the interwar period, it was not until it was put into operation during World War II that the difficulties in using it against a U-boat deep under water were discovered fully.
9 Herwig, *"Luxury" Fleet*, pp. 228–229.

navies was to winnow out not only the lessons of the last war but also the direction towards which the innovations of the last war years pointed.

The Germans

Germany was the first power to conduct unrestricted submarine warfare and the only major power in the twentieth century to resort twice to this form of warfare. Between 1914 and 1918, 320 U-boats sortied; 178 were lost, including 134 due to antisubmarine operations. From 1939 to 1945, 940 U-boats sortied; 784 were destroyed, including 593 due to antisubmarine measures. The Kaiser's "pirate ships" dispatched 11.9 million tons of Allied shipping; Hitler's "gray wolves" sank 14 million tons – a much reduced level of success in terms of the effort and losses. In neither case did underwater raiders accomplish their objective: the economic strangulation of Britain.

In both instances, German naval doctrine emphasized construction of a Mahanian surface fleet and not U-boats. Admirals Alfred von Tirpitz and Erich Raeder drafted extensive battle-fleet plans designed to achieve a measure of naval control in the North Sea and Atlantic. Neither believed that submersibles constituted true sea power. Yet both reluctantly came to the conclusion that U-boats constituted the only available weapon against "perfidious Albion." In 1919, Germany possessed no U-boats; it surrendered 176 boats to the British, and Article 191 of the Treaty of Versailles forbade the German navy from building or acquiring submarines. In addition, most wartime U-boat commanders left the navy, while participation in the right-wing Kapp *Putsch* of 1920 discredited the few who remained. The story of German U-boat development in the interwar period is one of skillful, clandestine attempts to stay abreast of foreign technological developments. Despite the Republic's fiscal straits, however, money was not a major obstacle. Apparently the *Reichsmarine* acquired 100 million *Reichsmarks* (RM) between 1919 and 1922 from sales of ships, scrapped by order of the Allies. From this windfall, it established two unbudgeted secret funds: a "Special Fund" under Captain Walter Lohmann, and the "Black Fund" under Captain Gottfried Hansen. Before exposure in parliament in 1927 for misappropriating 27 million RM, Lohmann built a military-commercial empire encompassing nearly thirty firms – as diverse as a torpedo factory at Cadiz, the Phoebus film corporation in Berlin, a Berlin bacon company, and a concern that proposed to raise sunken ships by surrounding them with ice.[10]

---

[10] See International Military Tribunal, Nürnberg, *Der Prozess gegen die Hauptkriegsverbrecher*, vol. 34 (Nürnberg, 1949), pp. 555–560, 562–563. Document C 156, Captain Wilhelm Schüssler, "Der Kampf der Marine gegen Versailles 1919–1935," dated

With the aid of these secret funds, Krupp Germaniawerft (Kiel), A.G. Weser (Bremen), and Vulkanwerft (Hamburg and Stettin) established a joint stock company in the Netherlands in July 1922, N.V. Ingenieurskantoor voor Scheepsbouw (I.v.S.). Acting as a submarine design and consulting bureau, I.v.S. officially opened for business in October 1925 and in time built eight U-boats with German machinery for the Kawasaki shipyard at Kobe, Japan, as well as submarines for Finland, Spain, Sweden, and Turkey. Attempts in 1926 to sell submarine technology to the Soviet Union proved fruitless, partly because naval leaders continued to blame the collapse of 1918 on "Bolshevik" machinations within the service. In time the German navy acquired a 28 percent interest in I.v.S. as well as chairmanship of its executive board. Two dummy corporations, Mentor Bilanz and (after 1927) Igewit (Ingenieurbüro für Wirtschaft und Technik) acted as liaison between navy and I.v.S. A Krupp-financed merger of Spanish yards at Barcelona, Tarrogona, and Valencia led to the development of swell-less torpedoes, and a similar joint venture with the Swedes after 1927 produced a trackless electric torpedo that became the basic German torpedo of World War II. The Germans conducted research and development covertly abroad; practical application, leading to diffusion and doctrine, however, awaited future political developments.

The first real breakthrough for I.v.S. came in 1924 when it sold designs for four U-boats to Finland. The following year a special subvention of 1 million RM from Lohmann helped place orders for two Turkish U-boats at the Rotterdam yard of I.v.S.[11] And in 1928 the Spanish government, through Commander Wilhelm Canaris, ordered a 600-ton boat from the Germans. In all cases, German engineers drew up plans at I.v.S.; motors came from Germany; I.v.S. at the Fijenoord Yard at Rotterdam assembled ship parts, while only final assembly occurred at the Ecchevarrieta Yard at Cadiz.[12] In fact, the German Navy assumed 80 percent of the expenses in constructing the Spanish boat through I.v.S. In 1930 Raeder under-

---

1937. (Hereafter cited as Schüssler, "Der Kampf.") The most recent analyses of U-boat warfare are by John Terraine, *The U-Boat Wars: 1916–1945* (New York, 1989); and V.E. Tarrant, *The U-Boat Offensive 1914–1945* (Annapolis, 1989).

[11] See the superb work by Allison W. Sackville, "The Development of the German U-Boat Arm 1919–1935," unpublished PhD dissertation, University of Washington, 1963, pp. 127. ff. The pros and cons of building U-boats abroad were examined by the navy in June 1927: BA/MA, Freiburg, RM20/993, Reichswehrministerium, Marineleitung, Schiffbauersatzplan, "Bau eines Ubootes."

[12] See Werner Rahn, *Reichsmarine und Landesverteidigung 1919–1928: Konzeption und Führung der Marine in der Weimarer Republik* (Munich, 1976), pp. 180 ff.

wrote the construction of a small (250 ton) U-boat in Finland at a cost of 1.5 million RM. This craft became the prototype for the U-1 through U-24 series boats. In addition, it served as a test craft for new torpedoes developed at Cadiz.

1925 marked a turning point in U-boat development. Buoyed by the financial windfall of the government diversion of $190 million from Dawes Plan loans to AEG, Krupp, Siemens-Schuckert, and Thyssen for rearmament, the German Navy established a special bureau to deal with "antisubmarine warfare questions," in reality, a thinly disguised U-boat development office. In July 1925 Admiral Hans Zenker, head of the *Reichsmarine*, blessed such clandestine machinations arguing that they "advance our plans and allow us to gather experience in the further development of weapons otherwise forbidden us."[13]

Perhaps most importantly, Zenker appointed the recently retired Rear Admiral Arno Spindler as the U-boat specialist at the naval archive. Spindler used his position to guide German submarine development over the next decade. The navy broadly defined his mission as follows: to collate wartime experiences of U-boat commanders, coordinate submarine activity among German firms operating abroad, establish a theoretical U-boat training program, establish an espionage network to gather information on submarine innovations occurring elsewhere, run a propaganda bureau to attack Versailles' submarine restrictions, and develop contingency plans for use of U-boats in a future war.[14]

Spindler made slow but steady progress. The naval academy began a regular U-boat training program for midshipmen, and the class of 1924 was the first to study the history of the U-boat campaign as part of its professional education. Nucleus crews helped work on the boats being built in Finland, Sweden, and Spain, thus providing German submariners with sea experience. Not surprisingly, the first submarine training officer at Flensburg was Lieutenant Commander Werner Fürbringer, who had gained sea experience in 1927–1928 during the shakedown trials of a U-boat constructed in Holland for Turkey. His successor, Lieutenant Commander Robert Bräutigam, likewise had gained realistic training delivering boats to Finland, Spain, and Turkey.

In developing submarine doctrine, Spindler was a traditionalist who ignored the possible use of U-boats for commerce raiding in favor of support for the battle fleet. In fact, when Raeder engaged Spindler in 1932 to

---

[13] BA-MA, N 239, Nachlaß Levetzow, Band 5, Box 18. Zenker to Rear Admiral Magnus von Levetzow, 22 July 1925.

[14] Sackville, "Development of the German U-Boat Arm," pp. 173–174.

write the official history of the unrestricted submarine campaigns of 1914–1918, he did so to show that a submarine campaign that conformed to "prize" or "cruiser" rules was perfectly feasible. By the time that Spindler completed the fourth volume, dealing with the U-boat campaign of 1917, however, his views were no longer in vogue, and the navy physically destroyed this controversial volume.[15] This blatant attempt to misuse history to prove particular "lessons" stands in sharp contrast to Seeckt's insistence that the army study the lessons of the Great War, however unpleasant.[16]

Spindler also collected wartime accounts of Imperial submarine commanders with an eye toward preparing the navy for a future conflict. He assigned theoretical exercises (*Winterarbeiten*) to the brightest officers, and the results provide an interesting insight into German thinking about submarines during the 1920s. Most such treatises were by experienced wartime U-boat commanders, including three winners of the *Orden pour le mérite*. These studies served to keep alive U-boat traditions, attacked the prevalent view that the U-boats had little significance, and established a small but enthusiastic band of submariners.

By 1922, former U-boat "ace" Lieutenant Commander (later Admiral) Kurt von Mellenthien had rigorously reviewed the technical shortcomings of wartime boats. He found them too slow, too lightly armed, incapable of diving more than seventy-five meters, and lacking adequate refit and repair bases. Lieutenant Commander (later Admiral) Wilhelm Marschall complemented Mellenthien's study that same year. The future German fleet commander demanded that the U-boats be able to sustain dives to 100 meters, more rapid dives to periscope depth, and a smaller turning radius. Unlike Mellenthien, Marschall also addressed issues of deployment and training. Any naval war, he argued, would immediately see the introduction of convoys from the start. Hence, German U-boats must train for night, surface attacks. This approach would provide surprise, allow greater speed with diesel motors, increase torpedo firings, and permit easier escape. While some argued that the advance in detection devices had rendered surface attacks obsolete,[17] Marschall's recommendation became German doctrine in 1935 – albeit with Captain Karl Dönitz taking full credit.

---

[15] See Rahn, *Reichsmarine*, p. 128. The volume was reprinted in 1964. The fifth volume, which took the U-boat campaign through 1918, did not appear until 1966.

[16] See James S. Corum, *The Roots of Blitzkrieg: Hans von Seeckt and German Military Reform* (Lawrence, KS, 1993).

[17] Dirk Horten, "Auswertung von Ubooterfahrungen im Ersten Weltkrieg: Eine Untersuchung anhand von Winterarbeiten in der Reichsmarine 1922–1933," unpubl. ms., Führungsakademie der Bundeswehr, Hamburg, 1972, pp. 2–16.

Two other *Wingerarbeiten* are of interest. In 1927 Lieutenant Commander Lothar von Heimburg addressed how U-boats could function in the face of improved detection. Accepting that antisubmarine warfare had outstripped U-boat capabilities, Heimburg argued that massed submarines alone could overcome qualitative disadvantages. To avoid detection, U-boats would have to attack "blind," that is, without raising periscopes. Acoustic sensing devices would guide the submarine during its final approach, while "hydraulic" mechanisms would curtail engine and screw noises. Above all, wartime experience suggested that joint services cooperation would be critical. "In the future, it is unthinkable that [navies] would conduct any major campaign or reconnaissance without the support of the air force." Thus, constant coordination between submerged U-boats and maritime strike aircraft offered the only chance for success.[18]

Lieutenant Commander Fürbringer echoed Heimburg's concerns in 1932. He argued that during the day enemy aircraft would force U-boats under water and thus greatly reduce their speed and radius. Antisubmarine detection further threatened the U-boat at sea. "The U-boat has lost its freedom of operation. Alone, it is no longer capable of dealing with its many adversaries." Fürbringer saw succor only in combined operations. Submarines must operate in groups and coordinate efforts with aircraft and surface vessels through a single command center. Above all, the "U-line" would be at the mercy of air support – both to fend off the enemy and spot potential targets.[19]

By the end of the 1920s Germany was well placed to develop a submarine force, should international passivity permit. In 1927–1928 the naval staff had drafted contingency plans for a future U-boat force of eighty-four boats: thirty-six small 360-ton "F" boats, twelve 800-ton "T" minelayers, and thirty-six 650-ton "G" boats. The planners rejected larger boats (up to 7,500 tons) as slow, unwieldy, vulnerable to attack as gun platforms, and useless in any war against commerce – as World War I had shown.

Such speculative musings took on a more realistic air in Defence Minister General Kurt von Schleicher's *Umbauprogramm* of July 1932. Under the plan, Germany would create a U-boat arm, a naval air arm, and three U-boat divisions of sixteen craft over the next five year – all in flagrant violation of the Treaty of Versailles. While Schleicher proceeded cautiously – for example, initially only storing the requisite U-boat materials until "the political situation" allowed actual construction – the en-

---

[18] Ibid., pp. 21–23.   [19] Ibid., pp. 24–27.

vironment altered drastically with Hitler's appointment as chancellor on 30 January 1933. The new regime removed all barriers to integrating national political objectives with military preparations, while providing a blank check to all defense funding. The Führer tripled the secret "B-Haushalt" for construction of forbidden weapons to 21 million RM and in the first three months in office added 115.7 million RM to the navy's official budget of 186 million RM. In November 1933 Hermann Göring informed Benito Mussolini that Germany intended to build 10,000 tons of submarines, twice the tonnage in Schleicher's *Umbauplan*. By 1938, Germany would possess not sixteen but thirty-six U-boats, with another eighty-two under construction.[20]

The German Navy won a critical victory on 18 June 1935. The Anglo-German Naval Agreement permitted it to build up to 35 percent of British submarine strength, or 18,445 tons. At that time German long-range planning for thirty-six U-boats encompassed only 12,500 tons, and even Raeder's most inflated demands had envisaged only 45 percent of British tonnage, or seventy-two submarines. Moreover, the 1935 accord permitted the Germans to achieve parity with Britain in submarine tonnage – 64,149 tons – should they see a need to do so.[21] In short, the Germans received *carte blanche* from London to proceed with U-boat development. They did not delay: by September 1938, the *Kriegsmarine* had seventy-two boats (45 percent of British submarine tonnage) either built or building; orders on the books aimed at reaching parity.[22] On 22 June 1935 a relieved Raeder ended "deception" measures with regard to U-boat development. The "antisubmarine" school now became the "submarine" school; its first official class included twenty-eight officers, nine engineers, and 580 enlisted men.[23] Realistic, rigorous, professional military training was now possible.

U-1 slid into the water eleven days after the signing of the Anglo-German Naval Agreement. For the remainder of 1935, Germany launched a U-boat every eight days. The clandestine research and development efforts paid off handsomely. I.v.S.'s branch office at Bremen had designs

---

20 BA-MA, N 328, Nachlaß Admiral Erich Förste, vol. 15. "Entwicklung: Denkschrift aus den Akten."
21 See Donald C. Watt, "The Anglo-German Naval Agreement of 1935: An Interim Judgment," *Journal of Modern History*, vol. 28, 1956, pp. 155–175.
22 BA-MA, RM 20/880 Oberkommando der Marine, Typenfragen-Schiffsbaupläne. Report of 19 September 1938 for Raeder. The British were informed of the parity option in December 1938.
23 BA-MA, RM20/870 Reichswehrminister, Verschiedenes. Raeder directive of 22 June 1935, "Enttarnung der Uboote."

on hand for U-27 to U-44; Igewit had secretly stored parts of twelve U-boats at Kiel and built a giant covered hall designed to build six U-boats at once, all safe from hostile observation, and state-of-the-art torpedoes were mass produced at Fürstenwalde.[24] Enterprising engineers already were hard at work on designing refueling boats (Type 111, or "*Milch Kuh*"), steam-turbine driven boats, and even an experimental U-boat (the V-80) powered by hydrogen peroxide turbines, capable of twenty-eight knots submerged – the so-called Walter boat. Late in 1935, Dönitz assumed command over the First U-Boat Flotilla; one year later, he became "Commander of U-boats."

If German submarine development had an Achilles' heel, it lay in its doctrinal conceptions and failure to grasp the larger operational and organizational issues that a future U-boat campaign might raise. Barred by international agreement from building either capital ships or submarines, and until 1934 concentrating on a possible hostile Franco-Polish coalition, German planners stressed the need for destroyers, mines, and nets – as well as patience and perseverance.[25]

When the German Navy introduced U-boats for the first time to its war-gaming in 1927, their missions were traditional: to attack enemy warships and military transports, to lay mines, and to scout. That same year, Spindler's "Handbook for U-boat Commanders" stipulated that they would conduct warfare against merchantmen only on the "stop, search, and seize" principle. Not surprisingly, fleet commanders in 1929 argued that superior British detection devices had "substantially reduced" the U-boat's military value. Nor could U-boat supporters have been enthusiastic when Raeder depicted the submarine in October 1930 as "one of the best *defensive* means of the weaker" naval power.[26]

The German-Polish Non-Aggression Pact of January 1934, which refocused the navy's planning to a possible war against a Franco-Soviet coalition, and the Anglo-German Naval Agreement forced the navy's planners to address submarine doctrine. Raeder and his staff produced a number of strategic studies in 1937 and 1938, all of which stressed a combined-arms approach featuring surface action groups, independent cruiser raiders, and naval air and submarine support groups. In February 1937 Raeder presented his "Basic Thoughts on the Conduct of Naval Warfare," wherein he confirmed his belief that the U-boats had failed to alter the last

[24] Schüssler, "Der Kampf," pp. 572–578.
[25] See the 1923–24 *Winterarbeit* by Captain Paul Wülfing von Ditten in BA-MA, RM20/1547 Flottenabteilung, Vortrag Nr. 32.
[26] Ibid., pp. 318 ff. My italics.

war's outcome. U-boats could achieve strategic success, he argued, only if they operated in conjunction with the battle fleet.[27]

Raeder based his views on a position paper prepared in October 1936 by Vice Admiral Günther Guse, head of the *Marinekommandoamt*. Guse summarized the new features of naval and legal developments since 1914: there would never again be an "unrestricted" U-boat campaign; "cruiser" rules would apply to both surface and submerged craft; and the primary U-boat mission would be to *"protect"* Germany's oceanic commerce. Rejecting attacks on commerce, Guse defined the U-boat functions as scouting, minelaying, and attacking enemy warships and military transports.[28]

The culmination of such strategic ruminations came in 1938 during the Czech crisis, when Hitler informed Raeder that the navy must consider Britain as a hostile power. Shaken, Raeder ordered Commander Hellmuth Heye of the operation's division to "game" a naval war with Britain. The results were sobering. Heye, like Guse, began by arguing that U-boats, "despite unexpectedly good results," had failed in World War I due to "improved antisubmarine measures." In addition, Imperial Germany's U-boat force had been too small. In any future war, Heye asserted, Germany would need at least 100 submarines for defensive patrols in "domestic" waters and the parts for at least a hundred more boats for series production on the outbreak of war. There would also be a major requirement for "supply ships" for fuel, food, and ammunition for boats on station. Heye argued that the U-boat was no longer the premier weapon of a *guerre de course*; the surface raider had taken its place. In conclusion he argued that "warfare against commercial shipping according to prize rules" would be "impossible." The only effective weapon remained the *"surface"* raider in the form of the "pocket-battleships." The war games of 1937–1938 came to similar conclusions. As in World War I, Germany would not possess sufficient U-boats *"alone* and in sufficient *time"* to have a decisive impact on the British position. Only the new surface raiders (*Panzerschiffe*) could be effective in any war on seaborne commerce.[29]

In the years immediately preceding the outbreak of the conflict the navy remained convinced that the British had solved the detection problem in antisubmarine warfare despite its own considerable problems in this area.

---

[27] BA-MA, RM 6/53, Raeder's treatise of 3 February 1937. Raeder was hardly a forward-thinking military leader; at one time he went so far as to characterize carriers as "only gasoline tankers." Michael Salewski, *Die deutsche Seekriegsleitung*, vol. 1, (Frankfurt am Main, 1970), p. 43.

[28] BA/MA, RM 20/1138, Reichswehrministerium, "Verwendung der Marine im Kriegsfalle," 28 October 1936.

[29] BA/MA, RM 20/1138, "Conclusions."

One of the papers for the *Winterarbeit* 1939 went so far as to state: "the importance of U-boats has considerably declined compared to 1915. One can assume that England has good detection gear, which makes torpedo attacks on a secured unit or convoy impossible."[30] This was obviously a major factor in the German underestimation of the U-boat weapon.

There were those, of course, who believed that the submarine possessed considerable potential for the next war. After World War II Dönitz and a number of his supporters argued that in the 1930s German submariners had waged a lonely effort against the "big ship" fanatics to develop the one weapon that could have won the war for Germany. Such claims, as with most postwar German apologias, need to be considered with a grain of salt. In reality, Dönitz shared much of Raeder's vision of naval warfare in the critical years of 1935 and 1936, while his junior rank hardly allowed him access to the navy's top decision-making levels before the war.[31]

Nevertheless, by November 1937 Dönitz was arguing that World War I had "brought the realization that the U-boat is well suited to menace the enemy's maritime life-lines, the enemy's trade."[32] He was now prepared to contemplate a second major campaign against merchant shipping. Accordingly, he began to drill his crews for night surface attacks, "wolf pack" tactics against convoys, and close-range (within 300 meters) firing. The winter exercises of 1938–1939, conducted against Atlantic convoys with an eye towards a possible Anglo-German war at sea, convinced Dönitz of the need for a fleet of at least 300 boats, mainly of the 517-ton and 740-ton variety.[33] Clearly Dönitz's conceptions had parted with those of the naval high command. Yet it would be inaccurate to predate the celebrated Dönitz-Raeder clash to the late 1930s. Indeed, not a single high-ranking officer shared Dönitz's concentration on a U-boat navy or an unrestricted campaign against Britain's sea lines of communications in the period before the war.[34]

---

[30] BA/MA, M 31/PG3458, Marine Kriegsakademie, Winterarbeit Kptlt. Haack, 1938/1939.

[31] See Jost Dülffer, *Weimar, Hitler, und die Marine: Reichspolitik und Flottenbau, 1920–1939* (Düsseldorf, 1973), pp. 387–388, 531 ff.

[32] BA/MA, PG 33970, "Die Verwendung von U-Booten im Rahmen des Flottenverbandes," 23 November 1937.

[33] Dönitz, *Zehn Jahre und zwanzig Tage*, pp. 18–19, 23, 36; BA/MA, N 328, Nachlaß Förste, Vol. 15, gives a figure of 249 U-boats under the "Z" Plan.

[34] It is interesting to note that Admiral Förste basically ghostwrote Raeder's memoirs in order to gloss over this clash. However, the disagreement resurfaced during the years of incarceration for both admirals at Spandau: Albert Speer, *Spandauer Tagebücher* (Frankfurt/Berlin/Vienna, 1975), pp. 176, 193, 331.

What was missing in Dönitz's arguments was a wider understanding of what the war on British commerce would require. Not surprisingly as a captain in command of submarines in the mid-1930s, he was thinking in terms of numbers of boats, tactics at the lowest level, and technological fixes. For example, Dönitz and his chief engineer urged experimentation with the Walter proposal as early as 1936.[35] The difficulty was that Dönitz's "fixes" were only relevant to the U-boat war *as it had existed in 1918*. Thus, Dönitz believed that the movement of convoys through Britain's western approaches would allow for concentrated attacks that would wreck British shipping and win the war. Nowhere in Dönitz's arguments before the war was there an analysis of possible weak links in the structure of trade and the support of trade for its war economy, as there was in the Air Corps Tactical School.

What Dönitz discovered in his conduct of the U-boat war was that the enemy also has options. The British could concentrate escorts and air power in the western approaches and inflict prohibitive losses on their tormentors. Then the war would move west to the central Atlantic, the shores of the United States, and perhaps even the Caribbean. In this wider arena U-boats would confront two great problems: range and intelligence. With the entire expanse of the Atlantic available, the British could select varied routing to the United Kingdom, while U-boats would have to travel greater distances to reach their concentration areas. In such circumstances, the Germans would require vast radio traffic to control the submarines and concentrate their attacks. Might not such traffic provide the enemy with an important intelligence advantage? And were the Germans building the correct U-boat types for a war over such distances? In retrospect, they were not. The new types of U-boats were most suitable for Britain's western approaches – not for a war off the coast of the United States and the Caribbean.

Since any submarine campaign would involve a number of complex issues, one would expect the Germans to have gamed at least some of the problems. Again they did not; such an approach was entirely foreign to the German naval culture. It is instructive to note that the war gaming of the United States Navy in the 1920s and 1930s not only gained insight into the problem of using air power to support the fleet, but worked out many of the larger logistic and support questions on which the great Central Pacific campaign of 1943–1945 would rest.

By not preparing their innovations for a submarine campaign on a broader base, the German Navy floundered from one battle in the Atlantic to the next with no clearer direction than the idea that U-boats must sink ships. In 1940 it achieved stunning success in a campaign that mirrored the war in 1918; similarly, against the inept operations of the United States Navy, it savaged shipping off the east coast off the United States in the first half of 1942. But as their enemies adapted, the Germans could only respond with technological and tactical changes; in the end such changes were too little and too late. The Germans developed excellent weapons and trained their forces within an ingenious tactical framework, but they failed to innovate in the larger sense. Their war lacked the operational endurance and intelligence base to attack Allied commerce in the open expanses of the Atlantic in a coherent fashion.

From Hitler to Raeder and the naval planners, the Germans remained wedded to a stolid belief that only a symmetrical Mahanian surface fleet – not U-boats – constituted sea power. After all, had that not been the key to British naval mastery? On 3 September 1939 Raeder noted in the *Seekriegsleitung*'s war diary that Germany had already lost the Battle of the Atlantic – a sad comment on the naval leadership's lack of imagination and failure to innovate.

> Today the war against England and France broke out. . . .It is self-evident that the navy is in no manner sufficiently equipped in the fall of 1939 to embark on a great struggle with England. It is true in the short time since 1935 . . . we have created a well-trained submarine force which at present time has twenty-six boats capable of use in the Atlantic, but which is, nevertheless, much too weak to be decisive in war. Surface forces, however, are still too few in numbers and strength compared to the English fleet that they . . . can only show that they know how to die with honor and thus, create the basis for the recreation of a future fleet.[36]

The British

What was the attitude of the Royal Navy toward the submarine during the interwar years? The British adopted a two-pronged policy. Officially they proposed abolition of underwater warships from the discussions at Versailles in 1919 to the London Conferences in 1935–1936; realistically they sought to define binding rules on the waging of submarine warfare. Thus, while publicly announcing a drastic reduction in submarine funding at Versailles, they restored the budget for submersibles in De-

[36] OKM, Berlin, 3.9.39, "Gedanken des Oberbefehlshabers der Kriegsmarine zum Kriegsausbruch," National Archives and Records Service, T-1022/2238/PG 33525.

cember 1919 out of fear that reduction would "seriously hamper" the Royal Navy's future development.

A basic division between the British, on one hand, and the French, on the other, continued throughout the interwar period over the submarine. At the Washington Conference in October 1921, Admiral Sir David Beatty, First Sea Lord, bluntly decreed that while abolition was still the official policy, "it is certainly not to be expected." King George V pressed for abolition, but the Committee of Imperial Defence thought the proposition unacceptable to the other powers. In December the First Lord of the Admiralty, Arthur H. Lee, offered to scrap "the largest, most modern, and most efficient submarine fleet in the world," namely the British, "if the other powers will do the same and desist from further building."

The one, limited gain of the Washington Conference was a resolution by Senator Elihu Root, decreeing that submarines were not "under any circumstances" exempt from international rules of "visit and search" before seizing merchant ships. A clause in the "Root resolutions" stating that violators were "liable to trial and punishment as if for an act of piracy," however, found disfavor with the Royal Navy, which feared that such a rule might also apply to its submarines. Ultimately, French obstructionism took the teeth out of even Root's moderate approach.[37]

President Warren G. Harding persuaded the naval powers at the Washington Naval Conference to declare a ten-year "naval holiday" regarding warship replacements. Thus, by 1932 the Royal Navy had 40,000 tons of submarines on the books, but unbuilt. Admiral W.W. Fischer, deputy head of the naval staff, proposed the construction of sixty large submarines, at the rate of six per annum, but the Admiralty's Permanent Secretary, Sir Oswyn Murray, informed the First Sea Lord that this goal was "politically and financially impossible." George V once again pressed for the abolition of "this terrible weapon."[38]

The final contractual agreement with regards to submarines was the Anglo-German Naval Agreement, which allowed Germany to build up to 45 percent of British submarine strength and achieve parity in submersibles if "in their own opinion" they deemed such action necessary. And while the 1937 British "New Standard of Naval Strength" called for a force of 198 destroyers and eighty-two submarines, the Committee of Imperial Defence in December 1937 expressed confidence that the submarine menace had declined precipitously. Perhaps the members of the

---

[37] See Lawrence H. Douglas, "The Submarine and the Washington Conference of 1921," *Naval War College Review*, vol. 26 (1974), pp. 86–100.

[38] Ibid., pp. 21–50.

committee took at face value the London Submarine Agreement of 1936, wherein Germany had declared that, in case of war, it would sink merchantmen only in accordance with international law.

The reason for this happy optimism is not difficult to explain. Ignorance and indifference led to a general failure to think through the lessons of the last war. "Virtually every . . . antisubmarine lesson of the first . . . war," in the words of Lieutenant Commander D.W. Waters of the Royal Navy, "had to be, and ultimately was, relearnt in the second at immense cost in blood, tears and treasure."[39] Success is not always a great teacher, and the Royal Navy had, after all, accomplished its major objectives during World War I.

Another reason for the failure to learn from the experiences of 1914–1917 was the general belief that both the public and naval professionals had overestimated the effects of Germany's unrestricted submarine warfare during the war. Over and over, the Admiralty belabored the fact that U-boats had sunk 11 million tons of merchant shipping. Yet in reality, hunger had plagued Germany's population with approximately 763,000 civilian deaths attributed to the blockade. While there had been hardships in Britain, the British had not even had to reduce daily rations of hay and oats for race horses.[40] Obviously, as these facts became general knowledge in the 1920s, the wartime panic occasioned by unrestricted submarine warfare receded from memory of not only civilians, but sailors as well.

The British basically forgot that convoys alone had played the crucial role in blunting the U-boat offensive in 1917. Through the end of July 1918, convoys had reduced the loss of merchant shipping to about one in 120 (ninety-nine out of 12,008), or 0.83 percent. Not a single American troop transport bound for Europe was ever torpedoed. And these results emerged despite the fact that only a mere fraction of British warships, about 5 percent of the Royal Navy's 5,018, had waged the convoy effort.[41]

Convoys fell by the wayside because naval professionals despised the

---

[39] Cited in Arthur Marder, "The Influence of History on Sea Power," in *From the Dardanelles to Oran: Studies of the Royal Navy in War and Peace, 1915–1940* (London, 1974), p. 38; and in reply, David MacGregor, "The Use, Misuse, and Non-Use of History: The Royal Navy and the Operational Lessons of the First World War," *The Journal of Military History*, vol. 56, October 1992, pp. 603–615.

[40] See especially C. Paul Vincent, *The Politics of Hunger: The Allied Blockade of Germany, 1915–1919* (Athens, OH, 1985).

[41] Arthur J. Marder, *From Dreadnought to Scapa Flow: The Royal Navy in the Fisher Era, 1904–1919*, vol. 5 (London, 1970), pp. 97–118.

dull and monotonous work of escorting convoys. They viewed convoys as a temporary expedient rather than a tool of war. Throughout the interwar period convoys represented merely one more antisubmarine device among a host of others such as the arming of merchant vessels, the use of smoke apparati, the Otter defense gear against mines, and the hydrophone. Navy planners refused to recognize that the convoy lay at the heart of any successful effort to control the submarine. Above all, they viewed convoys as a purely defensive measure and pointed out with relish that escorts had only managed to sink twenty submarines during the course of the war – out of a total German loss of 178 boats. They refused to recognize that what counted was not the number of submarine "kills," but rather the number of merchantmen that arrived safely in port. Not surprisingly Britain's naval doctrine remained geared to major fleet engagements and played down the threat of submarines.

One must also note that the Royal Navy in the late 1930s placed its faith in "Flower-"class corvettes, which featured anti-aircraft guns at the cost of limited antisubmarine equipment. British naval doctrine at this time stressed trade protection in the North Sea and along the home coast; hence the corvettes were only 180 feet long and lacked the speed required to pursue submarines in heavy Atlantic weather. The corvettes "remained the luxury end of the escort spectrum" and thus drew funds away from escort vessels that could stand the rigors of Atlantic operations.[42]

Traditional service beliefs, then, operated against learning the lessons of 1917–1918. The Royal Navy remained wedded to its obsession with battleships and major fleet engagements. Jellicoe may have eschewed the much-desired "second Trafalgar," but who could say what a future war might bring? Because the British once more viewed Germany as a future opponent, the Royal Navy's officer corps resurrected the rhetoric of prewar days in thinking about the potential threat. Indeed, Admiral Sir Casper John admitted that most officers viewed convoy protection "with martial antipathy." Command of a fleet destroyer rather than a convoy escort was "career enhancing." Advocacy for antisubmarine warfare often resulted in the termination of one's career.

Fleet actions dominated naval thinking. Vice Admiral Sir Peter Cazalet conceded that it was difficult during the interwar years to get anyone in the Royal Navy to take any interest in antisubmarine matters. As a result, rigorous professional antisubmarine training was hard to come by.

---

[42] See Andrew Lambert, "Seapower 1939–1940: Churchill and the Strategic Origins of the Battle of the Atlantic," *Journal of Strategic Studies*, Vol. 17, March 1994, pp. 93–94.

"[Antisubmarine] exercises were admittedly very dull and were generally regarded as time-wasters and an unmitigated bore."[43]

In fact, the drastic economizing of defense outlays reduced preparedness at sea. While the Royal Navy had been accorded one-quarter of governmental expenditures before 1914, on the eve of the Great Depression that figure had fallen to only 6 percent. And while the Royal Navy overall received adequate funding up to 1929, antisubmarine forces, especially, failed to receive sufficient financial and technological resources. The Admiralty eliminated the antisubmarine division of its staff, to cite but one example, on the grounds of economy. One scholar has called the "Ten-Years-No-Major-War Rule" a "wet blanket," inhibiting fresh thinking throughout the navy. "Funds for research and development were curtailed; new ideas and new weapons were sadly lacking."[44]

In an unfriendly fiscal environment, the Royal Navy placed its faith in a single (and largely untested) technological device: asdic (acronym for Anti-Submarine Detection Investigation Committee). The British had developed this ultrasonic sound-wave device in 1917 to locate submarines at ranges of up to 1,000 meters. Naval officers widely viewed it as *the* critical deterrent to another U-boat campaign. As early as December 1918, the Director of Research at Parkeston Quay, Professor A.S. Eve, had boldly proclaimed asdic as "the most probable antidote to unlicensed submarine warfare." Sir Maurice Hankey, Secretary of the Cabinet and Committee of Imperial Defence, was confident that asdic had neutralized the U-boats, while Churchill simply referred to asdic as "the sacred treasure of the Admiralty."[45]

Senior naval officers agreed. Admiral Sir Ernle Chatfield deemed asdic an "epoch-making achievement," one that would render the hitherto "undetectable craft" detectable and thereby able to "keep the sea open as in golden days." Beatty, as First Sea Lord, opined after the 1920 sea trials that asdic experiments were of "first importance." Admiral Sir Henry H. Oliver, the Second Sea Lord, optimistically prophesied that the navy could safely escort convoys across oceans "under the protection of an asdic

---

[43] Although one must note that Stephen Roskill in *Naval Policy between the Wars*, vol. 1, *The Period of Anglo-American Antagonism, 1919–1929* (London, 1968), p. 536, is off the mark when he states: "But not one exercise in the protection of a slow mercantile convoy against submarine or air attack took place between 1919 and 1939."

[44] Peter Silverman, "The Ten-Year Rule," *Journal of the Royal United Services Journal*, vol. 116, March 1971, p. 45.

[45] Winston S. Churchill, *The Second World War*, vol. 1, *The Gathering Storm* (Boston, 1948), pp. 163–164.

screen."[46] As late as July 1936 a Joint Planning Staff paper trumpeted that as a result of asdic, "the submarine should never again be able to present us with the problems which we faced in 1917," a position seconded in August 1936 by the Director of Naval Intelligence.[47] And in June 1938, virtually on the eve of a second major war, the Royal Navy rosily assessed asdic's chances of detecting a submarine at 80 percent.

The First Lord, Sir Bolton Monsell, shared the general confidence in asdic as the effective antidote to the U-boat and added, for good measure, that Germany would never wage another unrestricted submarine campaign. In the Admiralty's estimation the Germans would employ their U-boats in a future war chiefly in daylight for reconnaissance. As for attacks on commerce, it deemed German surface raiders of the *Deutschland* class as more dangerous than U-boats. All of this overconfidence had a disastrous impact on the British conduct of antisubmarine war in the first years of the war. In fact, British survival resulted to a great degree from a similar belief in the *Seekriegsleitung* that submarines could not survive in the face of current antisubmarine technology.

Such attitudes in the Admiralty were also responsible for one of the most disastrous mistakes made by the Chamberlain government in its appeasement policies: the surrender of Britain's treaty rights to use the west coast ports of the Irish Republic in case of a war. The Admiralty had plenty of opportunity to appeal the government's proposed surrender of these rights, but failed to do so. Whether the government would have listened to such objections is another issue. The point is that the Admiralty showed itself generally disinterested in whether it would possess these ports, which in the last war had proven essential to projecting escort forces out into the North Atlantic. Allied convoys would pay a terrible price for the fact that the Irish treaty ports would not be available throughout the entire course of the coming war.

British confidence, however, was misplaced, and the evidence was already available that asdic offered only a partial solution to the U-boat threat. When a suggestion arose in the 1920s that the leaking of asdic capabilities might persuade other governments to abolish the submarine, Alan Hotham, Director of Naval Intelligence, warned London that such a "lifting of the veil" would reveal asdic for what it was – a "gigantic bluff." Hotham cautioned that Britain had "more to fear from submarines than any other pow-

[46] See Willem Hackmann, *Seek & Strike: Sonar, Anti-submarine Warfare, and the Royal Navy, 1914–1954* (London, 1984), p. 127.

[47] Cited in Gerald Jordan, ed., *Naval Warfare in the Twentieth Century, 1900–1945* (London, 1977), p. 135.

er" and added that only "if this country had an antidote" could it "let other nations waste their resources on submarines." In 1929 Captain R. Bellairs, Director of Plans at the Admiralty, urged the naval staff to devote available resources to developing better antisubmarine detection devices.

Instead, an almost comical campaign took place to mislead other navies concerning asdic and to keep its design and technology secret. The Admiralty ordered that its officers were at all times to refer to asdic as "asdevite" to throw spies off the trail. The very term "asdic" remained secret until 1929. Even upon declassification, all information on asdic installation remained secret. Every ship was to lock its asdic offices at all times when not in use, and asdic instruments on the bridge remained covered the minute foreign visitors came on board. A confidential fleet order of 1934 – repeated in 1938 – warned submarine commanders not to use asdic sets near foreign warships for fear of rousing unwanted interest.[48] But these measures designed to protect the secrecy of asdic in the end prevented realistic testing of the technology.

A small band of antisubmarine officers at Portland were the only ones to know that nine times out of ten, submarines were able to avoid detection by asdic and to penetrate destroyer screens. They were also well aware that the asdic echo was usually smothered by the transmission during the critical final 500 yards of approach and that asdic's effective range of one kilometer measured poorly against the passive hydrophones of submarines, which ranged effectively up to ten kilometers. Yet ignorance provided security. Admiral Sir Manley Power put it simply: "The sad truth is that much of the navy disliked and feared submarines and was all too inclined to think that if they shut their eyes, the bogeyman would go away."[49] In short most of the British naval planners still shared Admiral Arthur K. Wilson's denunciation of the submarine in 1902 as "underhanded, unfair, and damned unEnglish."[50]

At another level, the Royal Navy dismissed the benefits of joint services planning in general and the role that aircraft had played – albeit a small one – in the antisubmarine campaign of the last war. According to official British figures, aircraft had destroyed seven U-boats at sea, but even more significantly, where the Allies had deployed a combination of surface and air patrols, they had lost only five merchantmen in convoys. In addition, the mere presence of aircraft had tended to thwart the submarine.[51]

---

[48] Hackmann, *Seek & Strike*, p. 126.
[49] Cited by Marder, "Influence of History on Sea Power," p. 40.
[50] See Herwig, *"Luxury" Fleet*, p. 86.
[51] See Ronald Dodds, *The Brave Young Wings* (Stittsville, Canada, 1980), p. 196.

What was the upshot of this experience? Air power advocates grossly exaggerated the airplane's role as a submarine killer, while the navy discounted the submarine menace and passed the threat on to the air staff for resolution. Confusion abounded as the air staff also refused to deal with the question. In December 1937 the Committee of Imperial Defence informed the services that in case of war with Germany, convoy escort requirements would total 165 aircraft altogether. The air staff adroitly fended off the estimate by arguing that convoys would provide easier targets to the enemy than independently routed ships. Hence, the airmen matched the Admiralty's own disinterest in the use of convoys.

The naval staff encouraged the complacency. In autumn 1935, it endorsed evasive routing of merchant ships and argued that it would employ convoys only if rerouting "should prove ineffective." That same year, Lord E.M.C. Stanley, the Admiralty's Parliamentary and Financial Secretary, informed the House of Commons that "the convoy system will be introduced" only "when sinkings are so great that the country no longer feels justified in allowing ships to sail by themselves."[52] Indeed Lord Stanley assured the House that antisubmarine systems were "so very much better than they were during the [Great] War that we should want fewer protective vessels in the convoy."[53] As late as 1938 Admiral Sir Dudley Pound, commander-in-chief Mediterranean, was still insisting the escort hunting groups were superior to convoys.

If the Germans, against all expectations, *did* resort in the future to unrestricted warfare, the Royal Navy was flatly unprepared. In December 1919 it closed the antisubmarine warfare training school that Captain William Fisher had founded in 1917. Even when reconstituted in October 1920, the entire staff consisted of three instructors. Moreover, the navy abandoned experiments with explosive paravanes and hydrophones and then left asdic as the sole ship-borne, antisubmarine device. It also jettisoned trials with 3.5-inch stick bomb throwers against U-boats in 1934 and never completed even more promising trials with 6 inch light guns. These guns were designed for merchant ships and light craft to fire depth charges at submarines.

The RAF, which had control of virtually all air assets throughout the interwar period, proved even more disinterested in antisubmarine warfare. The air staff, under the spell of Trenchard's "strategic bombing" mania, argued in 1936 that aerial attacks on enemy submarine bases and yards constituted "the most effective use of aircraft in meeting the men-

---

[52] Hansard, *Parliamentary Debates* (House of Commons), 5th Series, ccic, col. 674.
[53] Ibid., col. 677.

ace of submarine attack against shipping."[54] But when war came in 1939, few aircraft were available for antisubmarine warfare, and instructions to maritime aircraft did not include the protection of merchant shipping.

In political terms, a firm conviction that no civilized country would again resort to unrestricted submarine warfare lulled the Royal Navy, as well as the United States Navy, into complacency. After all, the last campaign had failed. Moreover, it had brought the United States into the war, thereby assuring Germany's defeat. In addition, it had aroused public opinion against Berlin because the submarine invited charges of callous disregard of human life and sheer barbarism. Finally, the chances of another such conflict seemed slim with the League of Nations to serve as an umbrella to ward off future wars of aggression. Britain's rosy perception of the likelihood of peace in Europe in the foreseeable future precluded realistic assessment.

Above all, the Royal Navy suffered from a sterile anti-intellectualism in the postwar era that prevented a thorough study of the "lessons" of 1914–1918. At the Royal Naval Staff College, refighting Jutland became the major obsession, a fixation that ran through the navy as a "deadening virus." As late as 1934, the staff college ran a three day examination of the 1916 battle. By contrast, it devoted a single hour to the study of submarines and completely ignored the U-boat campaign of 1917 and the antisubmarine efforts of the same period.[55] Additionally, the Admiralty devoted the greater part of its attention to protection of trade in the North Sea and British coastal waters. Few saw any reason to be concerned about a German threat in the Atlantic in the interwar period.

The antisubmarine division of the naval staff did analyze the experience gathered against submarines in World War I, and its findings appeared in a technical history series shortly after 1918. Yet the Admiralty classified the volumes and hence made them inaccessible to most officers. In 1939 it then declared them obsolete and destroyed them. At another level, officials censored the reform-minded *Naval Review* until 1926, in order, as Andrew Bonar Law, the Lord Privy Seal, put it, to avoid "washing dirty linen in public."[56]

A third example of the navy's anti-intellectualism concerns the naval staff's historical section. On 4 August 1914, Churchill, then First Lord,

---

[54] Cited in Stephen Roskill, *Naval Policy between the Wars*, vol. 2, *The Period of Reluctant Rearmament, 1930–1939* (London, 1976), p. 393.

[55] Marder, "Influence of History on Sea Power," p. 48.

[56] Arthur J. Marder, *Portrait of an Admiral, The Life and Papers of Sir Herbert Richmond* (London, 1952), p. 340. Richmond tartly commented: "that if the linen is not washed at all, it creates a pestilence."

created the section to gather experience for future study, as an "aid and a guide for all who were called upon to determine, in peace and war, the naval policy of the country." Its subsequent fate sheds much light on Admiralty attitudes. During the war, it remained buried within the Committee of Imperial Defence. Resurrected by Captain A.C. Dewar in November 1918 to examine the war's lessons and deduce principles for future guidance, the historical section consisted of only two officers in 1923 – on the grounds of economy. Thus, instead of completing its work by 1932, as planned, it had failed to produce the crucial volumes on the convoy battles of 1917–1918 by the time that the next war had broken out in 1939.

Antihistorical attitudes in the navy also throttled criticism of the Admiralty and individual combat commanders. A senior admiral accurately summed up the results:

> It would have needed men of intellectual stature far beyond that of Beatty, Madden, etc., to hoist in the fact that, from the view of the value of money, we would have done better to pay off a couple of battleships in the 1920s to insure that those 1914–1918 histories were truthfully and properly digested.[57]

In short, most naval officers thought that little could be (or should be) learned from historical study of the conduct of antisubmarine warfare in the last conflict.

In practical terms, the decision to ignore history proved costly. Whereas World War I had shown that armed craft designed as merchant vessels (the so-called Q ships) were ineffective – destroying but six U-boats at a cost of twenty-seven of their own number – the Admiralty immediately commissioned forty Q ships for service in the Atlantic in 1939–1940. In fact during 1940 these ships failed to bag a single U-boat. With regard to the fleet base at Scapa Flow, the Admiralty again ignored the warnings of 1914–1918: in 1939 it failed to protect the anchorage with more than a single line of antisubmarine nets across the main entrances. As a result, German submarines and mines sank one British battleship (*Royal Oak*), damaged another (*Nelson*), and damaged a heavy cruiser (*Belfast*).

Despite the experiences of World War I, the Admiralty clung until 1943 to the shibboleth that "the larger the convoy the greater the risk." In fact the reverse was true. Rollo Appleyard had mathematically analyzed "the

---

[57] Cited by Marder, "Influence of History on Sea Power," p. 61. See also Peter Kemp, "War Studies in the Royal Navy," *Journal of the Royal United Services Institute*, vol. 111, May 1966, pp. 151–155.

law of convoy size" for the Royal Navy in 1918 and concluded that whereas escorts protect the convoy's perimeter and not the individual ships within, and whereas large convoy's perimeter is only slightly larger than that of a small convoy, "the area occupied by the ships increases as the square, while the perimeter is directly proportional to the length of the radius."[58] In short, what counts is not the ratio of merchant ships to escorts, but the ratio of the attack area around a convoy to the number of close escorts. The larger the convoys the further the Admiralty could spread the small numbers of escort vessels then available.

At another level, up to World War II the Admiralty believed that U-boats would not make night surface attacks. In part, this estimate resulted from the Admiralty's ban up to 1937 on submarine operations at night during exercises for safety reasons. Mirror-imaging once again precluded serious historical analysis, for even a cursory glance at the last war would have revealed that by 1918 two out of every three U-boat attacks were occurring at night and on the surface.

The first German torpedoing of a merchant ship (*Athenia*) took place on 3 September 1939. The first British convoy sailed three days later. But years of neglect had taken their toll on British preparedness: insufficient escorts, unsuitable escort types, cruisers without effective antisubmarine devices, inadequately trained antisubmarine groups, lack of air power on convoy routes, and diversion of escort craft from the "defensive" convoy patrols to "offensive" hunting groups characterized British antisubmarine operations. The Royal Navy entered World War II with only 100 destroyers and thirty-eight submarines in service.

The one bright side of this dismal situation had to do with intelligence. Here the navy's misuse of "Room 40's" invaluable intelligence during Jutland alerted the Admiralty to the importance of integrating intelligence quickly and directly into operations. Thus, while the Operational Intelligence Center contained a staff of only thirty in 1939, it would soon find itself involved in an exponential expansion to cover the demands for intelligence in the war against the U-boat.

How to explain British behavior? On an organizational plane it is clear that the Royal Navy lacked a core of officers from the prior war with influence who believed that antisubmarine warfare represented a crucial role. Moreover, the navy proved dilatory in its efforts to examine the historical lessons of the last war; and that inattention provided most senior officers with the comfortable illusion that the success of the convoy in

---

[58] Commander D. W. Waters, quoted in Marder, "Influence of History on Sea Power," p. 46.

1917 indicated that the submarine was a relatively insignificant threat to British security.

It is also clear that the culture of the officer corps discouraged serious examination of disturbing challenges to the *status quo*. Finally, the financial constraints up to 1939 were so stringent as to discourage nearly all innovation. So desperate was the British situation in 1940, when the Germans gained the ports in western France, that only the most massive innovation – major investments in science and technology (that paid dividends in 1943), an infusion of civilians into both fighting units and support services, creation of armadas of escort vessels and supporting aircraft, and a willingness to support intelligence and its dissemination with unheard-of resources – allowed the British to survive the terrible threat to their sea lines of communications.

The Americans

The U.S. Navy was the only navy to participate in both major antisubmarine *and* submarine campaigns in its conduct of World War II. In the Atlantic, American escort forces played a considerable role in the convoy battles that won the Battle of the Atlantic; in the Pacific, U.S. submarines waged *and won* the war against Japanese commerce, in effect sundering Japan's sea lines of communications. Because the former campaign mirrored the worst aspects of Britain's failure to innovate in antisubmarine warfare, this section will focus on American preparations for the Pacific – a campaign that in the end proved successful after a fitful start.[59]

The United States Navy did not possess a commerce-raiding submarine strategy much before 1941. To be sure, on 7 December 1941 Franklin D. Roosevelt endorsed Admiral Harold R. Stark's proposal that the United States begin unrestricted submarine warfare against Japan in response to Pearl Harbor, and the American president informed Congressional leaders that same day that American policy in the Pacific was simply "stran-

---

[59] The less said about the American preparations the better; after all, not only did the United States Navy have the entire interwar period but also two years and three months of watching the British flounder in their efforts to master the threat. Nevertheless, American naval leaders responsible for the defense of merchant shipping along the East Coast of the United States and in the Caribbean resolutely set themselves against learning anything from the British. A British naval officer summed up the situation nicely in a comment directed at Rear Admiral Richard S. Edward of Admiral Ernest J. King's Staff: "The trouble is, admiral, it's not only your bloody ships you are losing. A lot of them are ours." Quoted in Eliot A. Cohen and John Gooch, *Military Misfortunes, The Anatomy of Failure in War* (New York, 1990), p. 60.

gulation of Japan – strangulation altogether."[60] The resulting campaign was a success: approximately 300 U.S. submarines conducted 1,500 sorties against Japanese shipping during World War II; they sank 4,779,902 tons of merchant shipping as well as 540,192 tons of warships, a total of 54.6 percent of all Japanese tonnage.[61] But in light of its strident opposition to German unrestricted warfare during World War I by what strange odyssey had the United States arrived at such a draconian policy?

The navy's General Board had defined two functions for submarines since 1911: coastal protection (defensive) and fleet operations (offensive). The failure of Imperial Germany's unrestricted U-boat offensive 1917–1918 confirmed established U.S. beliefs that commerce raiding was *not* the proper employment for submarines. Moreover, after 1919 the navy's civilian leadership argued that submersibles were violations of international law. Finally, the naval planning group for the London Conference advised the chief of naval operations that "the national conscience" would never permit use of submarines in "the destruction of enemy merchant shipping."[62] In short, the national strategic culture militated against the conduct of unrestricted submarine warfare.

A minority within the navy, however, was hesitant to forgo submarine warfare. Not surprisingly, Captain H.H. Bemis in the submarine section of the Office of Naval Operations favored retention of submersibles on account of the impossibility of outlawing any weapons system via "a scrap of paper." In November 1920 he argued that it would be "criminal" for the United States to "abolish submarines," especially with the potential threat posed by the Japanese in the Pacific. Theodore Roosevelt, Jr., then Assistant Secretary of the Navy, concurred, arguing that "against the two island empires – Great Britain and Japan," it was imperative that Washington "not permit our hands to be tied as regards submarines."[63]

Hence, from the earliest debates over submarines in the United States Navy, ambivalence characterized the American position. On the one hand, moral arguments existed against submarine warfare: the U-boats had failed in their mission, and the Kaiser's "pirate ships" had galvanized world opinion against Germany, while international law prohibited "un-

---

[60] Quoted in Samuel Flagg Bemis, "Submarine Warfare in the Strategy of American Defense and Diplomacy, 1915–1945," p. 34. Unpublished manuscript, Box 65, Folder 785, Manuscripts and Archives, Yale University Library.

[61] See Charles A. Lockwood, *Sink 'em All: Submarine Warfare in the Pacific* (New York, 1951).

[62] See Ernest Andrade, Jr., "Submarine Policy in the United States Navy, 1919–1941," *Military Affairs*, vol. 35, April 1971, pp. 50–55.

[63] Quoted in Douglas, "The Submarine and the Washington Conference," p. 87.

restricted" attacks against surface commerce. On the other hand, there were cogent operational reasons for retaining submarines: they were the only vessels that could effectively patrol the vast expanse of the Pacific. Captain Thomas C. Hart of the submarine section informed his colleagues that "the submarine will be an extremely valuable weapon for . . . operations against Japanese commerce."[64] Could the navy forgo the submarine when the nation's security was at stake?

Three major roadblocks faced the apostles of the "silent service." First, a Mahanian strategic culture dominated the United States Navy throughout the interwar period. Secretary of War Henry L. Stimson only slightly exaggerated the navy's cultural framework when he commented that it had a particular psychology, according to which "Neptune was God, Mahan his prophet, and the United States Navy the only true church."[65] In January 1919, the Atlantic Fleet's command confidently stated that "the best way to destroy commerce" was not to attack it directly, but rather to "destroy the forces that defend it." In other words, the primary objective in naval warfare remained "the destruction of the enemy's main force."[66] Moreover, as early as 1921 the General Board believed that the future lay with naval aviation and not submarines.

A lack of suitable submarines also affected priorities. U.S. boats during World War I were small, cramped, and unseaworthy; subsequent S-class craft were no significant improvement. Even the next generation of V-fleet submarines only had a top speed of eighteen knots, three knots short of the desired best speed.[67] Thus, unable to keep up with the fleet, submarines were incapable of fulfilling their assigned mission.

Navy planners remained ambivalent on submarines. Most opposed total abolition, and few saw utility in legal limitations. Admiral William S. Pratt, President of the Naval War College, spoke for many when he stated that any nation on the verge of defeat "would use the submarine to save itself," the "tremendous power of public opinion" notwithstanding. Admiral Pratt was equally forthright. Likening the Root resolutions

---

[64] Quoted in J. E. Talbott, "Weapons Development, War Planning, and Policy: The U.S. Navy and the Submarine, 1917–1941," *Naval War College Review*, vol. 37, May/June 1984, p. 56.

[65] Henry L. Stimson and McGeorge Bundy, *On Active Service in Peace and War* (New York, 1948), p. 506.

[66] Memorandum by Commander W. S. Pye, General Board Study Serial No. 894 (420–426).

[67] The most recent analysis of U.S. submarine construction is by Gary E. Weir, *Building American Submarines, 1914–1940* (Washington, DC, 1990); Weir has summarized his main points in "Search for an American Submarine Strategy and Design, 1916–1936," *Naval War College Review*, vol. 44, Winter 1991, pp. 34–39.

to Prohibition law, he argued that they were "made to be broken."[68] The General Board, for its part, failed to make up its collective mind concerning the efficacy of the submarine in a future war. As the board prepared for the abortive Geneva Conference of 1927, its senior member, Admiral Hillary P. Jones, avowed that the submarine was "not well adapted to lawful war on commerce." A host of technical problems worked against success: the crews of submarines were too small to allow for prize crews; submarines had insufficient room to take on crews of captured vessels; and public opinion in the United States would never permit unrestricted underwater warfare.[69]

Thus, the mission of American submarines remained the defense of Pearl Harbor, reconnaissance, and fleet operations, especially against large enemy warships. Tactical exercises reflected these stated missions. In part, this approach mirrored the hard reality that years of parsimony had created a shortage of building ways, berthing, yard capacity, draftsmen, and skilled laborers. The ultimate result was that the United States was eighteen submarines under the allowable treaty strength in 1935 (parity with Britain at 52,700 tons). While Japan laid down every ton to which it was entitled (two-thirds of British strength), the United States slipped to third place.[70] The General Board's recommendation that the country build six submarines per year from 1934 to 1940 remained a dead letter, while inadequate funding precluded design and engineering of new technologies.

To be sure, a general organizational framework existed that could remedy this situation. In 1926, 1930, and 1932, war games at the Naval War College explored the use of submarines on long-range patrol in enemy waters and as a scouting screen deployed in advance of the fleet. By and large, Newport war gamers were convinced that potential adversaries (namely Japan) would resort to unrestricted submarine warfare against the United States. That same year, a conference of submariners convened in Washington to report to the Chief of Naval Operations on the possible utilization of submarines in a future war. As early as May 1933, Roosevelt diverted funds from the National Recovery Administration for construction of four submarines. Under the provisions of the Vinson-Trammell Act

---

[68] Quoted in Douglas, "The Submarine and the Washington Conference," p. 97.

[69] See Bemis, "Submarine Warfare," p. 13.

[70] Gerald E. Wheeler, "National Policy Planning between the World Wars: Conflict between Ends and Means," *Naval War College Review*, vol. 21, February 1969, pp. 59–60; see also Janet M. Manson, *Diplomatic Ramifications of Unrestricted Submarine Warfare, 1939–1941* (Westport, CT, 1990), pp. 45 ff.

of 1934 and the Vinson Bill of 1938, Congress authorized the building of 117,450 tons of submarines.[71]

This support notwithstanding, the navy steadfastly clung to the notion that the "primary task of the submarines is to attack enemy heavy ships" such as battleships, battle-cruisers, and aircraft carriers. Planners stated that submarines could operate against heavy and light cruisers only by "special order." These instructions reflected the prevailing view that the navy should deploy submarines not in attacks on seaborne commerce, but as warship killers in tandem with the surface fleet.[72]

By the late 1930s, however, senior naval planners were challenging some of the assumptions on which War Plan Orange rested. The London Treaty had precluded the United States from constructing bases in the western Pacific, while Pearl Harbor, as every planner knew, was 4,850 miles from Manila. Could an American fleet realistically advance across the Pacific in search of a decisive engagement with its Japanese counterpart? Admiral Harry E. Yarnell, commander-in-chief Asiatic Fleet (1937–1938), thought not and argued for an alternative strategy. He identified Japan's economic trade as a more inviting target than its battle fleet and urged "the prosecution of [a] naval war of strangulation." This redirection of strategic thought resulted in the recasting of Orange in October 1940.

Admiral James O. Richardson, commander-in-chief U.S. Fleet, likewise favored the "long-range interdiction of enemy commerce." In January 1941, he informed Stark that the new "Plan Dog" had jettisoned the notion of an offensive in the western Pacific and instead dictated a wait-and-see attitude featuring defensive submarine patrols off Midway, Johnston, Wake, and Palmyra. In the event that Japan started a war, Richardson advocated that U.S. submarines "make an initial sweep for *Japanese merchantmen* and raiders in the Pacific."[73] Stark had already instructed Admiral Thomas C. Hart, commander-in-chief Asiatic Fleet, in November 1940 that "economic starvation" of "Japan's offensive power" would be the American objective in a Pacific war.[74] Doctrinal development thus came in line with more realistic national objectives.

Theoretical planning at Newport reflected these changes. In March 1941 Admiral Edward C. Kalbfus, president of the Naval War College,

---

[71] Andrade, "Submarine Policy in the United States Navy," pp. 53–54.   [72] Ibid., p. 55.
[73] See Richardson to Stark, 25 January 1941: National Archives: OA. Strategic Plans Division Records, Series III, Box 64, Plan O-1 Orange; and Box 147 J, Modern Military Records, U.S. National Archives.
[74] Talbott, "Weapons Development," p. 64.

passed along to the chief of naval operations (and eventually the General Board and Secretary of the Navy) a radical "solution" revising existing "tentative instructions" based on the Washington and London Accords. Arguing that "new weapons may well call for changes" in naval doctrine, Kalbfus and his staff endorsed the German establishment in World War I of "war zones" with the intention of "blockading" Japanese exports and imports. Citing the *real needs and interests* of the United States, he concluded: "it would seem unrealistic not to allow for and to plan definitely upon the declarations of war-zones." Although the board rejected these suggestions as having "no justification in international law," Admiral Richard K. Turner, director of war plans, alluded to the fact that the matters raised by Newport were already being "handled in a different manner."[75]

That other "manner" was undoubtedly war plan "Rainbow Five," which Stark approved in May 1941. "Rainbow Five" repeated the injunction to submarine commanders to obey existing "cruiser" laws regarding the conduct of underwater warfare, but the plan nevertheless stipulated that the Washington and London conventions should not apply to Japanese merchantmen because they were expected to be armed – and hence under command of the Imperial Japanese Navy. Echoing Kalbfus, Stark empowered submarine commanders to proclaim "strategic areas" in which enemy commerce would be considered legitimate targets.[76] Thus, to all intents and purposes, unrestricted submarine warfare was the *de facto* if not the *de jure* American naval strategy well before the attack on Pearl Harbor.[77]

It is hardly surprising then, that Stark instructed his naval forces to "execute against Japan unrestricted air and submarine warfare" on 7 December 1941.[78] And it seems fitting that Admiral Chester W. Nimitz assumed command of the Pacific Fleet on the deck of the *Grayling* on 31 December. Having lost his battle fleet, Nimitz, a submariner himself, appreciated that submarines – roughly fifty in the Pacific and Asian fleets – constituted his major striking force. Like German planners in both world wars, American naval strategists resorted to unrestricted subma-

---

[75] Bemis, "Submarine Warfare," pp. 22–27.

[76] See OA. Strategic Plans Division, Series III, Box 147F. Task Force Seven (Undersea Force) 352a Special Information.

[77] On 20 October 1941, the U.S. Navy informed the Admiralty in London that its own plans called for unrestricted submarine warfare – regardless of whether a potential adversary resorted to it. See Ronald Spector, *Eagle Against the Sun: The American War with Japan* (New York, 1985), p. 479.

[78] Bemis, "Submarine Warfare," p. 51.

rine warfare on the basis of hard logic; they had few other means of getting at their opponent.

How well the Americans had innovated on the technical and tactical sides in preparing for this conflict is open to question. In some respects they did very well; by 1941 they had come up with the best submarine of the war. Nevertheless, numerous setbacks plagued the development of an efficient and effective fleet submarine. Through 1930 the primary difficulty confronting American submariners was the lack of a suitable power plant. U-boats, surrendered at the end of World War I, had underlined how far behind U.S. technology in power plants was to what the Germans possessed. As late as 1931 the navy expended considerable sums to acquire German diesel technology despite the Depression's extraordinary financial restraints.[79]

But with the help of the Roosevelt administration the navy interested a number of American manufacturers in producing a small, high-speed diesel for submarines – an engine whose specifications matched those needed by locomotives.[80] The introduction of competition into developing a suitable power plant – helped by its utility in commercial markets – finally resulted in the production of a first-class engine.

To a considerable extent the improvement in American submarine design followed the improvement in engine performance. By the late 1930s the two came together to produce the best submarine of the war: the *Tambor* class of fleet submarines. The *Tambors* (1,500 tons) were nearly twice the size of German boats, with four diesels providing the power. With four engines the *Tambors* could cruise at the same time that they recharged their batteries; they possessed air-conditioning which made them livable even in the tropics and certainly kept up morale in the long cruises across the Pacific. Most important, the new class possessed speed and maneuverability on the surface and under water.

In 1937 the future Admiral Charles A. Lockwood, Jr., took over the "submarine desk" in Washington. With long experience in submarines and a first-class mind, Lockwood pushed the *Tambor* design against considerable opposition on the General Board. Those who opposed the new design favored smaller submarines (about the size of German U-boats) and denigrated frills such as air conditioning. They aimed to use submarines as defensive screens for Hawaii and the Philippines rather than as an offensive striking force.[81] Lockwood's arguments carried the day,

---

[79] Weir, *Building American Submarines*, p. 100–104.   [80] Blair, *Silent Victory*, p. 61.

[81] There was also some opposition to the idea of air conditioning the boats – obviously one did not want to pay too much attention to crew comfort.

but in winning the battle for construction of this new class, the submariners requested only six of these boats by 1940 – a total which one historian of the submarine war suggests represented "an absurdly low figure."[82] But they provided the prototype of the submarine class that would launch the devastating submarine offensive against the Japanese. In addition, the Americans had by this time developed effective computers that permitted submarine skippers to calculate their own relative position to that of the target and come up with accurate solutions for firing torpedoes.[83]

If the *Tambors* turned out to be an unqualified success, the provision of submarine commanders, a coherent and effective submarine doctrine, and torpedoes that worked for that force left much to be desired. On the level of personnel the navy is open to some criticism. At the start of the war there were too many submarine commanders who lacked the necessary initiative and drive. There were some, of course, who exhibited the necessary drive and courage but they were the exception rather than the rule.[84]

There was, moreover, a larger problem beyond the selection process for submarine commanders: the tactical framework within which the prewar navy had trained. Some American submariners had worked on wolf pack tactics in the 1930s.[85] Unfortunately, the emphasis in tactical preparations remained on fleet operations; moreover, skippers who "were caught during fleet exercises" more often than not received severe criticism from their superiors. Not surprisingly, the submarine force had too many commanders at the outbreak of the war who were unwilling to take risks in a war that very much depended on the decisiveness of the commander on the spot.

But the real scandal was the torpedo story. To begin with, for much of the period the navy depended on a sole provider – the torpedo factory at Newport that was under the direct patronage of the Rhode Island congressional delegation. In designing and producing new torpedoes, the Bureau of Ordnance (BuOrd) had developed a magnetic exploder that depended on the target ship's magnetic field to provide the stimulus to set it off. The design was flawed. Even more seriously, the Bureau of Ordnance had carried out its tests on torpedoes with light, dummy warheads. It refused to carry out any tests with live warheads *even in the initial months of the war* because of the expenses involved. When real explosive was

[82] Blair, *Silent Victory*, pp. 67–68. [83] Ibid., p. 65. [84] Ibid., pp. 198–200.
[85] Vice Admiral Charles A. Lockwood, *Down to the Sea in Subs* (New York, 1967), p. 201.

in the warhead, the torpedoes were considerable heavier and ran much deeper – so deep, in fact, that using proper settings, torpedoes ran eleven feet deeper than they were supposed to. As a result, many failed to work. Finally, to complete an extraordinary tale of incompetence, the contact exploder was so fragile that a direct hit was sufficient to destroy it without causing the torpedo to explode.

Nothing in three years of war caused the Bureau of Ordnance to lose faith in its torpedoes. When the British reported in 1940 that their submarines had great difficulty in utilizing a magnetic exploder and had deactivated the device, BuOrd refused to test its own torpedoes.[86] When reports from the Pacific indicated that something was wrong, again BuOrd refused to test its torpedoes; the failures obviously lay with the crews. The Bureau was finally forced to send an officer to the Pacific to examine crew practices in handling torpedoes. The officer charged with the task could "not point out a single fault in our preparations and maintenance procedures; nevertheless, [his] report, in summary, placed all the blame for the *Sargo* torpedo problems on *Sargo* personnel. As a result the Bureau of Ordnance reaffirmed their position that the Mark XIV torpedoes ran at their set depth."[87] Not until summer 1943 would the U.S. submarines have torpedoes that worked – a sad story indeed of innovation thwarted by arrogance and penny-pinching on the edge. By that time the Americans possessed a great fleet of submarines, all with well-trained crews, a magnificent support structure, and an intelligence system that was uncovering the movement of shipping on a day-to-day basis – a devastating combination.

There is, however, some danger in transposing that success to earlier in the war. Without a sufficient number of boats, the Americans might only have alerted the Japanese to the threat. While they might have done considerable damage to the Japanese, their initial attacks in 1942, even with torpedoes that worked, would probably not have sufficed to bring the Empire to collapse. With time to react the Japanese might have marshalled the resources necessary to master the submarine. However, when the onslaught hit their unprepared forces in 1943 the Japanese had no chance against a strong, well-prepared, and numerous U.S. submarine fleet.

CONCLUSION

The period after World War I witnessed significant declines in the levels of defense spending in Britain, Germany, and the United States. Popula-

---

[86] Blair, *Silent Victory*, p. 71.   [87] Quoted in Blair, *Silent Victory*, p. 170.

tions weary of exorbitant wartime outlays in the years 1914 to 1918 demanded vast reductions in military budgets and concentration instead on domestic spending. Funds that escaped the proverbial budget axes in London, Berlin, and Washington went to traditional force structures such as battle fleets rather than into dubious and still experimental service arms such as submarines and antisubmarine warfare. While many politicians and most naval officers recognized the need to stay abreast of technological developments in the "silent service," they also appreciated the hard reality that major programs would command major portions of defense allocations.

Inadequate access to national fiscal and technological resources hampered research and development. This, in turn, led to reduced experimentation in innovation, diffusion, and doctrine, thereby impeding the development of overarching operational and strategic frameworks. No coherent views of the next war developed and the navies involved established no strategic mission for the submarine. While potential enemies were readily discernible – Germany for Britain, Britain for Germany, and Japan for the United States – failure to establish clear national policy objectives obviated systematic force preparation.

Instead, ambivalence and confusion dominated the submarine policies of Germany, Britain, and the United States during the interwar period. On the surface, the three powers avoided open discussion of unrestricted submarine warfare for fear of accusations of condoning the "horrors and lawlessness" of Imperial Germany's U-boat campaign of 1917–1918. As a result, all three had publicly announced by 1936 their intention never to resort to unrestricted submarine warfare. Yet within the inner sanctums of the naval services, German and American planners, in particular, wondered whether they could eschew an offensive submarine strategy in case of national emergency; in both instances, the "realists" won the day. Britain, for its part, as the dominant Mahanian naval power, easily assumed the moral high ground and demanded the abolition of submarines.

A consensus existed in Berlin, London, and Washington regarding the lessons to be learned from the last war: Germany's unrestricted submarine campaign of 1917–1918 had failed. Despite the high losses to U-boats, Britain had not collapsed – as promised by German admirals. Indeed, Allied and American construction eventually outpaced losses, while the Germans failed to torpedo a single eastbound American troop transport. Above all, the political ramifications of unrestricted submarine warfare had been calamitous: Imperial Germany guaranteed its own defeat by bringing the vast manpower and material reserves of the United States into the war.

Technological considerations, while never a determining factor, also affected the doctrinal framework in which naval planners operated and weighed heavily against a repeat of 1917–1918. Submarines remained precariously prone to accidental sinking, abetted by crew fatigue and carelessness. Limited radius, slow speed especially when submerged, unreliable torpedoes, and lack of underwater communications further negated their effectiveness. If they operated according to "prize" rules, submarines could neither provide sufficient prize crews, nor could they take on board the crews of such prizes. Finally, what future would there be for admirals in a fleet of submarines? Imperial Germany had seriously toyed in the spring of 1917 with the idea of creating a "special cemetery for our existing submarines after the war" in order not to upset "organization and promotion" within the officer corps.[88]

In terms of an overarching doctrinal framework the major naval establishments were united in the belief that submarines could never constitute true sea power and exercise either sea control or sea denial. This staunch orthodoxy worked to block innovation. Mahan remained the prophet and *The Influence of Sea Power upon History* the Bible. Moreover, the lessons of World War I seemed clear: symmetrical battle fleets with 30,000-ton battleships and battle-cruisers as their locus of power alone could command the seas. To be sure, a few radicals both within and without the naval establishment were beginning to tout the coming dominance of air power, but none seriously contemplated a commanding role, be it sea control or sea denial, for submarines.

Fiscal considerations also operated against submarine (and antisubmarine) development. London as well as Washington sought to reduce military budgets drastically. The Germans had no choice. Parsimonious legislative assemblies were reluctant to provide funds even for highly visible and evidently powerful capital ships, much less a bewilderingly diffuse armada of submarines and destroyers and associated research and development costs. Germany was forbidden to build either capital ships or submarines and had to undertake clandestine research and development cautiously and in modest proportions.

Given the above conditions, it is not surprising that interwar naval doctrine in Germany, Britain, and the United States assigned underwater craft purely military roles such as scouting, minelaying, and operations against the enemy's heavy warships. While numerous officers on both sides of the Atlantic grumbled about legal restrictions ("cruiser" rules of search, seize, and destroy) placed on submarines, they eventually shared the belief that

---

[88] See Herwig, *"Luxury" Fleet*, p. 224.

unrestricted submarine warfare was a thing of the past. They could live even with Paragraph IV of the London Accord, which decreed the same rules for surface and underwater warfare.

The three navies quickly forgot, or in the case of the Royal Navy literally destroyed, the "lessons" of World War I with regard to antisubmarine warfare. Both convoy protection and antisubmarine work were tedious and dull, unlikely to attract either funds or recruits. For naval officers, appointment "plums" led to the main battle fleet – at worst, to its protective destroyer screen. It is interesting to note that the U.S. Navy failed to develop a new class of "destroyer escorts" until 1943. The British lulled themselves into a false sense of security concerning their ability to detect and destroy enemy submarines through asdic. Its American counterpart vaguely remembered the wartime mine barrage and President Woodrow Wilson's admonition to hunt not the hornets but rather their nests, yet it did little to perfect either convoy or antisubmarine warfare. In short, the "lessons" of 1917–1918 all had to be relearned at great cost during World War II.[89]

Of course, a small cadre of submarine enthusiasts existed, especially in the American and German navies, and they secretly harbored the conviction that any future war *might* witness a renewed underwater campaign against merchant shipping – either by the adversary or by one's own forces. Indeed, a power could not neglect this potent weapon if its security was in the balance. But these were voices in the wilderness, drowned by the belief in the supremacy of surface fleets. The small band of submariners who advocated offensive warfare against commercial shipping were more visionary than realistic even within their own services.

The fact that the submariners' predictions became reality after 1939 and 1941 was due less to their warnings – and certainly not to a change in official naval doctrinal development – than to the reality that submarines were all that remained to hurl against the enemy in the cases of Germany and the United States. Lacking sufficient capital ships with which to challenge British naval supremacy, Raeder conceded in September 1939 that his fleet could only "die gallantly." Dönitz posited a less negative solution, namely, a renewed submarine warfare *à outrance*

---

[89] One needs recall only what Rear Admiral Samuel E. Morison called the "merry massacre" that occurred in American waters in the spring of 1942 as 1.5 million tons of shipping were lost to the U-boats – at a cost of only one of the latter (U-85). Lighted channel markers helped the U-boats enter U.S. ports, lighted city skylines provided a perfect background to sight targets, and convoys were ignored until April. S. E. Morison, *History of United States Naval Operations in World War II*, vol, 1; *The Battle of the Atlantic, 1939–1943* (Boston, 1947–1962), p. 125.

against Britain's maritime lifeline. Stark and Nimitz, deprived of their main battle fleet at Pearl Harbor in December 1941, turned easily to submarine warfare as the only viable option against Imperial Japan – past American moral outrage and international law notwithstanding. In both the American and the German case, war was truly the engine of innovation.

7

FROM RADIO TO RADAR

Interwar military adaptation to technological change in Germany, the United Kingdom, and the United States

ALAN BEYERCHEN

BACKGROUND AND TERMINOLOGY

Given the way that electronic warfare characterizes today's battlefields, it is sometimes difficult to remember that scientists discovered the very existence of radio waves barely a century ago. Yet already by 1904 a young German named Christian Hülsmeyer claimed his patented "telemobiloscope" could transmit radio waves and receive their reflections off a passing object. He suggested that such a device could prevent collisions at sea or aid navigation. Representatives of shipping lines flocked to various demonstrations in Germany and Holland and were impressed that the device could detect objects up to a range of approximately five kilometers. But there were no buyers. At the time investment capital was scarce in the maritime industry, wireless telegraphy already offered a means of communication among ship's captains (as well as a form of direction-finding by means of cross-bearings from shore stations), and the legal and technical relationships between this use of radio waves and the patented wireless telegraphy of the Marconi monopoly were unclear to shipowners.[1] That is, budgets were tight, other devices that seemed to fulfill present needs already existed, and it took real imagination to sort out the nature and possibilities of something so new. The first two factors are nearly constants, for resources throughout history are usually limited and few human needs are totally unmet. The key to the timing that turns a discovery or invention into successful innovation lies in whether laymen can envision its possibilities. And in this case that would take more than the promise of safer navigation at sea.

---

[1] David Pritchard, *The Radar War: Germany's Pioneering Achievement 1904–45* (Wellingborough, UK, 1989), pp. 13–20.

Although radio became a significant feature of World War I, the idea that later became radar languished in obscurity. Radio waves form a portion of the same electromagnetic spectrum as visible light and travel at the same speed. The difference lies in the lower frequencies, and thus longer wavelengths, of radio waves. Like light they reflect well off most metallic surfaces, but longer wavelengths allow radio to penetrate fog, clouds, and many solid objects. In contrast to light, they are also partially trapped within the multi-layered canopy around the earth known today as the ionosphere. This initially attracted researchers and inventors to the low frequency end of the spectrum, where they could bounce waves of thousands to hundreds of meters long off the canopy for long-range communications. Into the period of the World War I, sparking devices generated such waves. (This accounts for the way the German word for spark, *Funk,* is still embedded in terms dealing with radio communications such as *Rundfunk* for broadcast radio and the corporate name *Telefunken.*) Such devices offered an on-off form of communication and were well-suited to Morse code at sea, on land as a backup to wire telegraphy, or even in airships with multiple crew members where someone could operate the wireless telegraph. The increasing importance of single-seater aircraft encouraged the development of electron tubes that amplified signal strength and made wireless voice telephony possible. These "thermionic valves," as the British called them, offered stable continuous frequencies, as well as ways of modulating a sound wave onto a carrier wave. The advantages of voice communication that would keep both hands of a pilot free were obvious. The naval requirement for radio voice communication was less pressing, even where wireless was useful, such as in the case of convoys or submarines.

In general the Germans made more extensive use of radio on both sea and land than the British in World War I; they believed that it offered a way to transmit commands simultaneously to as many units as had receivers without the bottlenecks and delays caused by telegraph relays or telephone switching stations. They relied on encryption rather than radio silence to mask their communications. What they did not realize was that the British were particularly effective in direction-finding and other signals intelligence techniques, as demonstrated by their success in deducing the movement of the German High Sea Fleet out of its ports; in May 1916 that British advantage resulted in the Battle of Jutland and almost in the destruction of the German fleet. On the other hand, the Germans effectively confused the tactical situation by transferring the call sign of their flagship to shore before sailing.[2] The basic electronic war-

---

[2] Tony Devereux, *Messenger Gods of Battle: Radio, Radar, Sonar: The Story of Electronics in War* (London, 1991), pp. 44, 60–68.

fare pattern of measure, countermeasure, and counter-countermeasure was already apparent. The Germans actually had the opportunity to obtain radar capability at this time, but when the German engineer Hans Dominik shared the reports of his tests of a *Strahlenzieler* ("ray-aimer") with the Imperial Navy in February 1916, the navy responded that the device still needed six months' development and therefore would not be useful in the war.[3]

When the Great War ended, radio development continued apace. Growing commercial possibilities tugged developments along despite the economic ups and downs of the postwar era. A wide range of manufacturers, products, and techniques for continuous wave propagation emerged. By 1930 the American technical lead in the immediate postwar years had begun to dissipate, but the Americans and British were ahead of the French, Italians, and Germans in market development.[4] When the Nazis came to power in 1933, they mass-produced simple and inexpensive sets, known as the *Volksempfänger* ("people's receiver") with which to hear the voices of Hitler and Goebbels. In all these countries, radio broadcasting and reception moved from the phase of innovation to that of diffusion.

It is important to pause briefly to explain terminology. Historians of technology often analyze patterns of change in phases: invention, research, development, and innovation follow one another in a complicated sequence with various feedback loops.[5] Innovation typically involves market considerations, which for present purposes one can replace with mission considerations. Furthermore, one should make the distinction between innovation as the spread of the new "best practice" and diffusion as a closely connected, later phase designating the spread of the new "average practice." Adaptation is primarily associated with the innovation phase, while the introduction of new military doctrine is in general closely associated with the diffusion phase.

Here it is also useful to distinguish technical, operational, and technological change during each of these phases as used in this essay. One can understand "technical change" as a matter of *equipment* or physical devices. When a new radar set or specifically designated system is men-

---

[3] Fritz Trenkle, *Die deutschen Funkmessverfahren bis 1945* (Stuttgart, 1979), p. 16. This presages a similar attitude evinced by German leaders in the early years of World War II.

[4] See Hugh G.J. Aitken, *The Continuous Wave: Technology and American Radio, 1900–1932* (Princeton, 1985); and Gordon Bussey, *Wireless: The Crucial Decade: History of the British Wireless Industry, 1924–34* (London, 1990).

[5] See Thomas P. Hughes, "The Development Phase of Technological Change," *Technology and Culture*, 17, No. 3, July 1976, pp. 423–431.

tioned, technical change is involved. "Operational change" designates the new function of sets or systems and the *procedures* for their collective employment. To envision radar as the technique of detecting targets by means of radio echoes, generating a range of devices and practices, is to focus on operational change. "Technological change" connotes the new set of parameters, that is, the new *context*, emerging from the interaction of technical and operational change with each other and with the environment. To understand radar as transforming the context of combat is to consider the emergence of a new logic. Technological innovations are thus understood as changes in the environment for determining the best practices involving mission considerations.

These distinctions are as arbitrary and overlapping – yet as useful – as the analogous distinctions among tactics, operations, and strategy in military theory. A simple schematic lays out some of the relationships:

Context	Technological change	Strategy
Procedures	Operational change	Operations
Equipment	Technical change	Tactics

The sequence of phases from invention through diffusion can appear at each level of change, which means there are three levels at which different forms of adaptation can occur.

It is also useful to note the role of strategic goals as perceptual filters. The Germans emerged from World War I with the desire to challenge the international *status quo*. Their policy was essentially a grand strategic offensive, in which aircraft, submarines, and tanks – the most potent offensive weaponry to emerge from the Great War – were key elements. Radio communication greatly enhanced the effectiveness of these weapons systems by providing a means of command and coordination of fast-moving or far-flung formations. Although the Germans maintained their First World War preference for radio telegraphy aboard bombers and ships, including submarines, they placed great emphasis on equipping their single-seat aircraft and armored units with voice radio sets and radio networks to link units and echelons. They also continued to place their faith in effective encryption techniques.[6] Radar, perceived as fundamentally defensive, received some technical, less operational, and even less technological attention until well into the Second World War. The Germans planned to be on the offensive and to know by radio where their forces were. They also expected to use signals intelligence to determine the locations of enemy forces.

The British, meanwhile, were on the grand strategic defensive after

---

[6] Devereux, *Messenger Gods*, pp. 58–71, 78–79.

World War I. They placed great emphasis on signals intelligence, which had already proven its worth. For this reason, they also expected their own and enemy policy to emphasize radio silence. Radar offered a form of intelligence of great defensive value and the British perceived it as a technical, operational, and technological response to German threats. British radar systems became technical countermeasures to aircraft and submarines. Radar employment patterns became operational counter-measures to German bomber or wolfpack deployment procedures. And radar's ultimate purpose was a technological countermeasure – a form of grand strategic response – to the technological changes in the mobility and velocity of combat made possible by the radio-coordination of enemy air and sea offensive formations.

At least until 1939, and in some ways until 1940, the Americans also had a stake in maintaining the international *status quo*, but acted as if they did not realize it. In effect, American policy was grand strategic indifference. The result was slow and unfocused adaptation to the changes wrought by both German and British radar developments, until it appeared in 1940–1941 that Germany would be successful in altering the international con-figuration of power. The Americans only then lurched toward a more de-finitive strategy and began to accelerate technological innovation. They were finally galvanized onto the grand strategic offensive by Pearl Harbor.

THE GERMANS

In this as in other fields, the victors usually write the history, so that the British are remembered as the great innovators of radar. The magnitude of the victory in the Battle of Britain overshadows other story lines. And when it became expedient in 1942 to respond to curiosity about British successes without revealing more highly secret signals intelligence tech-niques, Robert Watson Watt was promptly, semi-officially, and a bit pre-cipitously dubbed the "Father of Radar" for his pivotal role in the British research program. His self-congratulatory memoir ("By Radar's Greatest Pioneer") did nothing to dispel the notion that he was its inventor and certainly ignored both German and American precursors.[7] Winston Churchill's evocative images of the "Wizard War" and the "Battle of the Beams" have also contributed powerfully to postwar legends.[8]

---

[7] Robert Watson-Watt, *Three Steps to Victory: A Personal Account by Radar's Greatest Pioneer* (London, 1957).

[8] Winston Churchill, *The Second World War*, vol. 2, *Their Finest Hour* (Boston, 1949), pp. 381–397.

In terms of *technical* change, the claim to key innovation belongs more properly to the Germans than to the British. This should come as no surprise given the close connection of science and technology in radio research during the interwar period and German industrial and scientific prowess. The Germans produced the broadest range of radar sets, with finer resolution, better capabilities, more rugged construction, and greater versatility than anyone else before the outbreak of the war. They were also the first to explore systematically centimeter wavelengths (today usually called microwaves), which turned out to be the most crucial portion of the frequency spectrum for radar developments later in the war.

Yet clearly the Germans did something wrong. In fact, they did many things wrong, the most important of which was to overemphasize technical innovation and take operational and technological innovation for granted. They also produced a gaggle of competing agencies and research programs that did not communicate well with one another. And they became complacent, inferring their own superiority from the intelligence available. In some ways they were the victims of their own successes and positive reinforcement of their preconceptions. Once they awoke, they again forged ahead technically and began to catch up operationally and technologically, but it was too late to retrieve the desperate situation in the closing stages of the war.

The German mixture of technical advance and operational lag makes for an instructive case, even if we look only at the highlights. The early advances in equipment derived largely from the program run by Rudolf Kühnhold of the German Navy Signals Research Division in Kiel.[9] Using an analogy with an echo sounder he had worked with in 1929, Kühnhold suggested in April 1933 that centimetric waves might provide radiolocation of surface vessels, and perhaps even aircraft. The Germans then conducted experiments at a continuous wavelength of 13.5 centimeters, but the transmitted power was too weak. A magnetron valve (a vacuum electronic device controlled by a magnetic field) acquired from Philips Corporation of Holland generated greater power at 13 cm, but was unstable. By employing a Yagi directional antenna at 48 cm, promising enough results occurred for Kühnhold to approach the Telefunken Company in early 1934 to suggest ways to expand the research program.

Unfortunately for the Germans, a permanent split emerged from the effort to gather together the small number of experts in early radar research.

---

[9] The following paragraphs are based on Pritchard, *Radar War*, pp. 32–49; S.S. Swords, *Technical History of the Beginnings of Radar* (London, 1986), pp. 92–101; and Trenkle, *Funkmessverfahren*, passim.

(The Germans began to call such research *Funkmess*, or "radio measurement.") Wilhelm Runge ran the radio receiver laboratory at Telefunken; his interests lay in decimeter waves for relay or point-to-point working. Kühnhold approached this mid-level manager in person. Knowing that no one in the company was working on centimeter waves and that the firm was still laboring under the budgetary constraints of the depression, Runge responded that he had neither the staff nor the means to collaborate with Kühnhold. The latter took this as a rejection by the entire company, and with a younger co-worker and some others founded a firm to do the needed work for the navy, a company which became known by its acronym GEMA. This consumed considerable time. Runge soon began his own able investigations within Telefunken, but without significant backing.

The naval research group made reasonably good progress in two directions, surface vessel and aircraft detection. By October 1934 it had won initial financial support from the Torpedo Research Establishment and begun to conduct its work under military secrecy. One technical problem in early radar was that transmitter output tended to swamp the echo from the target, so that transmitters and receivers had to be kept apart. By May 1935, the Germans had achieved a pulsed transmission technique with pauses to "listen" for the faint reflected signal, which allowed them to place the receiver close enough to the transmitter to make on-board naval equipment practical. A January 1936 demonstration before the commander of the navy, Admiral Erich Raeder, led to funding on the order of 160,000 Reichsmark (approximately $40,000), primarily from the Naval Ordnance Office with the idea that radiolocation would enhance target acquisition for torpedoes. Added to this for the year 1935–1936 was an additional 70,000 Reichsmark from the Luftwaffe. These were certainly nontrivial sums, but they dwindled rapidly in comparison with the huge capital outlays for ships allowed under the June 1935 Anglo-German Naval Agreement. GEMA had actually requested nearly two and one-half million Reichsmark for its research program.[10] Its efforts puttered ahead.

A crucial development was that, in order to extend the range of the radar systems, research had moved in the direction of longer wavelengths. Thus the GEMA set, completed in February 1936, operated on 1.8 m (165 Mhz). When tested, however, the operation improperly aligned its Yagi antenna, and instead of finding the intended surface vessel at 1.5 km,

[10] Frank Reuter, *Funkmess: Die Entwicklung und der Einsatz des RADAR-Verfahrens in Deutschland bis zum Ende des Zweiten Weltkrieges* (Opladen, 1971), pp. 23–24.

it detected an aircraft at a range of twenty-eight km. This GEMA set was returned to Berlin and reconfigured to 2.4 m (125 Mhz) with the Yagi replaced by a rotatable array of dipoles in front of a wire mesh designed to concentrate the beam. This was the forerunner of the "Freya" (A-1) series of aircraft early-warning sets, eventually mounted on an anti-aircraft artillery platform that functioned something like a swivelling floodlight. Due to ground reflection it could not effectively determine the elevation of a target aircraft. But it could quite accurately fix the range and bearing, and was a rugged, mobile piece of equipment for the time. With the antennas aligned properly and with further refinements, the related *Seetakt* (naval tactical) systems, starting with the 60 cm version on board the *Graf Spee* and some smaller vessels in 1938, could detect capital ships at ranges of 15-20 km with a bearing accuracy of + or − 3 degrees.

Meanwhile, Runge was following his own course with funds squirreled away from other projects. In summer 1935 he was able to measure the reradiated power "reflected" from an aircraft. This led him to pursue fire-control radar. Using common screening wire, by 1938 he had managed to construct a parabolic reflector antenna with a diameter of 3 m and to devise a method of conical scanning needed for automatic tracking of an aerial target. This was the beginning of the "Würzburg" series at the remarkably high frequency of 565 MHz, or 53 cm. They provided excellent gun-laying radars for the Flak artillery (*Flugabwehrkanone*) that gave readings of elevation as well as range and bearing at intermediate distances. The Würzburg systems would also be used in the role of ground-controlled interception.[11] If the "Freya" could be compared with a floodlight, then the "Würzburg" was more like a searchlight.

It may seem astounding that Runge remained in obscurity until 1938, but one indication of the tenor of the time is suggested by Luftwaffe General Ernst Udet's alleged response to Runge's explanation of his work. When Udet learned that the radar would locate an aircraft in a fifty km area at night or through fog, he reportedly exclaimed: "If you introduce that thing you'll take all the fun out of flying!" Runge noted in his memoirs that, of course, anti-aircraft fire was not exactly there to protect aircraft; moreover it was under control of the army rather than the air force in those days, "so perhaps his reaction was understandable, if not excusable."[12] One can extract from this episode not only the distress of a van-

---

[11] The best source on the family of Würzburg devices is Arthur O. Bauer, *Deckname "Würzburg": Eine Beitrag zur Erhellung des Geshichte des geheimnisumwitterten deutschen Radargeräts 1937–1945* (Herten, 1992).

[12] Pritchard, *Radar War*, p. 64.

ishing breed of World War I pilots, but a sense of the pervasiveness of interservice rivalry and the visceral preference for the offensive spirit over technical advances in the defense in the Third Reich.

By 1938 the number of firms and individuals with radar research projects had begun to expand considerably. The Lorenz Corporation, a small vacuum tube and antenna manufacturer, had come up with another fire-control radar, the A-2 "Kurfürst," and Siemens and Telefunken became directly involved in radar set production. In October 1938, in the aftermath of the Czech crisis, the director of the Luftwaffe's Signals Research Establishment, General Wolfgang Martini, borrowed a "Freya" set from the navy and was impressed with its aircraft detection capabilities. Not to be left behind by the navy, the air force set out to acquire its own devices. In March 1939 GEMA completed the first of fifty production-model "Freya" sets; the air force had actually ordered 200, but GEMA did not have the necessary production capacity. From 1938 onward, civilian scientists were brought in under military contract to investigate specific problems, but not to set the agenda for research.

Despite all this activity, there were a number of critical developments that undermined larger efforts. One was the abandonment of centimetric radar research in favor of longer waves. The utility of such wavelengths was obvious if sufficient power could be generated – shorter wavelengths meant both smaller antennas and better target resolution. Kühnhold's halting efforts in 1933 to exploit this part of the frequency spectrum, however, had indicated the difficulty in achieving the requisite transmission power and thus range. Furthermore, the equipment was not as stable as the Germans thought desirable, leading one participant in the German programs later to observe that these negative results constituted the fateful "first step toward longer waves."[13]

Given difficulties in establishing GEMA, interservice rivalries, and sufficient success with other pieces of equipment, there was little incentive to persist with shorter wavelengths. Furthermore, some experts believed that not enough energy would be reradiated off the target with centimetric waves, suggesting that the waves would behave enough like light to be "mirrored away" due to the angles on ship superstructures and aircraft configurations. In January 1939, the official decision was promulgated that all efforts were to be concentrated on the longer wavelengths; thus the Germans restricted research to specific lines. With the "Freya" and "Würzburg" families of radars the German military complacently

---

[13] Leo Brandt, *Zur Geschichte der Radartechnik in Deutschland und Grossbritannien* (Genoa, 1967), p. 14. All translations are by the author unless otherwise indicated.

thought in 1939–1940 that neither dramatic improvement nor extensive research projects were necessary.[14] All German radar sets until the middle of the war operated on medium or long wavelengths (53 cm down to 2.6 m). The great Allied breakthrough was to push above 10 cm.

Another crucial development was the failure of the Germans to use technical advance to spur operational innovation. They simply retained older practices with innovative equipment long after they should have changed their procedures. For example, when defensive, long-range "Freya" radar sets emerged as an invention from the German laboratories and then as an innovation from the manufacturers, they were viewed as an enhancement or replacement for the ground observer corps. R.V. Jones was in charge of assessment of German technical innovations for the British during the war. He has astutely observed:

> German philosophy ran roughly along the lines that here was an equipment which was marvelous in the sense that it would enable a single station to cover a circle of radius 150 kilometers and detect every aircraft within that range. Thus it could replace a large number of Observer Corps posts on the ground, and so was a magnificent way of economizing in Observer Corps. Moreover, where we had realized that in order to make maximum use of the radar information the stations had to be backed by a communications network which could handle the information with the necessary speed, the Germans seemed simply to have grafted their radar stations on to their existing observer corps network which had neither the speed nor the handling capacity that the radar information merited.[15]

Thanks to some perceptive individuals in key positions and an overriding sense of urgency about the need for warning about numbers and direction of impending attacks from the air, the British conceptualized more effective operational ways to employ their new devices.

A third crucial development for the Germans was a series of events early in the war that reinforced the direction of their errors. A line consisting of eight partially overlapping "Freya" stations and three naval gunnery radars was under construction in fall 1939 to give approximately 100 km warning of approaching Allied aircraft. On the second day of the war, the British directed a bombing raid at the military port of Wilhelmshaven on the North Sea. A German radar picked up the attacking formation in timely fashion, but the defenders failed to act upon infor-

---

[14] Ibid., p. 35.
[15] R.V. Jones, *The Wizard War: British Scientific Intelligence, 1939–1945* (New York, 1978), p. 199.

mation quickly enough. The German fighters only rose, guided by a "Freya" operator, as the British were heading for home. Successful detection led the Germans to believe that their decision to invest effort in use of longer wavelengths was correct. It also led them to believe that ground-controlled interception would ultimately work, and that there was no reason to face the problems with shorter wavelengths that would have to be resolved if radar were to be placed on board aircraft for airborne interception. When the British returned on 18 December, the Germans were ready to heed the warning given by the "Freya" sets. Only ten of twenty-four Wellingtons made it home.[16] This success encouraged the Germans to fall in the direction they were already leaning, toward belief in the primacy of technical innovation. They continued to be pushed by further evidence such as the inferiority of a British 4 m set captured at Dunkirk. Their errors would cost them the Battle of Britain and much more.

THE BRITISH

A more extensive treatment of interwar British achievements than of German is necessary to explain the interwar basis for the British victory in the Battle of Britain. Despite strong financial support for radar research from its inception in England, the British lagged behind the Germans in technical change until 1940. They more than made up for the deficiencies of their systems, however, by the manner in which they employed them. Therefore, they can properly lay claim to the key innovations in terms of *operational* change. The distinction was understood and articulated by Churchill in his memoir of the Second World War. In describing the German efforts before the outbreak of war in 1939 to discover the existence of British radar, he explained:

> The Germans would not have been surprised to hear our radar pulses, for they had developed a technically efficient radar system which was in some respects ahead of our own. What would have surprised them, however, was the extent to which we had turned our discoveries to practical effect, and woven all into our general air defence system. In this we led the world, and it was operational efficiency rather than novelty of equipment that was the British achievement.[17]

This is an accurate assessment of the general situation and displays a keen perception of the contrast between technical and operational leads in tech-

---

[16] Brandt, *Geschichte der Radartechnik*, pp. 18–19; Devereux, *Messenger Gods*, p. 117.
[17] Winston Churchill, *The Second World War*, vol. 1, *The Gathering Storm* (Boston, 1948), p. 156.

nologies, namely novelty of equipment versus effective adaptation of military thinking to perhaps less advanced devices.

Unfortunately, as in many other portions of his account, Churchill offers a rather distorted sense of events as actually perceived at the time. In particular, the British did not know that the Germans had a radar system of their own, much less that it was more technically efficient, although less operationally advanced than the British system. Churchill does not say that the British mistakenly believed in 1939 that they were both technically and operationally ahead of the Germans. They, too, drew a wrong conclusion from the 5 September raid on Wilhelmshaven: they inferred from the slow response of Nazi fighters that the Germans had no radar at all.[18] This proved to be an expensive mistake in the December raid on the same target, and would be magnified in many other raids before the British recognized German capabilities in the terms Churchill so lucidly used after the fact.

By the end of World War I, British military planners had two interlocked strategic nightmares to confront: the threat of submarines past, and the threat of bombers future. The first specter threatened to reinforce the consequences of Britain's position as an island by isolating it from the resources of the remaining world. The second threatened to destroy Britain's isolation by invasion of the skies overhead, as well as the viability of naval power on which rested defense against the submarine threat. Ultimately the British would best both threats by employing radar.

The initial technical response to the submarine threat was to continue the line of development laid down in the victory over the German U-boats, namely asdic. This was a system, in Churchill's evocative words, "of groping for submarines below the surface by means of sound waves through the water echoed back from any steel structure [it] met" and which was "the sacred treasure of the Admiralty."[19] The cultivation of such devices and procedures for employing it had required large annual expenditures and the training of thousands of officers and men. Asdic would prove indispensable for meeting the submarine danger in the Second World War, but it would also prove insufficient. Technical advances in U-boat and torpedo construction gave the Germans operational capabilities that necessitated finding U-boats at even greater ranges than this "sacred treasure" could achieve. Furthermore, the use of sound aided the underwater vessel more than the surface ship, for the active sonar range even with technical advances of the interwar period was on the order of 1.2 km, while

[18] Devereux, *Messenger Gods of Battle*, p. 78; Brandt, *Geschichte der Radartechnik*, pp. 17–21.

[19] Churchill, *The Gathering Storm*, pp. 163–164.

the submarine's passive hydrophones could detect vessels at more than ten times that range. Ultimately, airborne radar would prove crucial for anti-submarine warfare, because the aircraft could make better use of advances in electronics than the submarine, and even such radar as was available before the advent of centrimetric wavelength devices made a difference.[20]

While the Royal Navy pursued sonar techniques to meet the submarine threat, there seemed to be no defense whatsoever against the bomber. The Germans had launched raids on Britain by means of both dirigibles and bomber aircraft in the Great War; these attacks had killed approximately 1,500 persons and injured approximately 3,400. The result was demoralization and hysteria that affected production and daily life. But defenses improved, so that the Germans abandoned daylight raids at the end of August 1917. Gradually blackout restrictions, better emergency responses by authorities, and a reasonably effective ground-observer warning system emerged by the time the Germans ceased the attacks in May 1918. But the psychological and political impact was enormous. The mere newspaper suggestion in 1922 that the French had a bomber force greatly outnumbering the Royal Air Force fighters was enough to provoke a serious panic in the public. A new report by the air staff concluded in 1925 that an air offensive by the French would cause more casualties in a matter of days than the Germans had inflicted in the entire war, with no prospect of an effective defense.[21]

It was no wonder that serving officers and civilians alike believed the conservative leader Stanley Baldwin, when he proclaimed in the House of Commons in November 1932 that "the bomber will always get through."[22] Baldwin was arguing for disarmament as the only feasible response, while the Royal Air Force was committed to deterrence by means of its own bomber force. The idea was not that air attacks would necessarily obliterate enemy cities, but that they could destroy the enemy air fleet on the ground in the home country, along with its bases. Disarmament negotiations in Geneva collapsed over many issues, including the dual use questions concerning civilian aviation and the resurrection of offensive attitudes in Germany. Although commitment to bombers constituted Britain's only real defense in the skies, not until 1936 did the British

---

[20] Devereux, *Messenger Gods of Battle*, pp. 196–199.

[21] John Terraine, *The Right of the Line: The Royal Air Force in the European War, 1939–1945* (London, 1985), pp. 9–11; Marian McKenna, "The Development of Air Raid Precautions in World War I," in *Men at War: Politics, Technology and Innovation in the Twentieth Century*, ed. by Timothy Travers and Christon Archer (Chicago, 1982), pp. 173–195.

[22] Terraine, *Right of the Line*, p. 13.

organize a separate Bomber Command (as part of a general reorganization also erecting Fighter Command, Coastal Command, and Training
Command). During the same period, however, the RAF's acquisition community let new specifications for fighter aircraft, from which would come
the Hawker Hurricane and Supermarine Spitfire.

In order to send the prospective new fighters against enemy bombers,
those moving targets would have to be located. The only proposed
method was a network of concave, acoustically molded walls or "mirrors" that would use microphones to amplify the noise of distant aircraft
for ground observers. Since ambient sounds were also amplified and since
the increasing speed of aircraft made sound-ranging impossibly obsolete,
this suggestion was obviously impractical. The abandoned 200-foot long,
25-foot high test wall – facing Paris – still stands in Southeastern England
as mute testament to the desperate resignation that gripped Britain in
those years.[23] A critically important boundary emerged between those
who believed that the bomber must always get through and those who believed that it must *not* get through. The former included the vast majority of the RAF, for, as one writer has noted, "Since its very earliest days the
belief in the offensive role of the service had possessed religious force, with
Bomber Command as the priesthood."[24] Those who sought to resist the
bomber included Air Vice-Marshall Hugh C.T. Dowding, who both headed Fighter Command and earlier served as Air Member for Research and
Development. The countercurrent around Dowding created the conditions for rapid change once a new idea emerged.

As a consequence of the tension between adherents of the bomber and
the fighter, the offensive and the defensive, the situation in Britain was
fundamentally different than that in Germany at the time. German fighters were designed to escort bombers in total commitment to the offensive,
and Göring's Luftwaffe and Four-Year Plan offices controlled much of the
pace of rearmament. Radar remained, in contrast to the use of radio, a
peripheral innovation until conditions forced Germany onto the strategic
defensive in 1942. In this sense the situation was insulated from change
and remained structurally stable. In contrast, Britain's fundamentally defensive grand strategy created a hearing for the defensive, however overshadowed by conventional wisdom and airpower doctrine.

A situation of instability thus existed within the British government
and military, so that a seemingly trivial event triggered a cascade that was

23 David E. Fisher, *A Race on the Edge of Time: Radar – the Decisive Weapon of World
    War II* (New York, 1988), p. 23.
24 Terraine, *Right of the Line*, p. 77.

quite remarkable and worth following in some detail. The British had created the Directorate of Scientific Research within the Air Ministry as a response to the bomber scare of the early 1920s. It comprised exactly two employees, and in 1934 the junior of them (accompanied by Dowding) went to see the acoustical wall that was supposed to offer air defense warning. A.P. Rowe came back to the ministry and on his own initiative looked up all the files offering suggestions for air defense. He wrote a memo to his superior that unless science could offer something new, any war in the next ten years was hopeless. His superior also exercised bureaucratic initiative. H.E. Wimperis sought advice over lunch from his biologist friend A.V. Hill about the rumors concerning "death rays" that might boil the blood of enemy aircrew members and then wrote to his immediate superiors (who included Dowding) suggesting the formation of a committee. They approved. The small committee of "outside" experts and the two bureaucrats, under the leadership of Henry Tizard, a civilian adviser to the Air Ministry and once a Royal Flying Corps pilot, was to meet on 28 January 1935.

How did the Tizard Committee come to be pivotal in the British development of radar? For one thing, it was obscure enough to be out of the political in-fighting that raged over budgets and doctrine and thus could concentrate on the task of thinking through the problem of air defense anew. For another, its members were extraordinary men. Tizard, trained as a chemist, had extensive experience in the postwar period working on the boundaries between government agencies and the academic and industrial communities. The other two members of the committee were academics: Wimperis's friend Hill was a Nobel Prize-winning biologist, and P.M.S. Blackett was a later Nobel laureate in physics. Hill had army anti-aircraft artillery experience and Blackett had navy experience. Their backgrounds and credentials give some sense of why these men were open to new thoughts, and the actions of Rowe and Wimperis had positioned them to make a difference. For a third, Dowding was anxious to have new, workable ideas, and the Tizard Committee provided them on more than one occasion.

In order to get the first committee meeting moving promptly, Wimperis had asked another key figure to prepare a report on the feasibility of electromagnetic "death rays." Robert Watson Watt (a hyphen between the last two names was added when he was later knighted) was Superintendent of the Radio Research Laboratory at Slough. After a meeting with Wimperis on 18 January 1935, Watson Watt had transposed the issue into a straightforward problem for his junior co-worker, Arnold F. Wilkins: "calculate the radio-frequency power which should be radiated to raise the temperature of eight pints of water from 98 degrees F to 105 degrees F at a distance of five km and at a height of one km."[25]

Wilkins was clever enough to guess what this was about and calculated that a huge amount of power would be needed just to generate a fever in an airman's body, even if it were not shielded by the metal fuselage of an aircraft. Then Watson Watt asked him the right question at the right moment: what else might be of use to the people at the Air Ministry? Wilkins remembered an incident back in 1931 when he had visited post office engineers at Colny Heath who were conducting communications experiments. In 1925 Gregory Breit and Merle Tuve in the United States had directed radio-frequency beams into the night sky to explore the ionosphere, measuring its altitude by the time it took at the speed of light for the beam to return to earth (.0007 seconds, for an altitude of 130 miles). The post office team was using this technique to test the properties of various frequencies, but sporadically they would get unexpected results of much shorter time for the return of the beam. Someone told Wilkins over a cup of coffee they had realized that the anomalies were due to echoes from aircraft operating out of de Havilland's nearby Hatfield aerodrome. One part of their research report of June 1932 was in fact titled "Interference by Aeroplanes."[26] Wilkins now suggested to Watson Watt that this phenomenon might be useful for detecting aircraft. In his report on the impracticality of electromagnetic death rays, Watson Watt pointed out that before one could destroy a target, one would have to detect it and that he might have a suggestion for how to accomplish this task.[27]

When the Tizard Committee read Watson Watt's report, their interest was immediately piqued by his tantalizing suggestion, and they asked him to elaborate. He turned once again to Wilkins, and the result was an historic document of 12 February 1935, entitled "Detection and Location of Aircraft by Radio Methods." This memorandum laid out the technical development and innovation of radio direction finding (RDF), as the British called radar for some time, and even indicated clearly some lines of operational innovation.[28] It specified how "zones of short-wave radio illumination" could be set up through which aircraft could be made to fly. A wing-span of 25 m would function as a linear oscillator with a resonant

[25] Fisher, *Race on the Edge*, p. 31.   [26] Ibid., pp. 44–49.

[27] The text of the report for the January 28 meeting is given in Swords, *Beginnings of Radar,* pp. 278–281.

[28] The February 27 text of a final draft of the report is printed in Swords, *Beginnings of Radar,* pp. 281–285; and Watson-Watt, *Three Steps to Victory,* pp. 470–474.

wavelength of 50 m, reradiating a small but measurable amount of energy back in the direction of the transmitter. A choice would have to be made between a weak floodlight effect and a more focused but narrow searchlight effect. The fields involved would be about 10,000 times the strength of commercial communication, raising safety issues. Pulsed signals such as those already used in echo-sounding techniques for the atmosphere could be projected on a cathode-ray oscillograph calibrated with a linear distance scale. A line of such senders over a long front could be erected, and a coordinated double line would offer even greater precision in the location of aircraft. Since the ionosphere reflects radio waves down to earth again, secrecy would be a problem, so perhaps experiments should be conducted at a station such as Watson Watt's that was already conducting ionospheric research to serve as a cover for new work. Furthermore, Watson Watt anticipated the value of a central control room for the position plotting of data from multiple stations and suggested the use of transponders on board friendly aircraft to identify friend versus foe.

All in all, Watson Watt's new report was a startling and prescient document. No wonder the members of the Tizard Committee were elated – here was the first practical idea for aiding in the air defense of Britain, complete with data, calculations, and suggestions for locations of research sites and programs. Watson Watt, however, had no idea of the boundaries within the military his suggestions would require to be crossed; this was perhaps a crucial element of the expansiveness of his thoughts and the interconnected vision of an operational system that they offered.

The cascade that had begun with Rowe's visit to the acoustical mirror, including the fortunate accident of Wilkins's memory of a chance visit to Colny Heath, soon swelled to an active stream of activity. Although radar could have been invented any time after the experiments of Breit and Tuve, it took the right question asked at a propitious moment to trigger the key steps to research and development of the new techniques. Nearly everyone brought into contact with the highly secret project immediately realized its value for air defense. One of the most crucial was Dowding, who was asked already in February for £10,000 by Wimperis to execute Watson Watt's plan. Dowding quite sensibly demanded a practical test, got it on 26 February and promptly authorized the funds from his limited resources. A small research center was immediately set up at Orfordness on the coast northeast of Colchester and a four-man team headed by Wilkins went to work.

The small number of persons involved in the crucial 1934–1935 period in Britain is typical of the early phases of major technological change. Resources were limited, but at no time was British radar delayed due to financial starvation. Money was not the only scarce resource, and often advance was blocked by lack of a specific type of tube or wire or circuit, which had to be found or invented. Wilkins and his team were ingenious and dedicated, a sign of the allure of the innovative nature of the problem. They were also goaded along by Watson Watt, who preached what he called the "cult of the imperfect," which he explained as the view that one should "Give them the third best to go with; the second best comes too late, the best never comes."[29] Many was the occasion on which the team escaped a bottleneck by acquiring off-the-shelf components, many of them made in the United States. They depended on a broad commercial-industrial base, including the beginnings of television technology.[30]

By all accounts there were four crucial personalities in the early stages of British radar development, three of them positive influences and one negative. Dowding was indisputably the pivotal military figure, providing the pull toward new operational developments and innovation. He took a strong personal interest in radar research and development and even flew in the research aircraft to see the project's progress for himself. He also insisted that military personnel be posted right with the "Boffins," as the civilian researchers became known. This insured that the RAF personnel actually understood what was happening and that the civilians could be kept aware of military constraints and needs.[31] Furthermore, the basic tactics and requirements for night airborne interception were his own ideas.[32] Dowding was no orthodox thinker, and perhaps understood the technological implications of radio better than any other figure in the RAF. When a few years earlier he had been in charge of fighter defenses in a summer air exercise, he quietly posted radio vans under the likely paths of incoming bombers. As the bombers passed overhead, his observers would radio the news, so he always managed to have his fighters in position to receive them (as he was to do using radar in 1940). When the umpires changed the rules on him, he decided to place a wireless on board one of his fighters (unusual for the time) and ordered it to follow the bombers back to base and radio their position – which he then attacked as the bombers were on the ground refueling, destroying the entire "enemy" force.[33] The central location that was set up for radar information and fighter control at Bentley Priory produced the kind of qualitative change that was typical of his innovative mind.

[29] Watson-Watt, *Three Steps to Victory*, p. 74.
[30] See E.G. Bowen, *Radar Days* (Bristol, 1987), pp. 33, 37, 50 and 77.
[31] Swords, *Beginnings of Radar*, p. 193.
[32] Bowen, *Radar Days*, pp. 69–74.   [33] Fisher, *Race on the Edge*, p. 60.

Where Dowding provided the mission pull for innovation and diffusion, Watson Watt generated the entrepreneurial push. He was the impresario of radar, hurrying from one research and development location to the next or negotiating among the entities of government and industry. In September 1935, as he reported on progress to date, he proposed the line of stations operating on 13 m that was to become the Chain Home radar defense system for the British Isles. This replaced the projected chain of sound mirrors, which as late as May 1935 was still in the works for construction in 1936, and which was canceled in September.[34] The British government allocated no less than £100,000 (approximately $300,000) in December 1935 for the immediate construction of an initial system of five stations covering the approaches to the Thames estuary. Their construction was to be supervised by Wilkins. In May 1936, the radar team moved to a larger site at nearby Bawdsey Manor, where the prototype Chain Home station was to be built. Watson Watt became superintendent of this station, but also moved to the Air Ministry as director of communications development. Less directly involved in radar development, he remained its tireless salesman and advocate in government circles. By the outbreak of the war, the total cost of radar contracts placed was approximately £10,000,000.[35]

Between Dowding and Watson Watt was Tizard, whose committee repeatedly thought ahead and made crucial recommendations. One of the most important of Tizard's contributions was the Biggin Hill experiment in August and September 1936, in which the station at Bawdsey Manor was incorporated into RAF exercises focused on the Biggin Hill airfield. The exercises were to determine the number of interceptions that could be expected in daytime by using radar location and how close to a bomber it was possible to direct a fighter by ground instructions. Coordination of air warning, a filtering process for cross-referencing and assessing available information, and an efficient communications network for alerting the fighters and guiding them to their targets were developed. Although the experiment was started with only tepid support, during 1936 and 1937 Tizard gained the respect of operational commanders and the experiment generated the basic procedures with which Fighter Command would fight the Battle of Britain.[36]

Tizard also realized that with the interlinked Chain Home radars in place enemy bombers were likely to be driven from the day sky into the night, with consequent urgent need for airborne interception. Ground

---

[34] Swords, *Beginnings of Radar,* pp. 179–180. [35] Bowen, *Radar Days,* p. 28.
[36] Clark, *Tizard,* pp. 149–156.

control could not "vector" fighters (a phrase that came out of the Biggin Hill Experiment) within the 330 meters necessary to see enemy bombers at night. What the RAF needed was a radar small enough to place on board a fighter, and the job of turning a room full of equipment weighing tons into a mobile rig weighing less than a man was turned over to E.G. "Taffy" Bowen, a young physicist who had been on Wilkins's initial four-man team at Orfordness. By September 1937, Bowen's research group had managed to invent a self-contained airborne set at 1.5 m (200MHz) that found the Fleet during exercises when Coastal Command was grounded due to weather, was able to detect other aircraft, and could be used for rudimentary navigation along the coast.[37]

The fourth crucial personality was one of the few disbelievers in the importance of radar, F.A. Lindemann, an Oxford physicist and Churchill's close confidant and scientific adviser, who had been added to the Tizard Committee in summer 1935 at Churchill's insistence. Lindemann thought the program was something of a fraud and firmly opposed the diversion of resources to aircraft detection research rather than aircraft destruction or, preferably, to enhance the capability of bombers. Enamored of exotic devices such as aerial mines, he never understood the kinds of operational innovations that animated Dowding and Tizard. He was a constant thorn in the side of the Tizard Committee and only grudgingly accepted the value of radar when it promised to aid the bombing of Germany directly later in the war, although Churchill writes nothing of this.[38] To get around Lindemann, the Tizard Committee dissolved itself and reconstituted itself without him in October 1936. Due to Churchill's importance for rearmament, Lindemann's influence was far-reaching. Due to the growing urgent need for practical defense, however, even Churchill could not be appeased at the cost of the effectiveness of radar.

As war approached, radar became an ever more effective component of British defenses. By the time of the Munich crisis in autumn 1938, five radar stations were on line with supplementary mobile sets, and in October Fighter Command opened the filter room at Bentley Priory, the nerve center of the air defense system. At the outbreak of war, land lines connected eighteen of the Chain Home stations stretching from the Channel Coast to the Scottish border directly to the filter room. All operated on a single frequency in the band 22 to 27 MHz, which meant only one system of antenna arrays was needed. Furthermore a new system, desig-

[37] Bowen, *Radar Days*, pp. 30–31, 42–46.
[38] See Fisher, *Race on the Edge*, pp. 64–83; Ronald W. Clark, *Tizard* (Cambridge, MA, 1965), pp. 120–148; Bowen, *Radar Days*, pp. 73–75.

nated Chain Home Low, was under development to detect low-flying aircraft.[39]

The technical innovation of radar devices had given way to diffusion. Growing numbers of civilian and military personnel became familiar with the new equipment and the procedures for employing it, and even more were affected while shielded from direct knowledge of this fact by the need for secrecy. Furthermore, as more and more aircraft needed to be fitted with radar sets in the winter of 1939–1940, bottlenecks formed and had to be broken with unlikely measures. One was a change in air fitness rules regarding eyesight that allowed more school teachers to become airborne radar operators. Another was a selective draft of ham radio operators to provide enough technicians to service and repair radar sets.[40] In effect, the British urgency led not only to the use of off-the-shelf components for equipment, but off-the-shelf personnel, too. Another sign of diffusion was that Bowen moved on from work concerning air-to-surface-vessel radars (ASV) for sighting surface ships to the more sophisticated problem of detecting submarines.[41]

The British had completed operational development of radar by the outbreak of war and operational innovation was in full stride. There were still various glitches, however, as evidenced by the "Battle of Barking Creek" on 6 September 1939. One of the technical features of the Chain Home radar stations was that their antennas radiated both forward and backward at the same time. This meant that an operator could not distinguish between an echo from the forward signal and an echo from the backward signal. The technical solution to this technical problem was to shield the reception from the rear to eliminate the ambiguity. During the previous days of the war, the Chain Home system had been detecting incoming objects that seemed to disappear.

On this morning a false warning had resulted in the scrambling of a Hurricane squadron. But when it rose, it appeared on the radar screens as an *incoming* formation, causing another squadron to be scrambled. As these aircraft took off from behind the radar stations, they, too, appeared as spikes on the screen in front of the stations and were therefore identified as enemy, causing yet more units to be scrambled. The rear echelon of a squadron of Spitfires attacked what it thought was Messerschmidts, which turned out later to have been Hurricanes, for there were no German planes at all in this "battle."[42] The operational system had proven

---

[39] Swords, *Beginnings of Radar*, pp. 194–195.   [40] Bowen, *Radar Days*, p. 118.

[41] Ibid., pp. 101–102.

[42] Fisher, *Race on the Edge*, pp. 84–91; Terraine, *Right of the Line*, p. 110; Gavin Lyall, *Freedom's Battle: The War in the Air, 1939–45* (London, 1968), pp. 7–8.

unstable with positive feedbacks driving it out of control, leading to the loss of two Hurricanes and one Spitfire. Fortunately for the British, the technical and operational problems could be corrected before the Germans actually arrived in force in 1940. Part of that solution was better shielding, but part was to depend on the low-tech ground observer corps to report German aircraft on their way home from their targets.

British technological innovation was the outcome of the interaction of technical and operational innovation, and the changed context meant new parameters for both in turn. Dowding and Tizard, a military man and a scientist with the same goals, were the two figures who understood earliest and perhaps best how radar was transforming the context of combat. Their actions blurred the boundaries between technical and operational change, as each pressed for operational development and innovation that would spur technical advance. As new devices became available, they availed themselves of the possibilities, but they did not rely on technical innovation to alter procedures and thinking. In pointed contrast to the Germans, who economized in ground observers, they perceived the full potential of radar for contesting control of the air. As R.V. Jones has reflected:

> The essential point of our radar philosophy was that it enabled us to overcome the fundamental problem of intercepting the enemy not by flying continuous patrols, which would have been prohibitively expensive, but by sending up our fighters so as to be at the right place and the right time for interception. In other words, we regarded the main contribution of radar as a means of economizing in fighters, one of our most precious commodities.[43]

It is important here not to read too lightly over the word "philosophy." Thinking through (or rather among) the implications of fundamental change requires an interaction of practical and philosophic bents of mind.

The new logic of the technological change Dowding and Tizard perceived was a function of the velocity of contemporary combat and the German use of radio to manage faster-moving formations than could be coordinated without radio communication. They sensed the need for a synergistic response to the increased tempo of attack and defense produced by newer aircraft and were together the innovators of the interconnected early warning system, filter room, and command system. As one observer has commented, the individual Chain Home station left

---

[43] Jones, *Wizard War*, p. 199.

much to be desired in comparison with a "Freya" station. Yet centralized control made all the difference:

> Experience has since shown that any chain of defensive radar stations gathers extra strength from being centrally linked. It is able to react more quickly and effectively. The Freya coastal chain reacted rather slowly until increasing defensive urgency compelled central linking.[44]

Soundranging through acoustical mirrors seems almost quaint to us today, but it was a response to bombers that most people could understand. It was a technical innovation within a technological response to an overwhelming weapon that "will always get through." Sound mirrors and the single-minded deterrent strategy of Bomber Command were parts of the same logic. Only Fighter Command's warning at the speed of light, timely assessment of incoming data from warning systems, and economical deployment of defenses could counteract a radio-coordinated threat with a different logic. This logic contributed a winning strategy to the Battle of Britain in 1940–1941.

THE AMERICANS

Lacking the offensive determination of the Germans and the defensive desperation of the British, the Americans lagged both technically and operationally. They were not *far* behind, due to the competitive thrust of American industry and the general technical proficiency of American researchers. Once a sense of urgency emerged, it was not long before the United States was at the forefront of innovation. But the political isolationism that produced a posture of grand strategic indifference also created a context in which early efforts to develop the reflective properties of radio waves wandered off into the void. In the words of one historian, in contrast to the development of radar in Britain as "a definite solution to a pressing problem," radar in the U.S. began only as "a vague answer to uncertain threats."[45]

A brief reference to military technical intelligence offers a sense of the implications of that context for innovation. From the end of the World War I into the 1930s, the Ordnance Department was responsible for the development of new weapons for the U.S. Army. In order to equip the field forces with competitive weaponry, Ordnance needed good reports on for-

---

[44] Devereux, *Messenger Gods of Battle*, p. 106.
[45] David Kite Allison, *New Eye for the Navy: The Origin of Radar at the Naval Research Laboratory* (Washington, DC, 1981), p. 141.

eign military technical and operational innovations. However, it was dependent for such information on military attachés who usually had little technical expertise. The number of ordnance officers with the necessary experience, command of a foreign language, and private income to be posted in a foreign capital was vanishingly small; only nine served in such capacities from 1920 to 1940, and between November 1930 and May 1940 there were only two, one in Moscow and one in London.[46] The army official history noted that as a consequence, "data on details of foreign design and manufacturing methods was so intertwined with military intelligence that what filtered through to the Department was casual and tended to leave research to proceed in a near vacuum."[47] Misinformation (particularly about Germany) abounded, and ignorance was the norm. For example, the army lacked a competitive tank engine, when it thought it had one. And it retained the 37-mm light antitank gun for the sake of its mobility and justified this decision by pointing out that the 37-mm was perfectly sufficient for penetrating American armor.[48] The situation in the navy was not much better. This was the technological environment in which radar would have to emerge.

Thus it is not surprising that the Americans passed up the opportunity to be first with a military radar project. In June 1922 the Italian radio pioneer Guglielmo Marconi spoke in New York and extrapolated from some of his current experiments that reflected radio waves could be used at sea to determine the presence and bearings of other ships.[49] In September of the same year, Albert Hoyt Taylor and Leo C. Young, two engineers working in Washington for the Naval Aircraft Radio Laboratory (later to become the Naval Research Laboratory, or NRL), were conducting 5 m high-frequency experiments. They accidentally found that ships could be detected by reflected radio waves, and, realizing the military possibilities, wrote a letter to the Bureau of Engineering to ask for support. The Bureau never responded and Taylor and Young dropped the idea for nearly ten years.[50]

After the 1925 work of Breit and Tuve on the ionosphere, suggestions

[46] Constance McLaughlin Green, Harry C. Thomson, and Peter C. Roots, *The Ordnance Department: Planning Munitions for War,* in the series *United States Army in World War II, The Technical Services* (Washington, DC, 1955), p. 208.

[47] Ibid., p. 260.

[48] Ibid., pp. 184–185, 203.

[49] Henry E. Guerlac, *RADAR in World War II* (New York, 1987), pp. 41–42.

[50] Allison, *New Eye for the Navy,* pp. 40–41, suggests that the navy may actually have been well served, since any equipment developed at that time would have been crude and cumbersome and perhaps hindered development of more sophisticated devices later.

of how to use radio waves to measure distances to various objects began to proliferate. In 1930, Taylor and another NRL engineer were able to detect the presence of aircraft, and in his detailed memorandum on this fact Taylor referred back to his 1922 suggestions. The naval officer in charge of the NRL endorsed the idea as having great naval value for aerial defense. Yet there were widespread doubts in the Bureau of Engineering that radio detection would be of use on moving vessels. In a January 1932 memorandum to the Secretary of War, the Secretary of the Navy offered the view that this project on airplane echoes might be more appropriate for the army.[51] The NRL managed to retain the project, but did not budget for it separately.

Really encouraging results did not begin to appear until NRL researchers followed a suggestion by Leo Young and shifted from the beat method to pulses in 1934. When the proposal arose in the 1935 congressional budget hearing (only a few months after the Tizard Committee had begun its work) that "death rays" ought to be studied as they were in Europe, the navy could simply ignore the matter in favor of a system that was beginning to work. By spring 1936 a 28.2 MHz rig developed under Robert M. Page had met with such success that a demonstration was given to the Chief of Naval Operations and Rear Admiral Harold Bowen, Chief of the Bureau of Engineering. On 12 June 1936, Bowen ordered the project classified secret; he became one of radar's strongest advocates and later served as director of the NRL.

The key developments now shifted to greater power with a 200 MHz set, and the next step was to place one of these sets on board ship. The commander-in-chief of the First Fleet was keenly interested, and tests were run on the destroyer *Leary* in April 1937. Further successes led to outfitting the battleship *New York* with a 200 MHz model XAF radar in December 1938 for operational use. It provided both detection of aerial targets (at ranges of 100 nautical miles) and surface vessels (at 15 nautical miles), with some capability of spotting the fall of shot and tracking the flight of projectiles.[52] During exercises in January and February 1939, this radar set proved extremely effective, leading the commander of the Atlantic squadron to the view that it was "one of the most important military developments since the advent of radio itself."[53] By May of 1940, the first of twenty warships of the fleet had RCA-manufactured copies of the NRL-designed 200 MHz systems installed.

Given the development of successful equipment, clearly the navy per-

---

[51] Guerlac, *RADAR*, p. 66–67.   [52] Swords, *Beginnings of Radar*, p. 110.
[53] Allison, *New Eye for the Navy*, p. 110.

ceived the value of radar, but there were still major pockets of doubters. One of those was in the Bureau of Ordnance, whose responsibilities included fire-control systems. Effective gun-laying radars would have to use higher frequencies, preferably in the centimetric range. Aside from a liaison officer to NRL, no enthusiasm for this work could be kindled among bureau personnel. Added to general skepticism were the concerns for immediate practicality over long-term research and development and for questions of bureaucratic turf.[54]

Meanwhile, the army had begun its own set of radar projects. Already in 1926 Major William Blair, Chief of Research and Engineering at the Office of the Chief Signal Officer in Washington, had called attention to the inadequacy of sound detection of increasingly fast-flying aircraft and had proposed the study of radio microwaves for the purpose of aircraft detection.[55] In 1931 a project concerned with finding targets by means of electromagnetic radiation was established at the Signal Corps Laboratories at Fort Monmouth, although it initially focused on the infrared portion of the spectrum. Although Blair, with a doctorate in physics, was correct in pushing for microwave research, it was actually too early to generate sufficient power at those frequencies and much effort was frustrated.

The lone Signal Corps researcher assigned to radio detection in the mid-1930s was W.D. Herschberger, a civilian who had once worked at the NRL. Blair was convinced that the navy project was fine for general area surveillance, but not for the kind of precise targeting needed for accurate anti-aircraft fire. This is why he was eager to get to the higher frequencies.[56] In fall 1933, Herschberger generated a proposal for a pulsed transmission with the interval between transmission approximately of the same duration as the pulses. This was rooted in the technique pioneered by Breit and Tuve and differed from the technique Page adopted of much longer listening intervals than pulse durations.[57] Early in 1936 Herschberger visited the NRL, whose reports on radio echo research he regularly read, and learned of Page's technique.

When the navy imposed secrecy on the NRL project in June 1936 Herschberger realized that Page must have hit pay dirt. A month later the Signal Corps redistributed its budget to make $80,000 available for the radar research. In the fall Herschberger left the project to complete his

---

[54] Ibid., pp. 114–117.   [55] Guerlac, *RADAR*, p. 93.
[56] Ibid., p. 95.
[57] At least one German author has pointed to Herschberger as the "inventor" of American radar, noting that one of Herschberger's key sources was a German survey of radio-frequencies for ranging purposes published a few years earlier. Reuter, *Funkmess*, p. 39.

doctorate in physics, but work on a practical prototype forged ahead. In spring 1937 top army figures including the assistant chief of the army air corps (Brigadier General Henry H. Arnold) and the Secretary of War were given demonstrations. Out of this project came several basic radar sets: the SCR-268 (at 205 MHz, for searchlight and gun direction), the SCR-270 (at 110 MHz, a mobile rig for long-range aircraft detection), and the SCR-271 (110 MHz, fixed version of the SCR-270). The priority had shifted from gun-laying to long-range detection with the outbreak of war in Europe. Contracts for all of these were in effect by August 1940.[58]

Before moving on to the decisive event in the American portion of the story of radar, a few words of perspective are in order. In general the American efforts were similar to the German ones, in that technical advances filtered upward from long-standing small bands of researchers rather from a specific administrative decision as in the British case. Another contrast was the way the British "old boy" system of acquaintanceships among men such as Wimperis and Hill or Watson Watt cut through the kind of institutional boundaries that compartmentalized American research. Lastly, like the Germans, the Americans had neither an airborne interception nor ASV program, and, although interested, they were not pressed by defensive necessity to explore more persistently the possibility of centimetric radar. Only with the events of 1940 did a greater sense of urgency appear.

The third stage of radar technical innovation was about to begin in early 1940; it would be responsible for the ubiquity of radar by war's end. The first stage lasted from Hülsmeyer's invention of 1904 through the 1920s and consisted of the recognition that one could use radio wave echoes for detection of large objects such as ships (with continuous beams of any wavelength). The second stage occurred in the mid-1930s as researchers generated focused, pulsed radio beams that one could use like floodlights or searchlights of specific wavelengths to warn of the approach of smaller objects such as aircraft. The third stage was introduction of centimetric (microwave) radar in 1940–1941, which radically shortened antenna size and vastly improved the precision of target location.

The qualitative threshold of centimetric radar was crossed with a single device, the resonant cavity magnetron, whose inventors were British, but whose great exploiters were American. The "magnetron" was a vacuum electronic device controlled by a magnetic field, which had been introduced at General Electric and given its "Greco-Schenectady" name in the 1920s.[59] The technical breakthrough was to construct a set of cavities within which

[58] Guerlac, RADAR, pp. 99–109.   [59] Sword, Beginnings of Radar, p. 259.

electrons could be made to resonate at higher frequencies than the input power would normally lead one to expect. In effect, the resonant cavity magnetron was an electronic whistle, with electrons bouncing around inside in contrast to air and electromagnetic waves coming out instead of sound waves. Its inventors at the University of Birmingham, Henry A.H. Boot and John T. Randall, later wrote that they fortunately did not have time over winter 1939–1940 to read the relevant – and conflicting – literature on magnetrons, or they would have become too confused to move ahead.[60] They therefore went back to first principles and designed the arrangement of cavities based on their precise physical dimensions (much as a long whistle generates a low pitch and a short one a high pitch). The very first laboratory device produced 9.8 cm waves at the incredibly high power level of 400 watts – the device may have been analogous to a whistle, but its output was more akin to a siren. With this device, the British for a time leapt ahead of the Germans in technical innovation.

The invention of the cavity magnetron came too late to play a role in the Battle of Britain that summer, but it was just in time to play a key role in America's preparations for war. In June, as the Germans were mopping up French forces and the British were gasping from their exertions at Dunkirk, the Americans established a new bureaucratic entity. The National Defense Research Committee (NDRC) was the brainchild of Vannevar Bush, president of the Carnegie Institution, but it spoke to the aspirations of many American scientists who wanted to contribute to the national defense in view of the growing Nazi threat. Earlier in the year, A.V. Hill of the Tizard Committee had visited the United States to sound out the political temper. He reported back to Tizard that American scientists were overwhelmingly ahead of the general population in their sympathy for Britain's plight. Tizard convinced the military and civilian leadership, including Churchill now as Prime Minister, that an exchange of technical ideas with the Americans was feasible and necessary. Churchill decided not upon an exchange, but a gift of Britain's greatest technical secrets as the best guarantee of American support.

The Tizard Mission in September 1940 was a brilliant diplomatic stroke that entailed massive technological consequences. The mission was part of a strategic gamble that already included the earlier sharing of designs for the Rolls-Royce Merlin engine that would transform the lackluster P-51 into a superb long-distance fighter, and the provision of crucial information about the splitting of atoms that would greatly stimulate

---

[60] Henry H.A. Boot and John T. Randall, "Historical Notes on the Cavity Magnetron," *IEEE Transactions on Electron Devices*, 23, no. 7, July 1976, p. 724.

American atomic research. In September Tizard's team carried across the Atlantic such items as information on jet propulsion, the proximity fuse, and the Bofors repeating cannon. And it brought along one resonant cavity magnetron.[61]

Tizard, Taffy Bowen, and others of the team met with members of the NDRC concerned with microwave radar research. On 29 and 30 September the British explained their work on radar and displayed the magnetron device they had brought. The response of the Americans was nearly instantaneous, and during the next week meeting after meeting occurred among civilian and military members of the NDRC. By 12 October the shape of American radar research and development for the rest of the war had emerged. Following the English model, the Americans erected a central laboratory staffed primarily by civilian scientists and engineers and devoted to three connected research goals. The greatest urgency was development of 10 cm airborne interception radar. Nearly as important was a precision gun-laying radar. And the third project would be a long-range aid to aerial navigation, in which no signals would be sent out by the aircraft, but by which a navigator could at a range of 500 miles from the aid tell his position within a quarter of a mile. Within the next two weeks, industrial development contracts of nearly $140,000 were approved, and by early February the Radiation Laboratory at MIT received its initial funding of $455,000.[62]

How did the American military personnel and institutions adapt to these changed circumstances, in which civilian scientists would take over direction of technical and in many cases even operational innovation in military technologies? In general, they responded readily to this technological transformation, perhaps because military leaders realized how ill-prepared they were to go to war. There were some objections, such as those voiced by Admiral Bowen of the NRL, who felt that the years of research and development of radar at the NRL were being swept aside, and that his efforts to elevate NRL to the navy's central research agency were being undermined.[63] But overall, the interaction of military personnel and civilians both at the Rad Lab and out in the field was by all accounts constructive and mutually beneficial – as it generally was in Britain and certainly was not in Germany.[64]

American innovation was not just a matter of following the British

[61] For an appraisal of British versus American radar at this time, and thus a sense of what the British were offering, see Allison, *New Eye for the Navy,* Appendix G, pp. 211–214.

[62] Guerlac, *RADAR,* pp. 253–259. See also Bowen, *Radar Days,* pp. 150–178.

[63] Allison, *New Eye for the Navy,* pp. 161–174.

[64] Guerlac, RADAR, *passim,* especially Sections D and E.

lead. The invention of the Plan Position Indicator, for example, was a major technical advance over the displays used by the British. Although the idea of the kind of sweep motion that most of us regard as natural today was first hinted by Taffy Bowen, the Americans generated their own specific versions and greatly improved the synchronizer unit that forms the central timing device.[65] There were many other technical innovations traceable to American initiative, yet the greatest American contribution was perhaps technological rather than technical or operational. This was a shift from the defensive attitude of the British to a clearer understanding of the offensive capabilities offered by radar. Some of the new navigation techniques needed for long-range bombing, the combination of ASV with operations research, and initiation of searches for radar-guided munitions all fit within an altered context for combat. That there were some holdover attitudes still in 1940–1941 is indisputable – the ignoring of radar warning from an SCR-270 rig in Hawaii on December 7, 1941, was proof enough of the persistence of obstacles to technological innovation. Yet even the Pearl Harbor episode was in a sense an artifact of the shift from a defensive to an offensive mind-set.

WARTIME DEVELOPMENT AND CONCLUSIONS

The incorporation of radar into World War II's military planning and operations is a vast subject. It makes sense, however, to touch upon a few examples briefly before offering some concluding thoughts. The examples include aspects of naval surface action, antisubmarine warfare, and strategic bombing.

From the time of Elizabeth I to 1940, the heart of Britain's defense had always been surface vessels. Radar had its impact there, too, even if the narrative of naval radar is less dramatic than the well-known story of Chain Home and Fighter Command. The British naval radar program began in October 1935 at the Admiralty Signal School in Portsmouth. By September 1938, radar sets operating on 43 Mhz had been developed and installed in *HMS Sheffield* and *HMS Rodney*. Tests were successful enough for the Royal Navy to order rigs using 100 Mhz in quantity, and delivery began in February 1941. The British also developed sets at 600 Mhz and higher for surface search and fire-control.[66]

Two episodes of radar employed against surface ships are particularly well-known, neither of which would likely have had the same outcome without excellent British radar. During a number of critical hours in May

[65] Ibid., pp. 269–271.   [66] Allison, *New Eye for the Navy*, p. 147.

1941 the cruiser *HMS Suffolk* tracked the German battleship *Bismarck* in the North Atlantic solely by means of gunnery-control radar on the quite high frequency of 600 Mhz. The plane that found the battleship heading for Brest was searching with ASV radar. And when torpedo planes from the carrier *HMS Ark Royal* came in for the crucially disabling attack, they were using radar to find their target (which is also why they only recognized the nearby *HMS Sheffield* at the last moment on their first run).

Although aircraft played no significant role in destruction of the battlecruiser *Scharnhorst* off the Norwegian coast in December 1943, surface radar determined the entire battle. Almost all contact in the gloom over 300 miles above the Arctic Circle was on radar screens. Almost all the victors' salvos were radar-controlled. And the early damage to *Scharnhorst*'s gunnery radar doomed the German ship to surprise and to hopelessly uncertain responsive fire once discovered.[67] The Royal Navy clearly realized that without sufficient radar, modern surface ships were effectively blind.

The Battle of the Atlantic offers a different example of the incorporation of radar into naval operations. The defensive phase of the war against the U-boat had gone extremely poorly in 1940, with ninety-six merchant ships sunk during the end of the year with not a single U-boat lost to British countermeasures. With the introduction of the relatively poor ASV Mark II radar, the ratio dropped to ten to one. The quadruple punch of cryptologic intelligence (Ultra), operations research, Leigh Lights, and the first 10 cm radar brought the figure down sharply in March-May 1943 to 0.8 ships lost for each submarine sunk. Signals intelligence (Ultra and/or direction-finding) would often provide a location, operations research would suggest the most efficient use of resources, the ASV Mark III or Mark IV on board would yield positive contact, and the Leigh Light once in the vicinity of a surfaced U-boat at night would insure a kill. Efforts by the Germans to use radar warning devices were hamstrung by the German assumption that centimetric radar was not technically possible, or, when the Germans finally realized it was possible, their assumption that the British were not employing it operationally.[68] Once again, the British technological innovation in the use of radar played a pivotal countermeasure role in dispelling one of those post-World War I nightmares. As the German Admiral Dönitz put it, "The enemy has deprived the U-Boat of its essential feature – namely the element of surprise – by means of radar. With these methods . . . he has conquered the U-Boat menace."[69]

---

[67] Niehaus, *Radarschlacht*, pp. 222–230.   [68] Bowen, *Radar Days*, pp. 107–116.
[69] Ibid., p. 116.

Lastly, the strategic bombing campaigns were technological watersheds for radar. Churchill's "Battle of the Beams" referred to German navigational aids and British countermeasures to them, but British bombers needed navigational aids just as much. The basic problem was the great inaccuracy of bombing runs in this era. "Gee" and "Oboe" were improvements over dead-reckoning above a dark landscape, but it was British H_2S (s-band, 10 cm) radar that first offered effective on-board navigation to the target. This time the German countermeasure was quick and obvious – their fighters followed the beam back to its source and downed many bombers. (This makes it even more astonishing that the Germans did not realize that the same centimetric wavelengths were being used in ASV radars.) The Germans did the same thing with British IFF systems. The fact that British technical leaders such as Watson Watt did not want to believe that the Germans were using the radar beams as homing beacons let the losses mount.[70]

For their part, the Germans showed considerable technical and operational resilience on the defensive. Their defensive radars were temporarily blinded by "window," but that was not due to ignorance of its possible existence. They had their own *Düppel*, but had been forbidden to research its properties for fear that doing so would lead to its use and then Allied counter-use.[71] The German countermeasure to chaff was to employ doppler radars to distinguish the faster moving aircraft from the falling strips of foil. (Such radars are used today at nearly all airports to tell ground clutter from aircraft.) Resourcefulness was also displayed in the quick reverse engineering of the H_2S radar captured from a British bomber downed over Rotterdam in early February 1943. Within less than six months a copy "Rotterdam" prototype was ready for testing. The Germans then quickly accelerated a myriad of technical development programs and vastly improved their coordination of technical and operational advances.[72]

However, a common feature of radar defense networks is that they are assessment-driven systems subject to positive feedbacks, as the British realized after the Battle of Barking Creek. Again R.V. Jones had a particularly penetrating insight into the situation, as related in his report of October 4, 1943, concerning a raid a few days before on Bochum:

> The present German system is unstable in that once the controller has formed a picture of the situation it becomes increasingly easy for him to convince himself that he is right. Having made his guess at the target

[70] Jones, *Wizard War*, p. 392.   [71] Brandt, *Geschichte der Radartechnik*, p. 53.
[72] Devereux, *Messenger Gods*, pp. 159–186; Reuter, *Funkmess*, pp. 113–185.

from the early track of the bombers, he sends his fighters to a convenient beacon. These fighters are then reported by sound observations and, unless the observers are extremely skilled, they may easily be misidentified. The controller then interprets the observations as referring to British aircraft, and is thus confirmed in his initial misjudgment, and so may order up more fighters which may again be misidentified. At Bremen the self-deception went even further; the Flak opened fire, possibly delighted by the absence of Window, and at least one fighter dropped a flare presumably to illuminate our bombers which were in fact at Bochum, 150 miles away. The flare probably convinced the JD 2 (Fighter Division 2) controller that the Pathfinders had arrived for even when JD 1 announced that bombs were falling at Bochum, JD 2 countermanded the JD 1 order for the fighters to concentrate on Bochum.[73]

The cross-reference and assessment procedures in the filter room at Bentley Priory had been designed precisely to limit the scope of such mistakes. Yet it was virtually impossible to eliminate entirely the amplification of small initial errors from the data-collection system.

The need for cross-checking with other forms of intelligence also makes it difficult to determine the real effectiveness of radar as distinguished from signals intelligence and other sources of information. This fact is perhaps more obvious when one introduces consideration of Ultra intercepts, but it was important even aside from cryptologic input. During the Battle of Britain, intercepts of German radio transmissions were often more valuable than radar contact.[74] Also, radar was later credited with influence in the war as a cover for the more highly secret signals intelligence sources. As good as its techniques were, the reputation of operations research also benefitted from these security priorities. Since part of the adaptation to a new technology involves learning how to gauge its effectiveness, this is a nontrivial open issue in understanding military adaptation to radar.

There are a few general observations and speculations that can be offered at this point. One set of thoughts has to do with the frequently small number of individuals at the sharp end of significant change. This fact enhances the prospect of the disproportionate influence of chance and personality. The miscommunication between Kühnhold and Runge or the ac-

---

[73] Jones, *Wizard War,* p. 381. The instability of assessment-driven systems does not merely relate to electronic networks or other high-tech phenomena. Over one hundred and fifty years ago Carl von Clausewitz described the general tendency in intelligence assessments to fall victim to positive feedbacks that amplify an initial error. See chapter 6 of Book One of Carl von Clausewitz, *On War* (Princeton, NJ, 1979).

[74] Devereux, *Messenger Gods of Battle,* p. 48.

cident of Wilkins's visit to Colny Heath were incalculably small factors that eventually cascaded into producing the macroeffects characterizing the entire German or British radar effort.

This is perhaps why the intermingling of scientists and military personnel worked as well as it did for the British and Americans. Unplanned insights and solutions emerged from their interaction. This implies the importance of both a cadre of technical talent inside military organizations and a reservoir of external talent that is accessible. Furthermore, it highlights the need for permeable boundaries between types of personnel as well as among their ideas, a situation difficult to maintain since bureaucracies are generally designed to seal off rather than facilitate contact across boundaries. Technological innovation emerges not just from an additive combination of technical and operational change, but from their interaction. It then forms a new context and feedback process for both.

This may mean that key actors perform their roles because they are capable of sensing the changing context and doing something about it. Some persons, such as Dowding or Tizard, were capable of looking at technical change (such as an invention or new research) without a focus so sharp that it cast the implications of that change into darkness. Sometimes a less intense focus offered them greater acuity for what was at the borders of their vision. This may be the best way to sense the direction of complex, nonlinear processes characterized by positive feedbacks and self-organizing criticality.[75]

Another set of thoughts has to do with the fact that technical change has become endemic in our world. There are multiple locations and occasions for technical change to occur in practically any field. There are too many interacting variables to isolate and control, yet too few at the front end of major change to treat purely statistically. Although technical developments ran roughly parallel in time in Germany, Britain, and the United States, the British jumped ahead operationally and technologically because they perceived a need to adapt to a situation they had not caused and could not control. The Germans thought that they *could* control events and the Americans saw *no need* to control distant events that did not concern them. Perhaps the desire for a sense of control leads to

---

[75] See the popular expositions of these notions by W. Brian Arthur, "Positive Feedbacks in the Economy," *Scientific American*, February 1990, pp. 92–99; and Per Bak and Kan Chen, "Self-Organized Criticality," *Scientific American*, January 1991, pp. 46–53. For a discussion of warfare as a complex, nonlinear phenomenon, see Alan Beyerchen, "Clausewitz, Nonlinearity, and the Unpredictability of War," *International Security*, 17, Winter 1992/93, pp. 59–90.

complacency and too mechanistic a mind-set. The fluidity of circumstances, the feedback processes with entrainment, and the general feel of the onset of technological change are more akin to weather systems or biological (and historical) processes than to mechanical structures. How many military line officers trying to discern the prospects for change are trained as meteorologists or biologists? Perhaps more should be.

8

INNOVATION
Past and future[1]

WILLIAMSON MURRAY

In 1991 the Soviet Union collapsed; with it went the military power and threat that had dominated so much of the history of the last half-century. As the "Wall" between occupation zones created in 1945 disappeared, so too paradigms for understanding the Cold War vanished in the echo of collapsing concrete. From a relatively simple and clear bi-polar contest between the United States and the Soviet Union, one sees the emergence of a multi-polar world of far more complexity. To add to uncertainties, in the late 1980s the military race between East and West had entered a period of considerable military innovation, occasioned by cascading changes in both civil and military technologies. What those changes mean or where they are going is not certain even in the aftermath of the Gulf War. Nothing, however, suggests that the rapid pace of innovation – underlined by the conduct of the war against Iraq – will not continue into the next century.

We appear therefore to be entering a time of political, strategic, and technological uncertainty; yet a period where the threats seem more indeterminate. Western military institutions confront a future in which they will not receive anything similar to the funding and resources they received throughout the Cold War despite rapid technological change and innovation. Thus, they must innovate with less money and greater ambiguities about potential opponents and the nature of the wars they will have to fight. The chapters in this study examined innovation during the 1920s and 1930s. Military organizations innovated during the 1920s and 1930s under circumstances similar to those that will influence military in-

[1] The author would like to thank Barry Watts of Northrop-Grumman Corporation and Professor Alan Beyerchen of The Ohio State University's Department of History for their generous help in preparing this chapter.

novation in the next century; consequently, the past offers useful perspectives on how to think about innovation in the future. Whether we like it or not military innovation and changes are inevitable given the technological developments occurring in civil society. Thus, how military institutions innovate will be a critical factor in their performance on the battlefields of the twenty-first century.

The purpose then of this chapter is to tease out perspectives to help in thinking about the future. Nevertheless, the reader must recognize, as Lenin suggested, that "history generally, and the history of revolution in particular, is always richer in content, more varied, more many-sided, more lively and subtle than even the best historian and the best methodologist can imagine."[2]

THE NATURE OF INNOVATION

Before we examine the past we need to look more closely at innovation itself and the military organizations in which it occurs. Unlike other organizations, military forces in peacetime must innovate and prepare for a war 1) that will occur at some indeterminate point in the future, 2) against an opponent who may not yet be identified, 3) in political conditions which one cannot accurately predict, and 4) in an arena of brutality and violence which one cannot replicate. Moreover, military institutions exist in a culture of disciplined obedience in which soldiers, sailors, and airmen must remain steadfast in the face of terrifying conditions, while their psychological instinct for self preservation urges them to flight. But disciplined organizations rarely place a high value on new and untried ideas, concepts, and innovations. As Michael Howard has suggested:

> There are two great difficulties with which the professional soldier, sailor, or airman has to contend in equipping himself as a commander. First, his profession is almost unique in that he may only have to exercise it once in a lifetime, if indeed that often. It is as if a surgeon had to practice throughout his life on dummies for one real operation; or a barrister only appeared once or twice in court towards the close of his career; or a professional swimmer had to spend his life practicing on dry land for an Olympic Championship on which the fortunes of his entire nation depended. Secondly the complex problem of running a [military service] at all is liable to occupy his mind and skill so completely that it is easy to forget what it is being run *for*.[3]

---

[2] Quoted in Paul Feyerabend, *Against Method* (New York, 1993), p. 9.
[3] Michael Howard, "The Use and Abuse of Military History," *Journal of the Royal United Services Institute*, vol. 107, February 1962, p. 6.

It is impossible to replicate the conditions of war in peacetime, while war itself is so permeated with fog and friction that it is difficult for military organizations to determine what has actually happened on the battlefield.[4] Since human beings conduct war in the real world, that activity reflects the complexities of that arena: one that, as we are discovering from science, chance and nonlinear factors can dominate. The problem for the historian or the analyst interested in innovation is that the complexities of the process make it extraordinarily difficult to recover the past in simple, easily digestible form. The relations among technological innovations, the fundamentals of effective military operations, and innovations in concepts, doctrine, and organizations that govern those operations are fundamentally nonlinear: changes in inputs like weapons systems, whether large or small, do not necessarily yield changes of proportionate magnitude in outputs or combat dynamics.[5]

The explanation has to do with the nature of how interactions work in the real world. Since the 1950s, research from fields as diverse as meteorology, ecology, physics, and mathematics has uncovered a menagerie of dynamic systems so *simple* as to be virtual paragons of deterministic, clockwork mechanisms which, can nonetheless, give rise to long-term behavior so *complex* as to be literally unpredictable or "chaotic."[6] Phenomena as disparate as the oscillation of convection rolls in low-temperature helium,[7] the rise and fall of many biological populations, the

---

[4] Of all the armies that experienced the World War I battlefield, only the Germans managed to find effective ways to deal with what had happened on the tactical level. Not surprisingly it has taken military experts until the 1980s to begin to fumble towards an understanding of the true complexities and dynamics of the Western Front. As Clausewitz notes with considerable understatement: "In the dreadful presence of suffering and danger, emotion can easily overwhelm intellectual conviction, and in this psychological fog it is so hard to form clear and concise insights that changes of view become more understandable and excusable. Action can never be based on anything firmer than instinct, a sensing of the truth." Carl von Clausewitz, *On War* (Princeton, NJ, 1976), p. 108.

[5] I am indebted to Barry Watts for leading me through some of the complexities of non-linearity over the past decade.

[6] Tien-Yien Li and James A. Yorke first introduced the term "chaos" in their 1975 paper "Period Three Implies Chaos" to denote the unpredictability observed in certain "deterministic" but nonlinear feedback systems (Hao Bai-Lin, *Chaos* [Singapore, 1984], pp. 3 and 244). Technically, scientists today use the term "chaos" to denote "those nonrandom complicated motions that exhibit a very rapid growth of errors that, despite perfect determinism, inhibits any pragmatic ability to render long-term predictions." See Mitchell J. Feigenbaum in Peitgen, Jürgens, and Saupe, *Chaos and Fractals*, p. 6.

[7] Ian Stewart, *Does God Play Dice?*, pp. 212–213 confirmed the existence of nonlinear dynamics not just in turbulent fluids, but in all kinds of physical systems: electronic, optical, even biological (ibid., p. 213).

behavior of a driven nonlinear semiconductor oscillator, the dripping of a water faucet, and the beating of a normal human heart all exhibit such nonlinear dynamics. It now appears that stable systems with regular, simple, predictable dynamics "are in fact exceptions in nature rather than the rule."[8] And most crucial for our purposes, the local randomness or unpredictability exhibited by nonlinear systems is fundamental: gathering and processing more information with better algorithms and computers cannot, even in principle, make the inherent local unpredictability go away.[9]

The implications of these developments in science and mathematics are indeed profound. They suggest that the world as a whole does not work in a mechanistic, deterministic fashion, that complex social interactions like military innovation or actual combat do not reduce to simple, linear processes, and that the study of human affairs, the interplay of literally hundreds, if not thousands of independent variables, is more of an art than a science. The process of innovation within military institutions and cultures, which involves numerous actors, complex technologies, the uncertainties of conflict and human relations, forms a part of this world and is no more open to reductionist solutions than any other aspects of human affairs.[10]

However, human organizations in general and military cultures in particular seek to bring order and linearity to a world governed by chaotic complexity.[11] The preceding chapters suggest how dangerous such a line of approach is. To understand innovation in the interwar period one must

[8] John Briggs and F. David Peat, *Turbulent Mirror: An Illustrated Guide to Chaos Theory and the Science of Wholeness* (New York, 1989), p. 110.

[9] That state of affairs explains why so few historians of and commentators on war possess the quality of timelessness. Thucydides with his extraordinarily heavy emphasis on chance and Clausewitz with his dynamic understanding of the imponderables and uncertainties in war have approached a real understanding of the underlying nature of war and human affairs.

[10] To suggest that social systems exhibit chaos is actually a claim that *even if* the relevant dynamics of such systems could be reduced to a few dominant variables, the interaction of those variables would still produce a degree of complexity that renders predictability very problematic. I am indebted to Alan Beyerchen for this point.

[11] Recently Alan Beyerchen has argued that the tendency of students of Clausewitz to interpret him in terms of linear dynamics "has made it difficult to assimilate and appreciate the intent and contribution of *On War*. . . . The concepts and sensibility recently emerging in nonlinear science can be used to clarify not *his* confusion, but *our* truncated expectations for a theory of war – namely that it should conform to the restrictions of linearity. At the very least, such a sensibility may help us explore the stubborn intractability of prediction in war." Alan Beyerchen, "Clausewitz, Nonlinearity, and the Unpredictability of War," *International Security*, Winter 1992–1993, pp. 87–88.

not lose track of the fact that the interplay among human factors, uncertain knowledge, misreadings of the past, political and strategic parameters placed innovation on a complex playing field in which not only were the players uncertain of the future but they were often more concerned with immediate problems than with long-range changes. Even to innovators their goals often appeared opaque. And for commentators at the end of the century, the story remains even more complex and difficult to untangle since so much of the evidence and the interplay among those involved in innovation have disappeared into the dustbins of history.

THE PROCESS

There are important influences on innovation which guide the emphasis and interest of military organizations and which largely lie beyond the realm of control for human institutions. The most obvious is geography. American and Japanese interest in amphibious warfare, for example, resulted from the nature of the Pacific theater. Similarly American and British airmen evolved more radical beliefs in "strategic bombing" and sold those conceptions to their political leaders because their island nations enjoyed an immunity from land threats that continental powers did not. Finally, the Germans, confronting land war in two directions, focused on ground war and innovated more aggressively in that arena, because any repetition of World War I's stalemate guaranteed defeat.

Innovation takes place at all levels of war; perhaps the most difficult to understand are innovations that nations confront when major changes occur in the balance of power or the political framework within which wars occur. Beginning in 1792 the French Revolution and its successor, Napoleonic France, fundamentally altered the rules and unleashed a period of nearly twenty-five years of constant war. As Clausewitz suggests:

> War, untrammeled by any conventional restraints, had broken loose in all its elemental fury. This was due to the people's new share in these great affairs of state; and their participation, in turn, resulted partly from the impact that the revolution had on the internal conditions of every state and partly from the danger that France posed to everyone.[12]

The twentieth century has seen equally dramatic strategic changes – the most recent being the collapse of the Soviet Union. In the 1930s defense planners had to innovate in an environment of dramatic and incalculable changes – a difficult task, particularly in the democracies, because it proved so difficult to move national preconceptions.

[12] Carl von Clausewitz, *On War* (Princeton, NJ, 1976), pp. 592–593.

National assessments of the international arena have a crucial influence on whether military organizations innovate successfully. The focus of this study is not on strategic innovation or adaptation, but the reader must remember that the strategic arena, as well as political and military assessments of the strategic framework, is an essential prerequisite to successful innovation.[13] The broad innovations either undertaken or neglected by military institutions often depend on the political guidance and strategic framework within which those institutions operate – especially in the case of Western democracies. For all of Fuller's and Liddell Hart's carping criticism about the British Army's failure to innovate with tanks, the army prepared for the war that its government's strategic policy decreed most likely. The failure to prepare for a mobile, high-density armored war in fact reflected the very strategy of "limited liability" that Liddell Hart spent so much of the 1930s propagandizing.[14]

There are a number of difficulties that military institutions confront in innovating beyond national strategic assessment. We might therefore turn to a general examination of innovation. Chapter 7 of this book presents a useful schematic for thinking about the process:[15]

RELATIONSHIPS IN INNOVATION

Context	Technological change	Strategy
Procedures	Operational change	Operations
Equipment	Technical change	Tactics

Military innovations that have the greatest influence are those that change the context within which war takes place. By utilizing technical changes – the development of the Spitfire and Hurricane as well as radar, both of which he played a critical role in pushing – Dowding provided Fighter Command with an operational level change that enabled it to alter the bomber/fighter relationship. Stanley Baldwin's claim that "the

---

[13] As the editors of this study have suggested in a summative essay on the military effectiveness study sponsored by the Office of Net Assessment: "No amount of operational virtuosity ... redeemed fundamental flaws in political judgement. Whether policy shaped strategy or strategic imperatives drove policy was irrelevant. Miscalculations in both led to defeat, and any combination of politico-strategic error had disastrous results. ... This is because it is more important to make correct decisions at the political and strategic level than it is at the operational and tactical level. Mistakes in operations and tactics can be corrected, but political and strategic mistakes live for ever." Allan R. Millett and Williamson Murray, "Lessons of War," *The National Interest*, Winter 1988/1989.

[14] Williamson Murray, *The Change in the European Balance of Power, The Path to Ruin* (Princeton, NJ, 1984), pp. 86–87.

[15] See p. 268 in the Beyerchen essay in this work.

bomber will always get through" was no longer valid.[16] Victory in the Battle of Britain – and Dowding's claim to be one of the few great captains of the twentieth century – resulted from the fact that he built an effective air defense *system* that altered the entire context within which air forces operated. What is crucial in this example, as Beyerchen suggests, is that while the Germans may have possessed better equipment[17] and even tactics,[18] the British operated within a broader framework of contextual change. By so doing they created a new logic within which the Luftwaffe was incapable of winning.[19]

Innovation is more than incorporation of equipment and technical change into doctrine, practices, and tactics. Innovation in tactics and operational concepts can prove as important on the battlefield as changes in equipment. Such innovation changes the basic operating framework that military organizations have about the relationships among weapons systems and those who use them against the enemy. In most cases such innovation is evolutionary rather than revolutionary in nature. In fact, as the case studies suggest, revolutionary innovations are the exception. But what both forms of innovation indicate about the process is instructive in fostering successful innovation in the future.

Revolutionary innovation

Revolutionary innovation appears largely as a phenomenon of top-down leadership – leadership that is well-informed about the technical *as well as conceptual* aspects of possible innovation. The only clear case of revolutionary innovation in this volume is that of Dowding and the changes that he wrought in British thinking about air power innovation. As head of RAF research and development, he pushed for high-speed fighters and radar; but it was in developing a new context for thinking about air war – one that flew in the face of RAF doctrine – that Dowding revolution-

---

[16] Keith Middlemas and J. Barnes, *Baldwin: A Biography* (London, 1969), pp. 731.

[17] German radar and the Bf 109 were superior to British equivalents. For those who wish to suggest that this was not true in regards to the Spitfire versus the Bf 109, the exchange ratios suggest otherwise.

[18] At the beginning of the Battle of Britain Fighter Command was for the most part still using obsolete tactics of tightly flown vics, while the Germans were using the finger-four formation developed in Spain.

[19] This carried over into the Blitz when the British, alerted by inspired scientific intelligence, were able to interfere with the blind bombing devices that the Germans had developed to solve the technical problems involved in strategic bombing at night. See in particular: R.V. Jones, *The Wizard War, British Scientific Intelligence, 1939–1945* (New York, 1978), esp. pp. 127–144.

ized the complex dynamics of the operational framework within which air war occurred.[20]

How Dowding found himself in this position was largely a matter of chance. Most RAF senior officers considered research and development as a dead end; in 1937 Dowding took over Fighter Command as a sop for losing out in the race to become chief of air staff. In the eyes of most senior RAF officers, Fighter Command was a backwater. But the new Chamberlain government made a number of critical defense decisions in summer and fall 1937 – virtually all of which had a disastrous impact. The one correct decision that it made, however, was to emphasize construction of fighters over bombers.[21] In the largest sense, much of history is chance; in this case the chances of promotion and interservice politics provided the right man, at the right place, at the right time. But it is worth underlining Dowding's personal qualities: the evidence suggests that there *was no other* senior officer in the RAF with the requisite imagination and drive to carry through the contextual innovation that Dowding executed.[22]

What then were the qualities that made Dowding an extraordinary innovator and allowed him to create at such complex levels? Perhaps most important was his ability as a generalist to seek help from a wide number of sources, including civilians and scientists. Equally important was his capacity to keep larger issues in view. His vision was also enhanced by the concrete, palpable threat that Luftwaffe bombers and strategic bombing posed to the existence of Britain as an independent nation.

Unlike most who either wished the problem away in nostrums of "appeasement," or hoped that the RAF could reply in kind, Dowding addressed the threat in its own terms, its vulnerabilities as well as its strengths. In fact, he was able to reverse his thinking and recognize as early as 1940 that to strike German targets, the RAF would require long-

---

[20] It is worth noting that in summer 1938 some within the air staff made a major effort to buy the two-seat Defiant instead of the Hurricane and Spitfire. Again Dowding's understanding of the issues kept the emphasis on the right technical means for the system that he was already developing. See in particular PRO AIR 2/2964, Headquarters Fighter Command, RAF Stanmore, 25.6.38.

[21] The decision was financial, not strategic. Fighters were cheaper than bombers: hence the government's decision in favor of Fighter Command. Murray, *The Change in the European Balance of Power*, p. 82.

[22] It is worth noting that Trenchard on the Western Front had a reputation for sacking subordinates who showed any signs of weakness. At the height of the Battle of the Somme, Dowding, then a squadron commander, went to Trenchard and requested that his squadron be relieved because of its very heavy casualties. Trenchard not only agreed, but then never thereafter penalized Dowding's career for the request.

range escort fighters that could defeat the Luftwaffe in its own skies.[23] Admittedly, the British government provided Fighter Command with the resources for successful innovation, and there were fighter pilots who rallied to the new conception. Finally, Fighter Command recovered rapidly from the fiasco of the "Battle of Barking Creek" in early September 1939 when operators misread the radar plots of British interceptors. But without Dowding the British would not have created a *system* of air defense.

It is, however, worth introducing a word of caution. The history of the interwar period suggests a number of other cases where similar top-down leadership had a *disastrous* impact on the process of innovation. Certainly, French generals maintained tight control of every aspect of their army's doctrine and development; those who refused to support the revealed wisdom of the "methodical battle" soon found themselves on the outside looking in.[24] Similarly, Bomber Command and the RAF exercised top-down control over development of strategic bombing capabilities. The almost complete inability of Bomber Command at the beginning of the war to execute any air power missions was a direct result. The lesson may well be that if you are right, top-down leadership will allow you to get it very, very right. If you get it wrong, however, you will get it very, very wrong.

Evolutionary innovation

Most innovation, however, suggests a long, complex process involving organizational cultures, strategic requirements, the international situation, and the capacity to learn realistic, honest lessons from past as well as present military experience. Evolutionary innovation takes place over extended periods during which tactics, equipment, and conceptions change on a gradual basis. A close comparison between successful and unsuc-

---

[23] PRO AIR 16/1024, Minutes of the 20th Meeting of the Air Fighting Committee, held at the Air Ministry, Whitehall, 12.3.40.

[24] For a close look at the development of French doctrine, which made some considerable sense within the context of recent French military experience see Robert Doughty, *The Seeds of Disaster, The Development of French Army Doctrine, 1919–1939* (Hamden, CT, 1986). The fault in the French approach lay in a desire to exclude all differing opinions on the subject of mobility from the discussions. André Beaufre records in his memoirs that Gamelin had established the high command as the sole arbiter of army doctrine; all articles, lectures, and books by French officers had to receive approval before publication. As a result, he notes "everyone got the message, and a profound silence reigned until the awakening of 1940." André Beaufre, *1940, The Fall of France* (New York, 1968), p. 19.

cessful case studies suggests that the degree of innovation on a year-to-year basis is relatively small, but that cumulative, gradual change can lead to dramatically different results. The contrast between the French and German tactical systems in May 1940 could not have been more dramatic, but the changes and innovations that led to this breaking point had occurred over the previous two decades. However incremental the innovations, by 1940 a chasm existed between how the two forces thought about, prepared for, and executed on the battlefield.[25]

Evolutionary innovation depends on organizational focus over a sustained period of time rather than on one particular individual's capacity to guide the path of innovation for a short period of time. Military leadership has the most influence on innovation by long-term cultural changes rather than immediate short-term decisions. The development of armored warfare offers useful comparisons. The two foremost leaders of the British and German armies in the 1920s were Generals Hans von Seeckt and Field Marshal Lord Milne. Of the two Milne was the more willing to see the "army of the future" – to use De Gaulle's phrase – in terms of armored forces. Milne not only supported a series of armored maneuvers with the scarce funds, but commented in an address to senior officers of the mechanized force:

> It is up to us to find some means of bringing war back to what it was when the art of generalship was possible. The only means of doing this is to increase mobility on the battlefield. Now that is the point of the initiation of the armored brigade – to revive the possibility of the art of generalship.[26]

Seeckt, although interested in the possibilities of motorized warfare, never went so far. In 1928 he even cautioned the *Reichswehr*'s officers that "many prophets already see the whole army equipped with armored ve-

[25] Robert Doughty depicts the following scene occurring at the Ecole Supérieure de Guerre: "It may have been a cool, sunshiny day in Paris in 1922 when Col. J. Roger addressed the assembled officers in a small, dimly lit amphitheater. . . . The hundred or so officers sitting around him were probably exhausted from late-night studying and from an overly full schedule of professional and extracurricular activities. Given a choice, they undoubtedly would have preferred to be elsewhere. But they had to attend Colonel Roger's lecture on artillery. After a long and tedious explanation of the doctrine for the artillery's support of attacks against a defense in depth, the lecturer reached his conclusion. . . . The students had listened to a familiar explanation of the importance of fire power and centralized control." Bit by bit the deadening hand of revealed wisdom tightened on the mind of the French Army. The results showed clearly on the Meuse. Doughty, *Seeds of Disaster*, p. ix.

[26] B. H. Liddell Hart, *The Memoirs of Captain Liddell Hart*, vol. I (London, 1965), p. 129.

hicles and the complete replacement of the horseman by the motorized soldier. We are not yet that far."[27]

But the crucial point is that Seeckt was able to create a culture of innovation by the kind of officer corps he created at the end of World War I. That officer corps not only studied the last war in a thorough and realistic fashion, but then through Seeckt's influence evolved doctrinal concepts on the basis of past and present experience.[28] Moreover, the culture that Seeckt inculcated in the *Reichswehr* placed a high value on study and analysis of changes in doctrine, tactics, and technology; in other words he created a climate ideally suited to innovation.

Milne on the other hand took over the British Army well after the war. The army had made no serious effort to examine its experiences in World War I (and Milne would not set in motion such an effort until his last year as CIGS), while the army's regimental system placed little value on serious, professional study of war.[29] Consequently, Milne's influence was wholly personal and dissipated rapidly in the 1930s after his retirement, as a series of unimaginative officers took charge. These officers, particularly Montgomery-Massingberd, effectively sabotaged Milne's initiatives.[30] Clearly, this comparison underlines that the long-term decisions that affect the culture and values of the officer corps play a crucial role in the process of innovation.

SUCCESS IN INNOVATION

We might profitably turn to examining the larger themes that have emerged during the interwar period. Despite the difficulties, a number of military institutions did innovate with success; others, however, failed.

---

[27] Bundesarchiv/Militärarchiv, W 10-1/9, Oberstleutnant Matzky, "Kritische Untersuchung der Lehren von Douhet, Hart, Fuller, and Seeckt," Wehrmachtakademie, Nr. 90/35 g.K., Berlin, November 1935, p. 44.

[28] See Williamson Murray, "Armored Warfare: The British, French, and German Experiences," this study, chapter 1.

[29] For an examination of these issues see Brian Bond, *British Military Policy between the Two World Wars* (Oxford, 1980); see also Brian Bond and Williamson Murray, "The British Armed Forces, 1918–39," in *Military Effectiveness*, vol. 2, *The Interwar Period* (London, 1988).

[30] Montgomery-Massingberd suppressed the committee's highly critical lessons-learned analysis of the army's performance in the last war as well as its recommendations for the future. Harold R. Winton, *To Change an Army, General Sir John Burnett-Stuart and British Armored Doctrine, 1927–1938* (Lawrence, KA, 1988), pp. 130–131. Montgomery-Massingberd commented to Liddell Hart in the 1920s on Fuller's *The Foundations of a Science of War* in the following terms: "No, I have not read Fuller's book! And don't expect I ever shall. It would only annoy me!" Liddell Hart, *Memoirs*, p. 102.

The factors that worked to further the path to success or failure in innovation in this period tell something about those elements conducive to innovation in future decades.

1) Specificity

A number of factors contributed to successful innovation. The one that occurred in virtually every case was the presence of specific military problems the solution of which offered significant advantages to furthering the achievement of national strategy. The most obvious had to do with development of carrier aviation by U.S. and Japanese Navies. Both confronted a similar problem – how to extend the reach and striking power of battle fleets in the vastness of the Pacific – within a strategic context that a war between the two nations was a distinct possibility. The existence of this concrete problem also proved of considerable help in focusing the U.S. Marine Corps on amphibious warfare.

On the other hand, major European navies faced an arena in which carrier aviation offered less clear advantages, particularly given access to land bases for possible operations in a European theater. These navies, particularly those of the Italians and Germans, considered naval war a matter of the European archipelago. Consequently, it was difficult to visualize what carrier aviation could provide a fleet that land-based air could not.[31] Admittedly, the British had interests in out-of-area operations against Japan, but they believed such a war had to be fought largely from the great naval facility at Singapore with its air bases. Thus, neither strategic environment nor the naval war envisaged by potential combatants provided a conceptual framework to encourage innovative thinking on carrier aviation.

Similar military problems prompted the development of German armored warfare and close air support. An examination of the 1918 battles suggests that little separated British and German armies in their ability to break down front line defenses at the end of the war; neither, however, possessed the means to achieve operational exploitation.[32] But after the war, the Germans continued to believe that renewed continental war re-

---

[31] Murray, *The Change in the European Balance of Power*, pp. 45–47 and 115–117.

[32] For the best studies see Timothy Lupfer, *The Dynamics of Doctrine: The Changes in German Tactical Doctrine During the First World War* (Leavenworth, KA, 1981); Timothy Travers, *The Killing Ground, The British Army, the Western Front, and the Emergence of Modern Warfare, 1900–1918* (London, 1987); and Timothy Travers, *How the War Was Won, Command and Technology in the British Army on the Western Front, 1917–1918* (London, 1992).

mained a distinct possibility. Consequently, development of armored war and close air support represented real progress towards concrete objectives: namely, victory on the continental European battlefield.[33] The British, however, while sponsoring extensive experiments with tanks in the late 1920s and early 1930s, never found a place for such developments in an army committed to defending colonies and forbidden by its political masters until February 1939 even to consider a continental war.

For their part, the French also faced a concrete problem, namely the German threat, on which they concentrated development of their military forces throughout the interwar period. That focus, however, was not sufficient to overcome a number of systemic defects and misperceptions in French doctrine drawn from their particular experiences in World War I.

What this suggests is that an additional factor is necessary to push a concrete problem toward significant innovation: there must be clear institutional conceptions and interest in developing a new form of war. The Germans aimed to overthrow the results of World War I from the moment they signed the Treaty of Versailles; the French merely aimed to hold on to their gains. Moreover, the terrible casualty bill of the last war combined with serious disasters that the French had suffered when on the offensive – Plan XVII, the 1915 attacks, the Nivelle Offensive – to make the French leery of developing new forms of war. They had tried one too many recipes for the offensive.

Our conclusion, therefore, is that one precondition for significant military innovation is a concrete problem which the military institutions involved have vital interests in solving. We cannot, without a much wider survey, insist that it is a necessary condition, but its presence in all the cases examined in this study is certainly suggestive.

2) Military culture

Another factor conducive to innovation has to do with the organizational cultures which have to undertake preparations for war during periods of minimal funding. Military culture might best be described as the sum

---

[33] It is significant that there was little connection between the development of concepts for armored war and those of close air support in the German military. Through the invasion of France, in fact, close air support remained a capability developed strictly to help the army in the breakthrough battle. It was not until after the French campaign that the Germans developed the procedures to provide support in the mobile exploitation battle. See Williamson Murray, "The *Luftwaffe* Experience, 1939–1941," in *Case Studies in the Development of Close Air Support*, ed. by Benjamin Franklin Cooling (Washington, DC, 1990).

of the intellectual, professional, and traditional values of an officer corps; it plays a central role in how that officer corps assesses the external environment and how it analyzes the possible response that it might make to "the threat."[34] The culture of officer corps also plays a crucial role in how military forces prepare themselves for combat. This represents an essential element in successful innovation.

There are a number of complex factors that influence military cultures. Rarely, if ever, do military organizations receive the opportunity to innovate with a clear slate. The past weighs in with a leaden hand of tradition that can often block innovation. And not without reason. The approaches that succeeded on earlier battlefields were often worked out at a considerable cost in blood. Consequently, military cultures tend to change slowly, particularly in peacetime.

In the interwar period, the only sudden change in culture came in the German Army in 1919, when the victorious powers demanded a drastic downsizing. This provided Seeckt with the chance to make dramatic changes in the composition and value system of the officer corps. There were three separate constituencies to which Seeckt could have catered: the general staff, front-line heroes, and those with connections, particularly the nobility.[35] He deliberately favored the general staff; the other constituencies received short shrift.[36] As a result there was a distinct change in the army's ethos.

This change furnished the Germans with a number of advantages throughout the interwar period. First, it predisposed them to make the most thorough, complete, and honest assessment of the last war. One of the most frequently quoted axioms of historians is that generals prepare for the last war and that is why military organizations have a difficult time in the next conflict. In fact, most armies do nothing of the kind, and be-

[34] Samuel Huntington, *The Soldier and the State, The Theory and Politics of Civil-Military Relations* (Cambridge, MA, 1957). For a more recent, but less successful, effort to lay out the culture of the United States Air Force see Carl Bilder, *The Icarus Syndrome* (New Brunswick, NJ, 1994).

[35] Of course these constituencies were not mutually exclusive. Some officers belonged to all three, although the membership in the general staff was limited to a select few – only the most talented at the profession of war.

[36] This helps to explain why so many former officers with outstanding combat records found the Nazi movement so attractive. Ernst Röhm clearly hoped to have the SA movement replace the army as the defender of the new Germany and the result was the drastic purge of 1934 in which Hitler sided with the professional army which Seeckt had so heavily influenced. James S. Corum, *The Roots of Blitzkrieg, Hans von Seeckt and German Military Reform* (Lawrence, KA, 1992), pp. 33–37 is particularly good on Seeckt's favoring of the general staff.

cause they have not distilled the lessons of the last war, they end up repeating most of the same mistakes. But the Germans, already the most innovative in developing defensive and offensive tactics in World War I, now insured that a thorough study of those developments provided the *Reichswehr* with a firm grasp of the direction that modern land war was taking.

Moreover, within German military culture there was honest study and reflection as to the possibilities open to further development. In Britain, the CIGS, Montgomery-Massingberd, in 1932 suppressed the unfavorable report on the army's performance in the last war. Such an act was inconceivable in the German Army of the interwar period. But this culture of critical examination transcended learning the lessons of the last war. In the late 1930s, the same pattern appeared as the Germans conducted exercises and then combat operations. In all cases they continued to carry out ruthless, critical examinations of what had occurred in the field and as a result, they *recognized and learned from* mistakes. Key to the German approach was their treatment of mistakes or problems in using new equipment or procedures; they treated such incidents as part of the learning experience, not as items deserving punishment or the end of an officer's career. Thus, they were able to see the forest for the trees and change *the context* of offensive war.

Throughout the interwar period the German Army's culture provided for a trust and honesty between different levels of command. German officers in command positions were not afraid to express their belief that their units confronted deficiencies. The *Anschluss* is a good example of this process, as that operation highlighted serious weaknesses throughout the army and mobilization system. After-action reports from battalion to army level became more and more critical of troop performance, training, discipline, and doctrine the higher the level of command.

In the case of the Polish campaign in 1939, by every measurement a brilliant victory, the army's high command carried out a thorough examination of operations, examined lessons-learned analyses, thoroughly established a massive training program to fix problem areas, and then created feedback mechanisms to insure that combat units carried out training and exercises in accordance with directives aimed at fixing those weaknesses.[37]

The result was a stunning improvement in the army's performance lev-

---

[37] See Williamson Murray, "The German Response to Victory in Poland: A Case Study in Professionalism," in *German Military Effectiveness* (Baltimore, MD, 1992), pp. 229–243.

el in the French campaign. This willingness to examine performance with a critical eye was one of the major factors that enabled the Germans to perform consistently well in tactical operations, and most of the time at the operational level as well, throughout World War II.[38]

This cultural framework also played a crucial role in the innovation and development of armored war in Germany. The Germans incorporated innovation in armored war into a comprehensive and realistic understanding of modern war. A long-term process of steady, incremental improvements in tactics and doctrine resulted in mechanized forces that possessed capabilities well beyond those of other European armies. An essential component in such successful innovation was the ability of the Germans to conceptualize the operational and tactical levels of war in doctrinal statements. The army's *Die Truppenführung* provided a coherent doctrine for thinking about future battlefields. So coherent was that framework that it not only provided a means to integrate traditional branches – artillery and infantry – but it also allowed space for inclusion of evolving concepts of armored war and close air support within a doctrine aimed at fighting mobile, decentralized battles. Since German officers took doctrine and doctrinal manuals seriously they could comprehend the larger picture of combined arms. Moreover, once exposed to the possibilities of armored war in the Polish campaign, a substantial number of skeptics were convinced by the new form of war.[39] Thus, the Germans created the maximum potential for exploitation.

The Luftwaffe showed similar interest in doctrine. Its basic manual attempted to fit doctrine to the capabilities and purposes of war; consequently, the Germans evolved a broader concept of air power, which allowed them to utilize air forces more effectively on the battlefields through 1942. Thereafter, luckily for its enemies, errors in strategy placed the Luftwaffe in a situation where it had little opportunity to innovate in the face of its opponents' overwhelming superiority.[40]

Other armies that paid less attention to doctrine as an integrative part in the process of innovation had more difficulty in peace *and* in war. The

---

[38] For the German examination of the Anschluss see in particular: Heeresgruppenkommando 3., 18.7.38, "Der Einsatz der 8 Armee in März 1938 zur Wiedervereinigung Österreichs mit dem deutschen Reich," National Archives and Records Service (NARS) T-79/14/447; and Generalkommando XIII A.K., "Erfahrungsbericht über den Eisatz Österreichs März/April38," 6.5.38. NARS T-314/525/000319.

[39] The result was that an infantryman like Rommel, a skeptic before the war, was able to take command of the 7th Panzer Division in late 1939 and make it the outstanding division of the Battle of France.

[40] See Murray, *Luftwaffe*, chapters 6 and 7 for further discussion of this point.

British Army is a particularly good example of how the lack of a doctrinal framework made it almost impossible for British ground forces to innovate with the tank, even during World War II. Thus, British armored experiments in the late 1920s and early 1930s ironically benefitted the Germans more than they did the British themselves.[41]

But there were problems with German military culture that robbed the Germans of their tactical and operational innovations. The most brilliant battlefield performance quite simply could not make up for logistic and intelligence systems that failed to function in the modern world. Given the contempt that the German officers exhibited for these crucial areas – and the Luftwaffe and the navy were as bad as the army – the Germans were incapable of standing up to their opponents in a prolonged struggle. If tactical battlefield innovations provided the Germans an initial advantage, these were not sufficient to overcome the gross mistakes they made in logistics and intelligence, largely as a result of their military culture.[42]

The development of U.S. carrier aviation reinforces a number of points. Here Congressional "interference" in decreeing that command of aircraft carriers could only go to flyers provided a vital twist to the culture within which naval aviation developed. From that point, even senior officers found it to their advantage to win their wings and "join the aviation community." But in fairness to the U.S. Navy there was already a substantial push behind aviation. Moreover, the navy's culture created a realistic relationship between yearly exercises, planning for those exercises, and education and war gaming that occurred at the Naval War College. Part of the navy's development of carrier aviation rested on a process in which the Naval War College designed summer fleet problems; the fleet executed the problems in as realistic a fashion as possible; and then a careful, honest evaluation funneled the results back to Newport to help in designing the next year's problems. Finally, the Naval War College, well-connected with the fleet, helped to keep the officer corps informed as to what was occurring in developments of naval aviation and the concepts underlying its utilization.[43]

What is particularly striking is the degree of realism and imagination

---

[41] For a fuller discussion of this see Murray, "British Military Effectiveness: the World War II Experience," in *Military Effectiveness*, ed. by Millett and Murray, vol. 3, pp. 90–136.

[42] For the logistic side see Martin van Creveld, *Supplying War, Logistics from Wallenstein to Patton* (Cambridge, 1977). For the abysmal failure of German intelligence on the Eastern Front which mirrored the catastrophic failures in other theaters see David M. Glantz, *Soviet Military Deception in the Second World War* (London, 1989).

[43] I am indebted to Dr. Thomas Hone for much of this view as to what was going on in the United States Navy in the development of carrier aviation.

that war games at Newport displayed. As early as 1923, one of the games involved a Blue fleet of five aircraft carriers against an opponent with four. While some of the games displayed carriers in the mundane role of spotting for battleships, the Blue forces launched a strike of two hundred aircraft, armed with bombs and torpedoes, that crippled the enemy's carriers and one of his battleships. What is impressive is how much the participants learned from the game. As Steven Rosen has pointed out:

> Most important, concepts essential in the conduct of carrier war were worked out. The necessity of massing aircraft for strikes was highlighted. Rather than assigning aircraft to each battleship to act as its eyes, they were launched and kept in the air until large numbers could be assembled for an independent strike. The need for a coherent air-defense plan to coordinate the use of defensive aircraft was emphasized, and the commander of the Red fleet was faulted for failing to come up with such a plan. Control of the air was established as the first goal of operations.[44]

The U.S. Navy's approach to war gaming was similar to that of the German Army.[45] Neither military force used exercises or war gaming as a device to justify current, "revealed" doctrine or as a means to exclude possibilities. In other words, exercises aimed at illuminating possible uses for military forces and at suggesting what questions one might ask; they did not aim at providing "solutions" or answers. In peacetime, they were an educational vehicle for the officer corps. In war, gaming aimed at illuminating possibilities. For example, the German game for the Meuse crossing, which occurred in March 1940, *came to no resolution* on the critical question of whether the German armor should make the breakthrough without waiting for supporting infantry divisions to come up.[46]

One of the important components in successful innovation in the interwar period had to do with the ability of officers to use their imaginations in examining potential innovations. The atmosphere that institu-

---

[44] Steven Peter Rosen, *Winning the Next War, Innovation and the Modern Military* (Ithaca, 1991), p. 69.

[45] The similarities are striking. Both used exercises as a divergent funnel of possibilities rather than a convergent funnel to establish correct solutions. Exercises were experiments rather than drills. In terms of current U.S. practices the differences in philosophical approach are considerable. Exercises served the German Army and the U.S. Navy in the 1920s and 1930s as exploration rather than problem sets.

[46] Characteristically at the end of the war game the general commented "Well, I don't think you'll cross the river in the first place!" Guderian, with Hitler watching closely, replied: "There's no need for you to do so in any case." Heinz Guderian, *Panzer Leader* (New York, 1957), p. 71.

tions of professional military education fostered was central to developing such imaginative powers and thus to success in innovation. Where schools were more interested in inculcating an absolute doctrine – the case of the French Army and the Air Corps Tactical School – or where the values of military education never formed a significant portion of the officer corps' world view, the result was less successful or flawed innovation.

In this case, despite a few officers who educated themselves widely during the interwar period, the British Army remained prisoner of the cultural mores of its regiments. As Michael Howard has suggested, "the evidence is strong that the army was still as firmly geared to the pace and perspective of regimental soldiering as it had been before 1914; that too many of its members looked on soldiering as an agreeable and honorable occupation rather than as a serious profession demanding no less intellectual dedication than that of the doctor, the lawyer or the engineer."[47] The explanation for British failures early in the war had much to do with other factors, particularly underfunding by successive governments right through spring 1939. But the consistent failure of the British to improve throughout the war suggests a military culture that placed intellectual activity low among its priorities.[48]

FAILURE TO INNOVATE

The most serious failures in innovation during the interwar period had to do with the Italians. Where Anglo-American and German historians once placed the burden of Italian military failures on manpower and racial characteristics, recent scholarship has moved the burden of failure away from such racist interpretations and placed it precisely where it belongs – on the backs of an officer corps that failed its nation, its people, and itself.[49] A remark of General Ubaldo Soddu suggests the pervasive culture

[47] Michael Howard, "The Liddell Hart Memoirs," *Journal of the Royal United Services Institute*, February 1966, p. 61.

[48] Murray, "British Military Effectiveness: The World War II Experience." Professor Alan Beyerchen in reading this chapter noted: "If tradition and innovation in the past (particularly in peacetime) have been contradictory, then the challenge today is clear – the need is to create a "tradition of innovation" that breaks with the past on a consistent basis. But isn't that what science has been about all along? Should military education involve considerable history of science and history of technology? *Yes!*" Alan Beyerchen, communication to the author, summer 1994.

[49] As a recent work has suggested: "Mussolini had to take his generals as he found them: even the replacement of obvious incompetents was fraught with potential risk, given the monarchy's jealous special relationship with the military. The rigidity of the armed forces' seniority system ensured that replacements could only come from the topmost ranks, and

in the Italian military: "when you have a fine plate of *pasta* guaranteed for life, and a little music, you don't need anything more."[50] The Italian Army also presents an interesting paradigm for Third World military organizations in which the political system – as well as other factors – aims at the regime's political security rather than at the professional competence of its military forces.[51]

But the evidence throughout the interwar period suggests a wide-scale pattern of failures to innovate – one which reflects the larger problem of military ineffectiveness. In an examination of military effectiveness, Lieutenant General Jack Cushman, USA ret., commented on the performance of military institutions from 1914–1945 in the following terms:

> Thus in the spheres of operations and tactics, where military competence would seem to be a nation's rightful due, the twenty-one auditors' reports suggest for the most part less than general professional military competence and sometimes abysmal incompetence. One can doubt whether any other profession in these seven nations during the same periods would have received such poor ratings by similarly competent outside observers.[52]

1) Misuse of history

A substantial part of the failure to innovate is more complex than simple incompetence. For some military organizations there may be compelling reasons not to innovate or circumstances that circumscribe the possibilities. For the development of British carrier aviation, as Geoffrey Till's essay makes clear, the argument over Fleet Air Arm and the loss of most naval airmen to the RAF in 1918 created a situation that made innova-

---

nothing guaranteed such men would improve on the incumbents. The fundamental problem was the Italian general staff tradition: Custoza, Lissa, Adua, Caporetto." MacGregor Knox, *Mussolini Unleashed, 1939–1941, Politics and Strategy in Fascist Italy's Last War* (Cambridge, 1982), p. 16.

[50] When he was in command of the disastrous Italian invasion of Greece, Soddu spent his evenings in composing music for the sound tracks of Italian films, while his army collapsed. MacGregor Knox, *Mussolini Unleashed*, p. 57.

[51] One of the bizarre aspects of the Gulf War was the enormous overestimate of the capabilities of Iraqi military forces that U.S. military intelligence agencies made before the war. Yet any close examination of the Iraqi political regime could only have led to the conclusion that the Iraqi regime's military forces would not put up significant resistance. For such a study, published a year before the Gulf War, see Samir al-Khalil, *Republic of Fear, The Politics of Modern Iraq* (Berkeley, 1989).

[52] Lieutenant General John H. Cushman, U.S. Army ret., "Challenge and Response at the Operational and Tactical Levels, 1914–1945," in *Military Effectiveness*, vol. 3, p. 322.

tion, at least comparable to that taking place in Japan and the United States, almost impossible.

There are distinct barriers to innovation that appear in a number of the case studies in this book. Perhaps the most obvious is a wilful desire to discard history or to twist its lessons to justify current doctrine and beliefs. In 1924 the British air staff made explicit their rejection of the past in a memorandum to the Chiefs of Staff Committee. This staff study argued that the forces employed in attacking an enemy nation

> can either bomb military objectives in populated areas from the beginning of the war, with the objective of obtaining a decision by moral effect which such attacks will produce, and by the serious dislocation of the normal life of the country, or, alternatively, they can be used in the first instance to attack enemy aerodromes with a view to gaining some measure of air superiority and, when this has been gained, can be changed over to the direct attack on the nation. The latter alternative is the method which the lessons of military history seem to recommend, but the air staff are convinced that the former is the correct one.[53]

While military organizations, including air forces are not generally so willing to voice their disdain for past lessons, there appears to have been a pervasive belief in military organizations that failed to innovate successfully that the past is something that one can ignore.[54]

The result of disregarding the past in the RAF and the army air corps was a belief that long-range escort fighters were not needed, not possible, and irrelevant to the conduct of strategic bombing – the exact opposite of World War I's experiences. The only conflict in which aircraft had participated in large numbers, World War I, had underlined again and again that air superiority was *essential* to all air operations and particularly bombing operations. Without fighters, other aircraft inevitably suffered *prohibitive casualties*. Particularly damaging was the persistent belief through to late 1943 that long-range escort fighters were not needed when *actual combat experience* had indicated the opposite from the very beginning of the war.

If military organizations ignore the past, they also misuse it. In some cases this was almost unavoidable. The French, looking at a series of dis-

---

[53] PRO, AIR 20/40, Air Staff Memorandum 11A, March 1924.

[54] The United States Air Force has never been so explicit in its statements about history, but its pattern of behavior after each one of the wars that it has fought has been almost immediately to disregard the past and move smartly "back to the future." For a discussion of this behavior see Williamson Murray, "The United States Air Force: The Past as Prologue," in *America's Defense*, ed. Michael Mandelbaum (New York, 1989).

asters resulting from the offensives of 1914, 1915, and 1917, wrote off anything other than the most stylized, tightly-controlled "methodical battle." Their defeat in 1940 had the quality and inevitability of a Greek tragedy; but it is hard to see how they could have developed any other attitude towards offensive operations. Nevertheless, their interpretation of World War I was fundamentally flawed and historically inaccurate.

More difficult to explain is the reaction of most navies to the unrestricted submarine campaign of World War I. In retrospect, the Germans came all too close to breaking Britain's sea lines of communication in 1917. Yet, when the war was over, the *Kriegsmarine* wrote the U-boat off as a major weapon and based its hopes entirely on rebuilding the High Sea Fleet with battleships (and virtually no carriers).[55] Ironically, in 1936 Dönitz and his chief engineer pushed the naval high command to support technological developments aimed at providing U-boats with higher underwater speed – what would eventually become the Walter U-boat; but Raeder and the senior admirals displayed no interest in pushing technology for a form of naval war whose history they had dismissed.[56]

But it was not just the Germans who wrote off the submarine. The British, on the basis of their victory over the submarine and the development of sonar gave up on antisubmarine war and threw themselves entirely into insuring that Jutland would never again happen. As the naval critic Russell Grenfell commented in 1939:

> The great amount of battleship and heavy cruiser tonnage that has been laid down since rearmament started has been much in excess of our requirements in European waters and was, therefore, clearly designed for Far Eastern use as part of Sir Samuel Hoare's two-hemisphere fleet. At the same time we have been left seriously short of small ships for anti-submarine and anti-aircraft work in home waters. The Admiralty, therefore, seem to have been committing the grave error of preparing for ambitious operations in a far distant theater without first taking steps to ensure the safety of the home base.[57]

But it was left to the Japanese to make the most astonishing misuse of submarines in the war despite the fact that in the "long lance" torpedo they possessed the finest undersea weapon of the war. In the face of World

---

[55] As Admiral Erich Raeder, commander of the German Navy, suggested, carriers were "only gasoline tankers." Michael Salewski, *Die deutsche Seekriegsleitung, 1935–1945*, vol. 1, *1935–1941* (München, 1975), p.29.

[56] John Terraine, *The U-Boat Wars, 1916–1945* (New York, 1989), p. 617.

[57] Russell Grenfell, "Our Naval Needs," *Journal of the Royal United Services Institute*, August 1939, p. 495.

War I's lessons, and the observable evidence of the Battle of the Atlantic in 1940 and 1941, the Japanese still failed to attack Allied commerce in the Pacific: at the same time they devoted virtually no resources to protecting their own sea lines of communication.[58] In the end, they lost their merchant shipping to U.S. submarines, while they inflicted hardly any damage on the shipping of their opponents – an astonishing disfunction.

2) Rigidity

Rigidity is undoubtedly a fact of life in many military organizations – one which has exercised a consistent and baleful influence over institutional capacity to innovate. There are a number of areas where rigidity appears: the most obvious are the cases of doctrinal rigidity. In the case of the French there are reasonable explanations for why their offensive doctrine remained rigid throughout the interwar period.[59] More difficult to explain in the French Army is why its doctrine remained so rigid in its understanding of defensive warfare. There, the lessons *and* the practices of combat in 1918 had underlined the absolute requirement for defense in depth and for reserves at both operational and tactical levels. In the debates in the early 1920s Pétain argued that the army should base the Maginot Line on the concept of a defense in depth.[60] By the late 1930s, however, the French had largely abandoned such understanding in favor of a rigidity in defensive tactics that mirrored their approach to offensive war.[61] What lay at the heart of the French approach was a belief that the "methodical" battle was the only sensible route in all cases to effective utilization of military force without heavy casualties. Thus, defensive war remained in the same cocoon as offensive doctrine except that the French ironically given national strategy, displayed less interest in it.

The French also believed that the Germans *could not and would not*, in the end, perform in a radically different fashion from their own

---

[58] As a result much of the commerce that sailed from the west coast of the United States to support the great military buildup and campaigns in the Pacific sailed singly and without escort throughout the war.

[59] Any visit to the ossuary at Verdun should explain a considerable portion of the French rigidity.

[60] Doughty, *Seeds of Disaster*, p. 54.

[61] As Churchill records his meeting with the French high command on 16 May 1940: "I then asked: 'Where is the strategic reserve?' and, breaking into French, which I used indifferently (in every sense): 'Où est la masse de manoeuvre?' General Gamelin turned to me and, with a shake of the head and a shrug, said: 'Aucune.'" Winston Churchill, *The Second World War, Their Finest Hour* (Boston, 1949), p. 46.

forces.[62] To a certain extent this rigidity reflected an inability and unwillingness to recognize not only that their opponent possessed alternative options and conceptions, but that he might exercise those options. This was mirror-imaging of the worst sort. Immediately after the 1940 defeat, the French historian, Marc Bloch (a reserve officer and observer of the collapse), put his finger on a major cause for the disaster: "our minds [were] too [in]elastic for us ever to admit the possibility that the enemy might move with the speed which he actually achieved."[63]

Such rigidity marked others besides the French. Both the RAF and the U.S. Army Air Forces remained wedded to their belief that the bomber would always get through well into World War II. Therefore, they mistakenly argued that long-range fighter escorts were technologically impossible and then drove their procurement programs in that direction.[64] Such rigidity persisted right into 1943 in the case of the Americans and into 1944 with the British, despite the evidence from the Battle of Britain where bomber formations had proven anything but "self defending." But even stronger evidence that the Combined Bomber Offensive would fail without long-range escort fighters had come in the night and day operations against Germany in 1942 and 1943.

Even then bomber commanders displayed little real interest in long-range fighters until defeat stared them directly in the face. The final irony of their rigidity came in the fact that the P-51 arrived at the end of 1943 largely by accident: its creation more the result of the capitalist impulses of the manufacturer and chance than a result of innovation by air commanders eager to protect their bombers. Even more ironically, Bomber Command's campaign against Germany's cities was saved only by the intervention of Allied ground armies which captured the German early warning system and robbed the enemy of the radar plots required to concentrate his night fighters against the bomber stream.

The rigidity of such military institutions was enhanced by institutional biases against feedback that contradicted doctrine, conceptions, or preparations for war. French exercises aimed at inculcating current doctrine, the "revealed" truth, into units.[65] There was little learning, because

---

[62] The Iraqis fundamentally repeated this same mistake in 1991.

[63] Marc Bloch, *Strange Defeat* (New York, 1968), p. 45.

[64] At least that is what the chief of air staff informed Churchill in 1941. Webster and Frankland, *The Strategic Air Offensive Against Germany*, vol. 1, p. 177.

[65] There may be a larger cultural pattern here caused by the basic philosophy of the French educational system. By their very nature drills obviate feedback, while they instill a particular kind of rigid learning. Much of French education in class emphasizes dictation which inculcates a large-scale, deep cultural pattern on those who go through the system.

the high command had the "answers." Consequently there was no need for feedback loops to learn "lessons" from either exercises or combat. The British Army displayed no greater willingness or interest in learning from its exercises and possessed no effective system to disseminate experience throughout its units. Even during the war there is little evidence that the British incorporated battle experience into the training of those preparing to go into combat for the first time.[66]

There was ample information flowing back from the Middle East theater, but the Home Forces appear to have paid virtually no attention to such after-action reports. Each division working up for combat for the first time had to innovate almost entirely on its own. Hence tactical innovation came on the battlefield – a most expensive school. Robert Crisp, an armored officer in North Africa, described the results most graphically:

> Other officers told me of how they had seen the Hussars charging into the Jerry tanks, sitting on top of their turrets more or less with their whips out. "It looked like the run-up to the first fence at a point-to-point," the adjutant described it. The first action was very typical of a number of those early encounters involving cavalry regiments. They had incredible enthusiasm and dash, and sheer exciting courage which was only curbed by the rapidly decreasing stock of dashing officers and tanks.[67]

Such rigidity led organizations to shut off alternative paths that might have eased the way for military operations. The belief that the bomber would always get through led airmen to minimize the potential of the Luftwaffe to interfere with bomber operations. In the case of the RAF and the U.S. Army Air Forces, it resulted in a minimization of the need for technical support to aid the accuracy of attacks at night and bad weather. The measure of air effectiveness thus became the number of tons dropped, sorties flown, and acres of cities damaged or destroyed; air war became an end in itself, and real measures of effectiveness simply failed to interest the Portals and the Harrises.[68]

Surely the most rigid military organization in the interwar period was the one that Stalin created. By his devastating purges of the Red Army –

---

[66] See Murray, "British Military Effectiveness in the Second World War."

[67] Robert Crisp, *Brazen Chariots* (New York, 1960), p. 32.

[68] In fairness, there were of course airmen like Dowding, Tedder, Spaatz, and Doolittle. But only Dowding was at center stage at the beginning of the war; Tedder was in a peripheral theater and Spaatz and Doolittle did not make their mark on operations against Germany until 1944. Harris remained in command of Bomber Command to the bitter end, despite his willful flaunting of Portal's instructions.

and its sister services – the Soviet dictator insured the absolute political loyalty of the Soviet Union's military institutions. Most military innovation ceased; much of the Red Army's officer corps grasped after mindless conceptions of "revolutionary" warfare which severely diminished its capacity to fight *and* made it incapable of understanding how its enemy would fight.[69] The results led directly to the most costly military defeats in history, the consequences of which the Soviets escaped only by the appalling strategic and political misjudgments of their opponent.

IMPLICATIONS FOR THE FUTURE

The chapters in this study suggest some rough and ready parameters for successful innovation. First of all, one must not think in terms of individuals – future Dowdings or Guderians – in furthering innovation in coming decades. The history of the interwar period suggests that one needs to think in terms of creating an officer corps educated and encouraged to innovate – a far larger problem than selecting one single innovative officer. The education and value system of that officer corps are essential to effective innovation. Professional military education was clearly a major player in the process of innovation in the interwar period; it will probably be even more important in the future, but only if it provides the broad conceptual framework that innovation requires.

In the larger picture, educational values among officers require a dedicated commitment to their profession. Only that willingness to think through the business of war will allow leaders to see the potential of long-term innovations. Moreover, it is also essential that officers have connections with, and an understanding of, the technologies in a civilian world dominated by innovation and technology. Dowding's example is particularly apt as a model for officers who are capable of furthering innovation in the future.

Innovation in the interwar period also demanded that military organizations judge the potential and possibilities of future war in a realistic fashion. Here the "muddy boots" business of exercises and realistic war games lay at the heart of effective innovation. The developments of

---

[69] As John Erickson has suggested: "*The failure to comprehend the essentials of German military doctrine in a tactical, operational sense and German 'war doctrine' in its widest context was the prime cause of the disaster; the effect of this was and had to be devastating, for such a failure impeded and inhibited effective operational planning*" [italics in the original]. John Erickson, "The Soviet Union, 1930–1941," in *Knowing One's Enemies: Intelligence Assessment before the Two World Wars*, ed. Ernest May (Princeton, NJ, 1984), p. 418.

German armor doctrine and close air support and of carrier air by the Japanese and Americans provide useful examples of the relationship among education, doctrine, war games, and exercises. Where military organizations and high commands "knew" the answers and drove the solutions, the results were less than satisfactory and stifled effective innovation.

What are the possible implications for those who will innovate during periods of low budgets, major technological changes, and an uncertain strategic environment? First, specific, detailed plans to enhance innovation probably represent a nonstarter. Courses on innovation, or offices of innovation, or even the creation of innovation specialties within the services will only draw individuals interested in a safe "career" niche, rather than driving, imaginative crusaders for innovation. If anything, such efforts to institutionalize innovation will inhibit rather than foster the process. Innovation demands officers in the mainstream of their profession, with some prospect of reaching the highest ranks, who have peer respect, and who are willing to take risks. The bureaucratization of innovation – particularly in the current framework of the U.S. military – guarantees its death.

How then to encourage innovation in current and prospective climates? The best route appears to be one that fosters change in service cultures. There are a number of areas where one could push cultural changes to encourage rather than discourage the process of innovation. The difficulty with such an approach is that one can only achieve cultural change over the long haul, not an approach that Americans traditionally have favored.

Here then are possible areas where the military might ease the process:

1. The services must focus innovation within a realistic framework. They need to think in terms of fighting wars against real opponents, with real capabilities, and with real strategic and political objectives. Exercises and war gaming must take place in concrete scenarios against realistic opponents who have the potential to give "Blue" forces a difficult time. Such scenarios must examine all three levels of war: strategic, operational, and tactical.

2. The services must rethink their operational tempo and the large numbers of exercises run over the course of each year. The value of exercises, particularly in periods when resources are short, lies not just in their conduct, but in their planning, and especially in the "lessons-learned" analysis. The latter process must involve more than writing reports that no one reads, but rethinking doctrine, training, and edu-

cation at every level. The value of exercises in the end depends entirely on the willingness and ability of their participants to think through the implications of what has gone well and what has gone badly.

3. The services must ensure that "lessons learned" analyses aim at more than merely validating current doctrine and processes. During the interwar period the French made serious efforts to examine the lessons of the last war and learn from the innumerable exercises conducted in the 1920s and 1930s. But they also created a system in which exercises and study occurred within narrowly constrained limits that insured the sanctioned approach would again prove out. They learned what made generals and staffs happy, a clear case of self-fulfilling prophecy, at least until the Germans arrived on the banks of the Meuse in May 1940.

4. At every level the services need to think in clear measures of effectiveness. They need to consider in a realistic fashion exactly what it is they wish to do to potential opponents. And, as changes occur in war, new measures and methods will be needed. Moreover, it is crucial that military organizations maintain their standards of professionalism.

5. One needs to rethink professional military education in fundamental ways. A significant portion of successful innovation in the interwar period depended on close relationships between schools of professional military innovation and the world of operations. The U.S. military lost its belief in professional military education after World War II despite the fact that individuals, like Eisenhower and Spruance, emphasized the connection between success in World War II and education at Leavenworth and Newport. But any approach to military education that encourages changes in cultural values and fosters intellectual curiosity would demand more than a better school system: it demands that professional military education remains a central concern *throughout the entire career of an officer.* One may not create another Dowding and manage his career to the top ranks, but one can foster a military culture where those promoted to the highest ranks possess the imagination and intellectual framework to support innovation.

6. Finally, the services must encourage familiarity with nonlinear analyses to a greater degree than is presently the case. With the heavy emphasis on engineering in all the services, an emphasis that the officer acquisition programs encourage and in some cases mandate, the U.S. military has created a mind set that is not conducive to innovation. The officer corps in general thinks of innovation in a fashion similar to the way the Luftwaffe thought about innovation in World War II, in quantitative and qualitative terms of equipment and techniques rather than in conceptual terms.

Innovation in the next century demands extensive cultural changes in how the services do business and even in the moral parameters within which they view the world. Some recent U.S. military leaders have recognized the need for major changes in the cultural frameworks, but unfortunately such general officers remain in the minority. Until, however, there is a wider recognition of the difficulties involved in innovation, the services will not see significant change.

9

PATTERNS OF MILITARY INNOVATION IN THE INTERWAR PERIOD[1]

ALLAN R. MILLETT

Agitated by the American treatment of Japanese immigrants in 1907, the diplomats of *Nihon Teikoku* fired off a series of protests, and Theodore Roosevelt volleyed back in even more bellicose language. Before the diplomatic climate improved, the United States and Japan had survived their first war scare, and Roosevelt had sent a "Great White Fleet" of sixteen battleships on a leisurely round-the-world cruise. In the Philippines, the likely theater for a war with Japan, a new ensign in the U.S. Navy, Chester W. Nimitz, took command of the coal burning, 92-foot island gunboat *Panay*, most noted for its low speed, lack of electricity, and sluggish handling. Nimitz could not have imagined that his modest command would become the focus of an international incident with Japan thirty years later, nor that thirty-seven years later as commander-in-chief Pacific Fleet he would destroy the Imperial Japanese Navy with his own fleet of seventeen fleet carriers, six fast battleships, and more than 800 other warships, submarines, amphibious ships, and service vessels.

In the same year the Quartermaster Department of the U.S. Army tested twelve automobiles for military missions and reported that they were unsatisfactory to replace "the standard means of army transportation." The army of 1907, which numbered only 64,000 officers and men of its statutory strength of 88,000, had thirty regiments of infantry, fifteen of cavalry, and thirty batteries of field artillery that moved by foot and hoof, all standard means, and 126 companies of coastal artillery that didn't move at all. At the United States Military Academy a theatrical third-year cadet from California struggled to keep up with his classmates in Ro-

[1] In addition to the participants in the Military Innovation Project who read this essay, the author thanks Dr. John F. Guilmartin, The Ohio State University, for his expert review.

mance languages and mathematics. Thirty-seven years later Lieutenant General George S. Patton, Jr., found himself in command of the Third Army, which in autumn 1944 numbered approximately 225,000 men and 30,000 vehicles in two corps of nine divisions and two cavalry groups. Although stalled by the Germans on the Moselle River, Third Army's biggest problem was gasoline (400,000 gallons a day) for its tanks, self-propelled howitzers, mechanized infantry half-tracks, cavalry scout cars, and convoys of ammunition-bearing trucks.

In December 1907, the U.S. Army Board of Ordnance and Fortification directed the reluctant Chief Signal Officer to solicit bids for a heavier-than-air flying machine, a delicate contraption flown for the first time four years earlier by the Wright brothers, visionary bicycle-makers from Dayton, Ohio. In Nome, Alaska, eleven-year old James H. Doolittle went to grammar school and played boyhood games, all at ground level. Thirty-seven years later Lieutenant General James H. Doolittle, a wartime celebrity for his 1942 bomber raid on Tokyo, commanded Eighth Air Force, U.S. Army Air Forces, the principal American instrument for the destruction of urban-industrial Germany. On 24 December 1944, Doolittle sent 1,884 bombers and 813 escort fighters to bomb Germany as a Christmas Eve "maximum effort," just another workday in Operation "Pointblank." Although Allied bomber forces failed to bomb the Germans into submission as their staunchest champions had hoped, they eliminated the Luftwaffe as a serious threat, limited the Wehrmacht's mobility, and reduced German morale and living conditions to survival levels.

Only reckless historians make absolute claims for special periods of military innovation, usually labeled a "revolution" of some sort, technological or social. The "Military Revolution of the Sixteenth Century" now covers enough centuries to qualify as several revolutions. The advent of nuclear weapons brought contemporary proclamations of revolutionary change that predicted everything from Armageddon to world peace. Neither occurred despite little improvement in statesmanship and another twenty million war-related deaths. Nevertheless, the two decades between the two world wars produced many examples of military innovation, subsequently tested with few exceptions (e.g., the invention of more deadly poison gases) in World War II. The interwar period thus merits its special status as a period of importance in the development of military capabilities and strategic thought.[2]

---

[2] This essay is based on the chapters in this work as well as earlier work for the Office of Net Assessment, OSD: Allan R. Millett and Williamson Murray, ed., *Military Effectiveness*, vol. 2: *The Interwar Period*, 3 vols. (London and New York, 1988). The notes are designed to guide the reader to additional readings on interwar military history.

The interwar military innovators, of course, had the experience of World War I upon which to draw, and the operational concepts and military technology they developed had their roots in wartime experiments. By the end of World War I, for example, Chester Nimitz had commanded a submarine and directed the adoption of diesel engines in the navy's submarine force; George Patton had commanded a tank brigade in the Meuse-Argonne offensive; and Jimmy Doolittle became a crack flying instructor for the air service. With World War I as a shining example of the need to return *rapid* decisiveness to warfare, the armed forces of the victors and vanquished turned to the lessons of the Great War.

Only one innovation of the interwar period represented a truly new approach to war-waging: strategic aerial bombardment. If urbanized, industrialized nations fought again, the war could (should?) be decided without a great clash of navies and armies. Instead, one or another of the belligerents would bludgeon the other into submission with a bombing offensive against enemy population centers and economic infrastructure. Bombs would ruin the enemy's capacity to field military forces and to feed and house its population, who had sacrificed their relative immunity by performing war-work. The shock of bombing might even stir the cowardly masses to revolt: a new government would then sue for peace. Only surrender would bring relief from the inexorable destruction that fell from the sky through all conceivable defenses. However, other military innovations offered different alternatives to conventional ways of warfare, not a new departure such as strategic bombing. Strategic considerations played a role, but only one school of thought in naval warfare, for example, approached a strategic revolution. Some submarine theorists saw unrestricted commerce raiding as the maritime equivalent to strategic bombardment, a way to strike at political will and economic capability without forcing the battle on land. Instead, the cascade of military innovation in the interwar period focused on well-known problems of distance, time, weather, terrain, firepower, strategic and operational mobility, force structure, and the character of the enemy's armed forces.[3]

In one form or another, the concept of sea power had drifted through the history of maritime nations since the appearance of warships in the Mediterranean of antiquity. (Thucydides did not have to read Mahan.) To exercise command of the sea in wartime in the twentieth century, naval

---

[3] For a review of military developments in the interwar period, see Theodore Ropp, *War in the Modern World* (rev. ed., New York, 1971) chapter 9, and William A. Preston, Alex Roland, and Sydney F. Wise, *Men in Arms: A History of Warfare and Its Relationships with Western Society* (5th ed., Fort Worth, TX, 1991), chapter 17. For a recent analysis of World War I innovation, see Hubert C. Johnson, *Breakthrough: Tactics, Technology, and the Search for Victory on the Western Front in World War I* (Novato, CA, 1994).

planners could, through new technology, create concepts of a naval campaign that incorporated aircraft, surface vessels, and submarines in tridimensional task forces directed against the enemy's fleet. Strategic considerations drove four major navies to develop aircraft carriers, and two more to begin building carriers when war came. All major navies, however, incorporated long-range reconnaissance and bombing aviation into their plans. They simply chose the different risks of basing such forces on land rather than accepting the challenge of building and operating aircraft carriers. Naval forces sought to improve virtually every operational capability with radio communications and direction-finding, various types of optics and electronics for target identification and munitions guidance, improved marine engines powered by diesel oil, advanced electrification of warships to reduce crew size and increase efficiency, more destructive and varied forms of ordnance like torpedoes and specially fuzed shells, and more effective armoring and compartmentalized construction to reduce combat damage. Some developments simply proceeded as extensions of general theories of naval warfare. Others rested on specific contingencies. Although the United States and Japan had not yet clashed, their military planners believed that their inevitable conflict would be an air-sea war involving fleet actions and the defense and seizure of forward operating air and naval bases throughout the Pacific. Such conceptions in turn influenced naval development.

For postwar armies, the basic challenge was to link operational mobility and effective firepower through some form of mechanization and motorization. All the major armies of 1939 had tanks and trucks and even some kind of armored force of combined arms, however different their tanks, force size, organization, and tactics. In addition, army planners foresaw many uses for tactical aviation and battled the political and military allure of strategic bombing. Just how different major land battles in the future might be remained a puzzle, but no one wanted to create another Western Front. However, military innovations based on improved technology and refined doctrine did not constitute a revolution. If such there was, it came from political revolution, the ability of the one-party totalitarian state to mold its society for war. Germany, Italy, and the Soviet Union showed the way, but they had imitators in Spain and the Eastern European states. National conscription demonstrated this resolve to produce mass armies and hard-working, long-suffering civilian work forces, but, after all, republican France also had compulsory military service. Many European states had used conscription before World War I, so the real novelty of fascism and communism was their dedication to the concept that *all* people had an assigned role in the mobilized nation.

Military innovation in the interwar period proceeded within an international geopolitical environment of great uncertainty and strategic ambiguity. Three of the major belligerents of World War I – Britain, the United States, and France – showed no inclination to alter the *status quo*: nonaggression toward others, a pale commitment to international cooperation represented by the League of Nations, and equivocal support for the national self-determination that followed the collapse of the Austro-Hungarian, Russian, and Ottoman empires. Although they held their overseas empires, the three postwar democracies pursued foreign policies most responsive to economic and domestic social issues, not security concerns. Most strikingly these countries failed to form any sort of coalition, formal or tacit, in the interwar period, which reversed a pattern that characterized their relations from the 1890s until the outbreak of World War I. The military implication of the democracies' unilateralism was that each country pursued a separate strategic vision that drove it to create forces that were not articulated with any ally. The single exception was the French link to Poland and Czechoslovakia, middle-sized powers of *Mitteleuropa* capable of bedeviling the new Soviet Union and post-imperial Germany. This pattern received further reinforcement from the fact that Britain, France, and the United States had overseas possessions to defend, a mission that drew resources from the metropolitan armed forces, especially the armies.

The last major influence, hardly inconsequential, was financial. The three democracies staggered under postwar debt and deferred industrial modernization and public spending. Democratic governments showed little inclination to increase revenues by postwar taxation, and the portion of the money they did collect and spend for defense represented a minuscule portion of national incomes. France made the largest investments, between 4 and 5 percent of gross domestic product, and increased spending during the rearmament period of the late 1930s. Britain averaged around 2 percent, and the United States spent about the same portion of national wealth on defense. However, the democracies still made sizable investments in their armed forces. In terms of the actual dollar value of their defense investments, the democracies stood second (the United States), third (France), and sixth (Britain) among the seven great powers in 1933. By 1938 despite nearly doubling military expenditures in the 1930s, the three democracies had slipped to fourth (Britain), fifth (United States), and sixth (France) in military spending.[4]

---

[4] Paul Kennedy, *The Rise and Fall of the Great Powers* (New York, 1987), pp. 275–343, provides a revealing discussion of economics and military spending in the interwar period.

The four remaining major military powers – Germany, the Soviet Union, Italy, and Japan – are not easily categorized by type of regime and foreign policy for the entire interwar period, but by the mid-1930s all were dominated by authoritarian governments led (with the exception of Japan) by a charismatic leader who gloried in military might, if not military values. From its birth the Soviet Union armed itself to the teeth to repulse a world of real and potential invaders, to spread revolution, and to destroy internal opponents. Italy developed armed forces to enhance its power in the Mediterranean and Africa at the expense of France and Britain and to support another fascist cause in Spain, but its military spending, tied to a weak economy, remained low until 1935–1940.

In Weimar Germany civilian diplomats and politicians tried to reverse the Versailles agreement, but the Reichswehr already behaved as if it knew a new Supreme War Leader would one day return the Reich to greatness. In 1930 the German armed forces ranked last of the seven powers in defense spending; in 1933 they moved to fourth place, and in 1938 Germany became the world leader.

Although allied with the victors of World War I, Japan had not reaped the full benefits of victory its military leadership coveted, especially army generals who held territorial designs on Manchuria and China. With the militarists in political control, Japan jumped from last in military spending (1933) to third (1938), lagging behind only Germany and the Soviet Union. Even before a European war became inevitable, Japan began its military domination of Manchuria (1931) and conquest of China (1937). By the end of the 1930s a common thread linked the policies of revisionist states (Germany, Japan, the Soviet Union, and Italy): they would use military power in an aggressive manner (and fight if necessary) beyond the borders they held in 1930, and they would seek alliances or neutrality pacts with one another as long as they could find some mutual benefit. They aimed to wrest control of the international system away from the democracies, their empires and commonwealths, and their middle-power collaborators and regional surrogates. Military innovation found a fertile environment in the ambitions of the Soviet Union, Germany, Italy, and Japan.

None of the military powers escaped the effects of the Great Depression, but the three democracies, sensitive to popular will communicated through legislative bodies, faced constraints on military spending that authoritarian regimes did not. The barrier of political opinion, elite and mass, worked in varied ways. The most obvious was an unwillingness to authorize defense spending at the levels recommended by the defense ministries, but the official recommendations often reflected pessimistic assumptions about political acceptability even before estimates left defense ministries. Another barrier to military innovation came from those portions of the politically aware public who believed that increased military spending and new operational capabilities of any sort would make their nation's foreign policy more bellicose. World War I had done little to improve public confidence in the judgment of senior military leaders.

Britain may have been the home of the most articulate appeasers and pacifists, but France and the United States had their share of isolationists, anti-imperialists, accommodationists, Right and Left radicals, and economic isolationists. An ingrained distrust of government corruption and secrecy made it especially difficult for military innovators to work without some sort of external scrutiny, a condition barely tolerable to the technological entrepreneurs of the democracies. The only countervailing influence came from civilian, commercial innovation, an advantage that applied to research and developments in transportation and communications, but not in ordnance.

For all the restraints military powers confronted in the interwar period – political, technological, financial, conceptual and intellectual, and operational – innovation swept forward on a transnational basis. All seven of the world's leading military powers shared a common investigation of the principal innovations of the era. The "invented here first" syndrome that swept the world of military analysis after World War II had a basis in fact, but not in quite the way claimants intended. The insights of science and engineering that made military innovation possible – when linked with political will and the capacity of the armed forces to change – developed on an international basis. An *idea* of innovation (for example, the explosive potential of splitting atoms) might be held in common among scientists of every major military power, but the operational capability of the atomic bomb came first to the Anglo-American alliance. Knowing how and why innovation flourished or lagged from one nation to another is an essential step toward understanding the enduring dynamics of military innovation and the challenges of military reform.

There are four central problems in assessing interwar military innovation. The first is determining the influence of strategic context. For the rationalists who make military plans or explain them later as reasoned careful analyses, the principal influence on organizing armed forces should be the requirements of deterring or waging war against a knowable enemy in identifiable theaters of operations. The essence of justifiable innovation (and a barrier against frivolous change) should be strategic calculation and the analysis of perceived threat, now known as "mission area assessment and analysis." The second consideration is the influence of technological discovery upon the development of military capabilities. One persistent assumption about military technology (sometimes even championed by technologists) is that it creates requirements on its own; innovation, therefore, occurs when the tools of war demand that military organizations adopt and use them. This "technological imperative" argument often includes a coda: if we do not exploit this weapon, others will, so at least we must develop it in order to study it. Certainly technological possibilities shaped interwar innovation, but never provided *dei ex machina*. Instead, the third consideration, the political behavior of military organizations, played a pivotal role in encouraging, retarding, and channeling technology-based innovations. Military innovation, however, is more than a product of bureaucratic competition or collaboration, even if such relationships provide powerful explanations. Instead, the fourth factor one must investigate is critical parallel technological developments in the civilian commercial sector and the interaction of defense and civil leaders and technological professionals. Regardless of their relative weight, one must discuss all four considerations in explaining military innovation.

THE IMPORTANCE OF STRATEGIC CALCULATIONS

However incomplete an explanation for innovation, the history of the interwar period does demonstrate a relationship between strategic net assessment and changes in military capability. The influence of strategic calculations can nevertheless cover a range of factors: the anticipated enemy, anticipated theaters of operations, the immediacy in distance and time from the possible outbreak of war, the balance between deterring war or simply preparing to fight it, the likely length of a potential conflict, the role of allies, the "lessons of the last war" that shape perceptions of how or how not to wage future wars, and the anticipated requirement for joint air-sea-land operations. Even within these complex sets of interacting considerations, there are inherent barriers to complete rationalism. For example, the armed forces of the same nation may have conflicting visions on which enemy they are most likely to fight, which can make substantial differences in every other strategic calculation. Another inherent limitation is whether a nation intends simply to destroy an enemy's armed forces (and presumably regime) or whether it intends to annex or dominate a conquered territory; forces designed simply to destroy can be far different than those required for extended occupation. Last, strategic planning may run afoul of internal political debate about the wisdom of war and the legitimacy of waging it with certain forces like submarines or strategic bombers.[5]

Naval development in the interwar era showed the influence of strategic calculations. The rise of the aircraft carrier reflected such considerations, for the United States and Japan envisioned a naval war fought in the Central Pacific where land-based air would be scarce and bases vulnerable. Japan also considered carriers a complement to the battleships denied it by treaty. Britain, the third of the great carrier-aviation pioneers, did not have the same compelling rationale for investment in carriers for many reasons: 1) the lack of an enemy carrier-based naval air threat in Europe; 2) the presumed availability of land-based air for maritime missions; and 3) the requirement for large surface forces to control the approaches to the British Isles and the exits from the Baltic, the English Channel, and the Mediterranean. British naval planners, in addition, avoided shaping the Royal Navy for battle against one specific foe; instead, they followed a generalized theory of sea control, which tended to keep the Royal Navy rooted in its battleship past. The U.S. Navy and the Imperial Japanese Navy, the only navies that could actually threaten Britain, were far and away more interested in fighting each other.

Many of the same calculations influenced amphibious warfare development, again with Japan and the United States emerging as world specialists. The Japanese Army and Navy, however, developed much different amphibious capabilities, both designed for each service's principal strategic interest, China for the army and the Central Pacific for the navy. The Americans had the advantage of two services, the navy and marine corps, with a shared requirement to seize forward operating naval and air bases in the execution of War Plan Orange (the plan for war with Japan). The U.S. Army had no pressing amphibious mission and thus conceded the field until World War II. Although the other major military powers had naval infantry, none of them had progressed beyond the landing party and colonial infantry missions of the nineteenth century. The Royal Navy did not create any amphibious capability despite the fact that it had contingency plans in Asia that required such forces.[6]

---

[5] This section is drawn principally from the essays on the seven military powers in Williamson Murray and Allan R. Millett, eds., *Calculations: Net Assessment and the Coming of World War II* (New York, 1992). See also Michael Geyer, "German Strategy in the Age of Machine Warfare," in Peter Paret, ed., *Makers of Modern Strategy* (Princeton, NJ, 1986), pp. 527–597.

[6] For a review of interwar naval developments, see Clark G. Reynolds, *Command of the Sea: The History and Strategy of Maritime Empires* (New York, 1974), pp. 473–500. For

The development of interwar submarine forces reflected strategic assessment. Every major navy had submarines, for their potential as an economy-of-force weapon for coast defense had general appeal. The great naval powers – Britain, Japan, and the United States – regarded submarines as useful fleet elements in an extended naval campaign since they could conduct reconnaissance missions, skirmish with an advancing enemy fleet, and cut off a fleet from its bases or floating supply trains. Even Germany did not envision a major role for submarines as commerce raiders, despite some obvious advantages for cruiser warfare. Yet Japan, Great Britain, France, and the United States were vulnerable to commerce raiding. Even the Royal Navy's potential continental enemies could be hurt by submarines, for the Germans required control of the Baltic to reach their Scandinavian suppliers, while Italy depended upon materials from abroad it could not find on the continent. Only the Soviet Union was more or less immune to economic strangulation, but even it required safe maritime passage in order to use imported weapons.

The Japanese and American navies faced a common problem: their submarines had to have the range to cover the Pacific Ocean, but they could not develop the speed to keep up with the surface fleet. Therefore, submarine planners had to consider employment concepts not tied to direct fleet support. For the U.S. Navy this strategic appreciation focused upon Japan's vulnerability to commerce raiding. The United States did not have an equal exploitable weakness, at least in the Pacific, but the United States certainly had a problem with merchant shipping between Latin

other general perspectives on interwar naval rivalry, see especially Kenneth J. Hagan, *This People's Navy: The Making of American Sea Power* (New York, 1991), pp. 259–304; Edward S. Miller, *War Plan Orange: The U.S. Strategy to Defeat Japan, 1897–1945* (Annapolis, MD, 1991); Charles M. Melhorn, *Two-Block Fox: The Rise of the Aircraft Carrier* (Annapolis, MD, 1974); Stephen Roskill, *Naval Policy between the Wars*, 2 vols., (London, 1968, 1976); Kenneth J. Clifford, *Amphibious Warfare Development in Britain and America from 1920–1940* (Laurens, NY, 1983); H. P. Willmott, *Empires in Balance: Japanese and Allied Pacific Strategies to April 1942* (Annapolis, MD, 1982); James J. Sadkovich, ed., *Reevaluating Major Naval Combatants of World War II* (Westport, CT, 1990); Rear Adm. Yōichi Hirama, JMSDF (Ret.), "Japanese Naval Preparations for World War II," *Naval War College Review*, Spring 1991, pp. 63–81; Jost Dülffer, "Determinants of German Naval Policy, 1920–1939," in Wilhelm Deist, ed., *The German Military in the Age of Total War* (Dover, NH, 1985), pp. 152–170; Mark P. Parillo, *The Japanese Merchant Marine in World War II* (Annapolis, MD, 1993); Paolo Coletta, "Prelude to War: Japan, the United States, and the Aircraft Carrier, 1919–1945," *Prologue*, 23, Winter 1991, pp. 343–359; David MacGregor, "Innovation in Naval Warfare in Britain and the United States between the First and Second World Wars," PhD Dissertation, University of Rochester, 1990; David C. Evans and Mark R. Peattie, "Kaigun: Doctrine and Technology in the Imperial Japanese Navy, 1887–1941," in press.

American ports and its own in the Gulf of Mexico as well as commerce from Europe across the Atlantic Ocean. Based on its World War I experience, the *Reichsmarine* led the world in submarine tactics in the 1930s, but other navies were more advanced in the technical sense. No navy showed any special expertise or interest in antisubmarine warfare, even Germany. For the United States, Britain, and Japan the strategic reason for worrying about the safety of their merchant fleets would seem obvious, but their navies did not stress the mission, at least, in part, because of an unjustified faith in the active and passive undersea sound-detection systems of the era. As the experts in undersea warfare predicted, commerce protection would not be any nation's strong suit when the next naval war came.

The development of combined arms mobile land warfare also had roots in strategic realities, especially those of the Soviet Union and Germany. Continental nations whose fate had rested historically in the hands of their armies, Germany and Russia, both defeated in World War I, turned with a vengeance to reform their armies. Within a decade they had surpassed the best 1918 standards, probably held by the British Army. In the 1920s neither nation considered the other an inevitable enemy; instead they shared the distinction of being defeated pariahs. If any one enemy dominated their concerns, it was Poland. The "army of the future," as Charles de Gaulle labelled it, took shape in the 1920s in Germany and Russia. Although tank forces gave this new model army its novelty, it gained its operational strength through a combination of armored units, motorized infantry, mobile artillery, radio communications, and the support of tactical aviation. The Germans and Russians also experimented with air-landed forces in order to give the battlefield additional depth. They correctly calculated that the command and logistical systems of modern armies offered tempting opportunities for demoralization and disorganization as well as destruction. All one needed was some sort of combined shock action, speed, and firepower to make a penetration into the bowels of an enemy army and then exploit the breakthrough with rampaging mobile forces.[7]

Strategic imperatives drove Germany and Russia into mobile land warfare since neither had decisive natural barriers, and neither could count on allies to distract their enemies. Of the other military powers,

---

[7] Kenneth Macksey, *Tank Warfare: A History of Tanks in Battle* (London, 1971), pp. 70–107; Richard Ogorkiewicz, *Armored Forces* (New York, 1970); John T. Hendrix, "The Interwar Army and Mechanization: The American Experience," *Journal of Strategic Studies*, 16, March 1993, pp. 75–108.

only Japan and France had a compelling reason to develop armored forces. Japan shared a common border with the Soviet Union in Korea and Manchuria and wanted to eliminate Russian influence in Asia. France, however, unlike Japan, did not emphasize mobile forces since it chose to defend its border with Germany with a belt of forts and heavy artillery, supported by infantry and automatic weapons. If it had not had an open border with Belgium, its armored forces (three light mechanized divisions and one real armored division by 1939) would have been even smaller.

The United States, Britain, and Italy also developed armored units, but all three armies saw no reason to stress mobile forces since geography and politics combined to give them safe or defensible borders; Italy could be invaded by land only by limited routes through the Alps, Britain by sea, and the United States through its barren and roadless southern and northern borders or by the lakes and two oceans. In the case of the U.S. Army, strategic considerations and tactical concepts combined to create light armored forces that it could rapidly deploy and employ in North America and the insular possessions against non-European enemies; cavalry tactics stressed the use of such forces in the exploitation phase of an attack (which favored speed) while infantry tactics dictated that heavier tanks spearhead the advance of foot-soldiers – at their speed. The logistical limitations of armored forces also weighed heavily upon American thinkers like Daniel Van Voorhis and Adna R. Chaffee, Jr. Armored forces, moreover, added little capability to the armies deployed for colonial peacekeeping, strikebreaking, or imperial defense. Given the utility of cavalry for the constabulary missions of the American, British, and Italian armies, it is something of a marvel that they had any tank forces at all.

Strategic considerations also shaped development of tactical aviation. The pioneers of armored warfare also led the way in integrating air and ground attacks in land operations. The Soviet Union and German armies kept their aviators focused on well-identified missions: air control over the battlefield, interdiction, offensive close air support, reconnaissance, and artillery fire direction. The Japanese and French army air arms followed suit, although they tended to seek and gain more autonomy for air defense and offensive air operations against an enemy's aviation base structure. The U.S. Army Air Corps and the Royal Air Force provided only minimal tactical air support for their national ground armies; they raised token participation to a high art form. The independent Italian Air Force also drifted away from close cooperation with the army since it argued that it should devote its scarce resources to air defense and selective bomber operations against either military or civilian targets.[8]

The tactical support of field armies, even the interdiction of enemy supply lines, did not appeal to airmen of several major military powers. Airmen wanted to win wars by themselves or at least contribute to victory on the basis of equality. The instruments of air power would be long-range bombers or pursuit aircraft, which would protect the homeland (air installations first, cities second) from bomber attack. Logical inconsistency did not bother air power missionaries: how could one defend against an unstoppable bomber force? With the possible exception of the Soviet Union, army pleas for tactical aviation, especially for ground attack missions, were crowded from the hearing. The airmen of the interwar period (with the exception of naval aviators) envisioned an independent role, which would provide a strategic alternative to the conventional naval and ground campaigns of the past. All the air forces required was independence and public funds.

The strategic uncertainty of the interwar period influenced development of radio communications security systems (and their compromise) as well as the exploitation of radio beams (radar) for navigation and target acquisition. World War I had demonstrated that if speed were the major consideration, the radio provided a method of transmitting orders and information that no other means could duplicate. (Clarity and security could be quite different matters.) Using sound and radio beams to identify objects under the sea or in the air had also had small tests in World War I. The German commitment, nurtured in the armed forces in the 1920s, to reverse the decisions of Versailles enhanced a national determination to pioneer in communications, radar, and cryptology. None of the *status quo* powers (Britain, the United States, and France) showed the same sense of urgency. In fact, the British, leaders in the field in 1920, let their edge in radio intelligence fade and focused their communications intelligence operations on the Soviet Union rather than Germany and Japan until well into the 1930s. The American military radio intelligence specialists concentrated on a potential enemy, Japan, from the start, but confronted many other problems that strategic vision could not address.

---

[8] Robin Higham, *Air Power: A Concise History* (New York, 1972), pp. 59–87; Lee Kennett, *The First Air War, 1914–1918* (New York, 1991); Richard P. Hallion, *Strike from the Sky: The History of Battlefield Air Attack, 1911–1945* (Washington, DC, 1989), pp. 45–75; Lee Kennett, *A History of Strategic Bombing* (New York, 1982); Richard R. Muller, "Close Air Support in Germany, Great Britain, and the United States, 1918–1941," chapter 4 in this study.

Japan, in turn, focused on American and Russian diplomatic and military radio traffic. In fact, the Poles and Dutch, menaced on several frontiers, did much of the pioneering work in cryptology, and their work influenced advances in Eastern European countries and France.[9]

The strategic evolution of the interwar period – from the flirtation with collective security in the 1920s to the hostile coalitions of 1939 – influenced military innovation, perhaps more than any other factor. The best-known wishful thinking of the era, like the British "Ten Year Rule" or the Kellogg-Briand Pact, gave the period's politicians a *post hoc* image as spineless romantics or, in the case of the Axis, crazed warmongers. Such stereotypes have some element of truth, but the relationship of force development and strategic planning displayed much cool consideration of each nation's likely future military needs. When Adolf Hitler pushed the Third Reich toward war, he placed primary emphasis upon the land forces of the Wehrmacht; when Franklin Roosevelt and the Congress collaborated to modernize and enlarge the U.S. armed forces, they invested in the navy and in army aviation. Even when a nation did not create forces adequate to its future needs – most clearly in the case of France and Italy – military planners, nevertheless, made their estimates on reasonable strategic assessments. "Fog" and "friction" may have shrouded the international politics of the potential belligerents, but those politics did define strategic planning and force structure.

TECHNOLOGY AND INNOVATION

The military technology of the interwar period provided ample promise for innovation, but it did not determine the process of change. In many cases, state-of-the-art technology would not support operational concepts developed by military visionaries. The range of technical challenges for scientists, engineers, designers, and technicians mounted from the demands of military officers. One trend that applied to almost every major military power was the redistribution of funds from armies to navies and air forces; only Britain and Germany did not fall into this pattern as both

---

[9] Ronald Lewin, *ULTRA Goes to War* (New York, 1978); David Kahn, *The Codebreakers* (New York, 1967); Ernest R. May, ed., *Knowing One's Enemies: Intelligence Assessment before the Two World Wars* (Princeton, NJ, 1984); M. Postan, D. Hay, and J.D. Scott, *Design and Development of Weapons: Studies in Government and Industrial Organization* (London, 1954); James Phinney Baxter, *Scientists against Time* (Boston, 1946); Robert Morris Page, *The Origin of Radar* (New York, 1962); Alan Beyerchen, "From Radio to Radar: Interwar Military Adaptation to Technological Change in Germany, Great Britain, and the United States," chapter 7 in this study.

the Royal Navy and German Army retained their primacy, but both surrendered some assets to their national air forces. The RAF, in fact, replaced the Royal Navy as most-favored service in 1938. Aviation development in every country, even the Soviet Union, absorbed a growing
budget share, much of it for aircraft and base structure. The movement
of armed forces from labor-intensive to capital-intensive organizations
continued apace, even though the continental European powers and
Japan maintained large armies. The emphasis on air and naval forces dramatized the parameters of military requirements: weapons had to travel
farther and faster, arrive when and where planned, and destroy enemy
forces with increased efficiency. Strategic bombing and maritime commerce destruction created new technical requirements. Not only was this
period rich in its menu of technical needs, but it is striking how close all
of the major powers remained in their technological advances.[10]

The demands for improved military technology fell into several broad
categories, all of which, with the possible exception of the exploitation of
the electro-magnetic spectrum, required greater progress in engineering
rather than scientific breakthroughs. In ordnance technology, the great
advances predated World War I and found validation in that conflict, and
postwar development in infantry and artillery firepower simply carried
the established lines of innovation along known paths. For example, the
infantry of the major powers fought World War II with advanced models
of the bolt-action, magazine-fed rifles used in World War I; the United
States made the only major improvement by adopting the semi-automat-
ic Garand M-1 rifle. In terms of close combat weapons, all the nations
soon had a submachine gun or machine pistol of some sort, but used long-
service World War I automatic pistols, which had replaced revolvers except in the British Army. (The United States did not replace the Colt
M1911 automatic until the 1980s.) The Germans led the way in developing lighter machine guns, but the British and Japanese nearly kept pace
with the Bren gun and Nambu. Although the Italians had many shortcomings, adequate firearms from Beretta and Mannlicher-Carcano was
not one of them. Russian technology might not thrill the engineers of
Browning of Belgium, but Russian firearms met the test of rugged de-
pendability, high rates of fire, and reasonable accuracy. Similar developments characterized artillery, where the major powers had shifted from

---

[10] For perspectives on technology, see Bernard and Fawn M. Brodie, *From Crossbow to H-
Bomb* (rev. ed., Bloomington, 1973) and Martin van Creveld, *Technology and War* (New
York, 1989). For a case study, see Gary E. Weir, *Building American Submarines,
1914–1940* (Washington, DC, 1991).

guns to howitzers during the World War, which improved shell-weight and accuracy in indirect fire. The chemistry of explosive charges, fuzing, and shell design improved incrementally to enhance destructiveness.

The mechanization and motorization of weapons and crews brought the most striking changes in land war. The development of mechanized vehicles and whole families of military wheeled vehicles required designs that had no analog in commercial manufacturing (armoring, weapons stations), but tracked and wheeled vehicles did have civilian variants, especially for agricultural production and longhaul trucking. Military planners understood the basic trade-offs: speed, range and fuel efficiency, weight, serviceability, weapons platform suitability, crew conditions, and engine power. Most of these engineering parameters could be altered if designers received guidance on what capabilities military users valued most; major engineering breakthroughs did not determine superiority, but represented optimum combinations of technology focused on a clear set of performance criteria that correctly anticipated future operational requirements, e.g., will this tank outgun its likely opposite number?

Other questions eluded engineering answers. When should a particular tank (or truck or reconnaissance vehicle) be adopted and produced? In what numbers? What was its likely service life? Should we wait for the next model? The U.S. Army, for example, started a major motorization program in 1926, but could not decide upon a family of standard vehicles until 1939, largely because the Quartermaster Department wanted technical standardization to determine which models it purchased. In the meantime the army acquired 360 different types of vehicles and a spare parts problem that baffled commercial manufacturers as well as army users. Even for the vaunted German Army, procurement decisions eluded rationalization, and only Hitler's insistence and profligate funding in the late 1930s gave the German Army some technological advantages, and even these advantages were neither universal, nor long-term.[11]

The technology of the interwar period also demonstrated the interna-

---

[11] Jac Weller, *Weapons and Tactics* (London, 1966), pp. 86–160; A.J. Barker, *British and American Infantry Weapons of World War II* (New York, 1978); Martin van Creveld, *Fighting Power: German and U.S. Army Performance, 1939–1945* (Westport, CT, 1982); U.S. War Department, *Handbook on German Military Forces 1945*, introduction by Stephen E. Ambrose, reprint (Baton Rouge, 1990); Daniel Beaver, "Deuce and a Half: Selecting Army Trucks, 1920–1945," in John Lynn, ed., *Feeding Mars: Logistics in Western Warfare from the Middle Ages to the Present* (Denver, 1993), pp. 251–270; Mark K. Blackburn, "A New Form of Transportation, the Quartermaster Corps and Standardization of the United States Army's Motor Trucks, 1907–1939," PhD Dissertation, Temple University, 1992.

tionalization and civilianization of military research and development, indeed the mechanization of Western Civilization. Largely influenced by German and British manufacturing, Japan created its own motorized and mechanized army units, more than adequate against the Chinese, inferior to the Russians, and largely sacrificial against the Americans. Japan's major limitation was not its ability to produce machine tools or finished products, but its access to raw materials and its thin base of national research and development for mass-producing industrial enterprises. In Europe even nations not noted for their industrial prowess could produce quality weapons because they shared a common cultural commitment to engineering progress. Some weapons even showed national characteristics: German optics were universally admired and German direct fire cannons (mounted or not) spread terror throughout the continent, but German drive trains for vehicles lacked staying power.

On the other hand, American armored vehicles did not match the same superb performance of army jeeps and trucks; the Americans seemed predisposed to carry, not fight. The Japanese favored designs of grace and lightness in fighter design, which maximized maneuverability but not survivability. The awesome Russian T-34 tank moved on a reengineered version of an American-designed chassis, a J. Walter Christie invention rejected by the U.S. Army for its weight and complexity, but Russia took great pride in the ruggedness of its home-built artillery and rocket-launchers. The British initially produced high-quality tanks and vehicles, but lost their edge in both design and production. In the British case, the lack of approved doctrine and high-command advocacy and understanding allowed tank development to become a victim of industrial competition (Carden vs. Vickers), decision making by committee, an institutional obsession with engine power and vehicle speed that reflected the influence of cavalry officers and civilian motorcar designers, and the lack of any standing experimental force. Even the creation of a Director of Mechanization did not cut the Gordian knot or free the one insight that eventually dominated German tank development: the tank gun should dictate the design of the vehicle that carried it. The British and Germans, in the meantime, marveled at the rugged reliability and sheer numbers of American wheeled vehicles. The industries that gave the world Daimler-Benz and Rolls Royce engines and automobiles failed to adjust fully to a Chevrolet war.[12]

The pattern of technological diffusion also occurred in aviation and

[12] Duncan Crow, *Tanks of World War II* (New York, 1979); Col. G. MacLeod Ross, *The Business of Tanks 1933 to 1945* (Ilfracombe, UK, 1976).

naval force development. In aviation strong ties between the air forces and commercial aviation moved research and development along at a breathtaking pace. In the United States, the army air corps worked in concert with such aviation companies as Boeing, Douglas, Martin, and North American to develop long-range, multi-engine aircraft, suitable for carrying cargo or dropping bombs. Junkers and Messerschmitt enjoyed similar relations with the Luftwaffe, and in Japan the great *zaibatsu* like Mitsubishi, Nakajima, and Kawasaki served the military aviation community. Aviation designers made one major leap forward when improved engines and materials made the metal-sheathed, closed-cockpit monoplane the standard world design, but only the development of the jet engine (well advanced in 1939 and operational in the Luftwaffe in 1944) freed aviation from the limitations of the internal-combustion aerial engine. American engineers, however, pushed the engine performance development to obtain long-range cruising capability and reliability. They exceeded their limits in the design of the Cyclone R-3350 engine for the B-29, an engine with flammable magnesium components for weight reduction. High performance, however, had to be modified to fit real-world operating conditions: simplicity for repair and maintenance in wartime conditions, durability, and dependable ordnance delivery. None of these requirements proved easy to meet. The Germans and Japanese took the high performance road for their fighters, but could not match American bombers, more specifically powerful, multi-row radial engines designed by the Pratt & Whitney and Wright corporations. The RAF produced superior interceptors, but not attack aircraft. The Soviets introduced superior ground attack aircraft, but nothing else of note, and the French and Italian air forces left the war early as identifiable national forces, but not just because they had poor aircraft. In fact, one Italian fighter, the Macchi 200/202, had performance characteristics like the Hawker Hurricane, and the French Dewoitine D-520 proved competitive with the Bf 109 in 1940.

The new global competitiveness in aviation had long been a feature of naval development. The establishment of carrier forces proved no different. By the end of World War I, the Royal Navy enjoyed a clear lead in carrier technology and concepts, but forfeited that advantage to the Americans and Japanese within twenty years. The Royal Navy at one time had eighteen vessels that could carry several aircraft (not just the floatplanes of battleship and cruiser gunfire spotters), and it knew as early as 1918 that carrier deck landings could be made with the help of arresting cables. Building and operating carriers required no great engineering feats with the possible exception of a later development, the catapult. Arguments about carrier design did not focus on technical barriers, but operational concerns. What anti-aircraft armament should a carrier carry? What armoring should be required to retard flight deck and hanger deck damage and fires, the latter much worse than a bomb hole or two. Should the flight deck be clear or provide space for an island superstructure from which officers could conduct flight operations and maneuver the carrier? What was the optimal design for carrier aircraft and what ordnance should they carry? What kinds of aircraft should have priority in the carrier air group: fighters for fleet protection, torpedo bombers, scout-dive bombers? Although these issues had technological components, they were essentially operational matters.

In the strictest technological terms, the all-round leader was probably the Imperial Japanese Navy, whose officers and engineers pursued technical excellence with a single-mindedness that amazed Europeans. The Japanese produced the best naval torpedo, the Americans the worst, but neither fully exploited radar for fire-control purposes as quickly as they might have. The Germans improved their capability with the development of snorkels for the U-boats' air-breathing diesel engines, but the Americans (not the Japanese or the Germans) waged the most successful submarine campaign in World War II. The Japanese and the Germans produced state-of-the-art battleships and lost most of them to Allied naval forces that enjoyed no special technical advantages. No submarine could operate with sufficient speed to participate effectively in fleet operations until the development of nuclear-powered engines. Nevertheless, relying on stealth and positioned at maritime chokepoints, submarines could wreak havoc on a battlefleet weak in antisubmarine warfare – as the Imperial Japanese Navy learned.[13]

Radio communications, communications interception, cryptology, and radar probably represent the most dramatic technological changes from one world war to the next. They all demanded cutting-edge engineering and design expertise, especially to make transmitters and receivers suitable for ships and aircraft. Two-way radios in tanks and fighters made effective *Blitzkrieg* and air defense possible. Although electronic warfare had its origins in World War I, it did not emerge until after the war as a major element of modern warfare with measures and countermeasures multiplying like *amoebea*. The communications technology and

---

[13] Geoffrey Till, "Adopting the Aircraft Carrier: The British, American, and Japanese Case Studies," chapter 5; Carl Boyd, "Experimentation and Innovation: Japanese Submarine Technology, 1904–1945," paper for the Southwestern Social Science Association meeting, March 1993.

the know-how for electronic warfare had developed as an international activity, so that all the World War II belligerents were capable of intercepting and decoding at least some of their enemies' transmissions. The Anglo-American Ultra operation is best known, but the Japanese and Germans both broke some Allied naval codes, and all the participants attempted to understand the enemy order-of-battle by plotting the location of radio transmissions. The most important mechanical innovation in communications penetration was the primitive computer, which could speed and simplify the mathematical unraveling of the complicated German Enigma code system. Electronic data-processing also made it possible to conduct operational analysis, both applied with good effect to air and naval warfare.

The dominant role of civilian scientists and technicians in the development of communications technology suggests a pattern that applies to aviation and motor vehicle development as well. Where technology promised dual-use military and commercial applications, it tended to advance more rapidly than "arsenal model" development (as in ordnance, for example) in which research and engineering development remained the domain of military managers and technicians. The influence of commercial manufacturers upon military procurement nevertheless could be counterproductive. The lobbying power of Detroit truck-builders prevented the U.S. Army from buying standardized components for trucks it would assemble itself to field requirements; Congress mandated that the army had to buy fully assembled trucks from civilian sources and hope that its contract process alone would insure satisfactory performance. Yet on the whole, the integration of civilian and military development in the transportation and communications industries worked well in all the industrialized nations. The opposite model – of innovation controlled by bureaucratized military technicians – offers miscalculations of mythic proportions. The worst example is the U.S. Navy's Mark XIV torpedo, designed with such secrecy and wishful thinking that its many defects required two years of wartime remedial engineering to correct – while Japanese vessels steamed along unharmed.

The key to technological exploitation became not so much the revolutionary character of inventions and processes, but creation of a management and logistical system that made the *application* of technological advantage possible. Marvellous German tanks and Japanese fighters could not run on air or even octane-depleted gasoline. The German submarine menace faded not only under the application of airborne radar, but under the fleets of Liberty ships that replaced sunken merchantmen. The Imperial Japanese Navy could not maintain the operational tempo of U.S. carrier groups because it did not have service ships capable of underway replenishment; an appropriate symbol for victory in the Pacific would be a black hose between a carrier and a fleet oiler. The proximity fuse that ruined Axis forces in 1944 had to be manufactured in the millions. The United States made 2.6 million trucks in World War II, more than the combined output of the Axis powers. Only amateurs discuss military technology without considering logistics.[14]

THE ORGANIZATIONAL POLITICS OF INNOVATION

Critics of the pace and scope of military innovation in the interwar period blame the general staffs and service organizations for not appreciating the potential of new forms of operations. Their contempt for high-ranking conservatives exceeds only their unhappiness with political leaders who failed to provide wise guidance and sufficient money, although the reformers' insatiability knew hardly any bounds. The history of interwar innovation actually provides a more ambiguous picture. If the developments were not enough to win World War II immediately, they certainly represented giant steps beyond the capabilities of 1918. In fact, the military politics of innovation, that is, interservice and intraservice struggles of the era, both advanced and retarded reform. Although general and service staff conservatism and interservice rivalry certainly slowed innovation, these organizational barriers to change were not immutable. When reformers found a way to institutionalize their new forms of warfare, they met with success. Messiahs are not enough; they need disciples. One way to further the idea of innovation is to institutionalize it in the services' school systems. This ploy, exploited in every nation, proved inadequate. Another battleground for the hearts and minds of the officer corps has always been the writing of doctrinal manuals. Again, the reformers met with considerable success, but these victories of words on paper did not suffice. In the end, the only sign of victory for reformers were real operational units that could perform wartime missions.

The pioneers of armored warfare, Germany and the Soviet Union, reflected the organizational realities that their ground forces dominated their military culture and their general staff organization. Thus, air force and naval opposition played a less consequential role in retarding devel-

---

[14] For an introduction to strategic and operational logistics, see Kenneth Macksey, *For Want of a Nail: The Impact on War of Logistics and Communications* (London, 1989); Maj. Gen. Julian Thompson, *The Lifeblood of War: Logistics in Armed Conflict* (London, 1991); and Martin Van Creveld, *Supplying War: Logistics from Wallenstein to Patton* (London, 1977). See also Thomas Wildenberg, "Chester Nimitz and the Development of Fueling at Sea," *Naval War College Review,* pp. 52–62.

opment than did intra-service conflict – disputes that pitted armored warfare missionaries against their brethren in the infantry, artillery, and cavalry. Yet, the concept of armored warfare (*Blitzkrieg*) involved the active participation of all three traditional branches and played to their own interest in motorization and limited mechanization. Moreover, Russo-German experimental exercises in the 1920s, held in Russia, demonstrated a potential for mobile warfare that appealed to both nations' general staffs. By the 1930s the concept of armored warfare received official sanction in the school systems of both nations and had entered the doctrinal publications on land warfare. The critical development, however, was an army commitment to further operational experimentation. Germany created its first armored division in 1935, the Soviet Union its first in 1931.

In France the officer education system provided instruction in the new concepts of mobile ground warfare and the varied uses of aviation in support of ground operations. Instructors at the *École Supérieure de Guerre* waxed rhapsodic on the potential of tank forces, wedded to mounted infantry and mobile artillery, for operations of high intensity, heavy firepower, and surprise. French theorists, led by Lieutenant Colonel Robert-August Touchon, advocated a type of warfare not unlike the armored warfare concepts associated with German and Russian reformers. Their ideas found no favor outside their classrooms. The French general staff and army field commanders remained champions of fixed fortifications, heavy artillery, and the deliberate battle. Moreover, the cavalry dominated mechanization policy, which meant fitting new *chars de combat* to traditional light cavalry roles. The French army formed its first true armored force, the *Division Cuirasée de Reserve*, in 1939, but did not change its operational doctrine. The Polish Army followed the same pattern of development, but was caught in 1939 changing its light mechanized battalions to larger, heavier armored forces, but artillery immobility slowed the conversion.[15]

Of the remaining military powers Japan, Italy, and the United States took the same course as France, mechanizing both cavalry and infantry regiments without fundamentally changing those arms' missions. The U.S. Army came closest to breaking into the armored warfare community with the creation of the 7th Cavalry Brigade (Mechanized) in 1938, a

[15] General Lucien Robineau, "De 1919 à 1939, La Pensée Militaire Française a l'Ecolé de la Grande Guerre des Armes Nouvelles," and Tadeusz Panecki, "Armée Polonaise, 1918–1939," papers delivered at the XIX Colloquy, International Commission of Military History, Istanbul, Turkey, 17–24 July 1993; E.C. Kiesling, "Reform – Why?: Military Doctrine in Interwar France," paper, Society for Military History, 8 April 1994.

unit designed to experiment with *blitzkrieg* techniques. Britain, however, offers the best example of innovation lost. When World War I ended, the British Army stood astride the military world in every aspect of ground warfare: theory, operational experience, advanced technology, and trained personnel. The compounding impact of many factors wiped out this advantage, including a lack of strategic rationale and paltry funding. The British Army, however, bore much of the responsibility for its withered expertise. The tradition of regimental soldiering meant that the officer corps could find little incentive to look toward a major change; the Royal Tank Corps attracted good officers, but they remained mavericks within the service as a whole. The army staff did not encourage experimentation after the exercises of 1934, but the army command and staff college at Camberley did, indeed, examine the mechanization problem in the abstract. The army's aversion, however, to published doctrine denied the officer corps the chance to debate armored warfare except in its classrooms, messes, and (gently) in military publications. To some degree J.F.C. Fuller and Basil H. Liddell Hart provided the public forum for such a discussion, but their thirst for celebrity status, book royalties, and international influence as well as their vitriolic attacks on the army's leadership made their writings suspect with the political and military establishment. To the degree that the British Army retained any interest in armored warfare, that interest reflected the work of little-known serving officers such as Percy Hobart and others.[16]

---

[16] For a transnational discussion of the changes in mobile land warfare and tactical aviation, see Williamson Murray, *The Change in the European Balance of Power, 1938–1939* (Princeton, NJ, 1984) and "Armored Warfare: The British, French, and German Experiences," this study, chapter 1. For important national discussions, see Robert A. Doughty, *The Seeds of Disaster: The Development of French Army Doctrine, 1919–1939* (Hamden, CT, 1985); Edgar O'Ballance, *The Red Army* (London, 1964); Herbert Rosinski, *The German Army* (New York, 1966); General André Beaufre, "Liddell Hart and the French Army, 1919–1939" and Robert J. O'Neill, "Doctrine and Training in the German Army, 1919–1939" in Michael Howard, ed., *The Theory and Practice of War* (New York, 1966), pp. 129–165; Correlli Barnett, *Britain and Her Army, 1509–1970* (New York, 1970), pp. 410–423; Field Marshal Lord Carver, *The Apostles of Mobility* (New York, 1979); Shelford Bidwell and Dominick Graham, *Fire-Power: British Army Weapons and Theories of War, 1904–1945* (London, 1982); Gen. Sir Frederick Pile, "Liddell Hart and the British Army, 1919–1939," in Howard, *Theory and Practice of War*, pp. 167–183; Tim Travers, *How the War Was Won: Command and Technology in the British Army of the Western Front, 1917–1918* (London, 1992); John Sweet, *Iron Arm: The Mechanization of Mussolini's Army, 1920–1940* (Westport, CT, 1980); David E. Johnson, "Fast Tanks and Heavy Bombers: The United States Army and the Development of Armor and Aviation Doctrines and Technologies, 1917 to 1945," PhD Dissertation, Duke University, 1990; and Harold R. Winton, "Tanks, Votes, and Budgets: The Politics of Mechanization and Armored Warfare in Britain, 1919–1939," paper, Society for Military History, 9 April 1994.

The advance of mechanized warfare from theory to operational capability could cause an institutional ripple-effect that could slow other worthwhile innovations. In the U.S. Army, for example, the mechanization of selected cavalry regiments complicated the problem of creating mobile field and anti-aircraft artillery. Motorization and mechanization policy as an administrative and logistical problem rested in the hands of the Ordnance and Quartermaster Departments, and their struggles over standardization caused much difficulty, but the conflicting operational requirements of the combat arms (each of which had a Washington office and chief of arm to bedevil the general staff) provided another set of barriers to motorization. The field artillery could buy adequate trucks to pull howitzers along roads to support the standard infantry division, but it could not find a commercial truck that it could modify for the speed and off-road operations of mechanized cavalry. It investigated self-propelled, tracked vehicles, but the available chassis could not accommodate even a 75 mm. gun or howitzer, let alone carry sufficient ammunition or the 105 mm. howitzer that artillerymen wanted to make standard throughout the army. Horse-drawn artillery could support horse cavalry regiments or infantry operating in rough terrain; as late as 1939 one of three divisional direct support artillery regiments was horse-drawn for this reason. But the army still had left unresolved the matter of mechanizing or motorizing the field artillery that supported tank regiments. The provision of mobile anti-aircraft artillery followed a similar course. The coast artillery corps, which assumed the mission of air defense in 1920, found the air corps a strong ally for fixed base defenses, but the ground army wanted mobile anti-aircraft batteries with weapons it could employ on existing vehicles. The 3-inch gun, favored by the chief of coast artillery, could meet mobility requirements (it was one-third the weight of a 105 mm. gun), but it could not fire fast enough or use advanced shells. The anti-aircraft planner filled the gap with rapid-fire 37 mm. cannon and the Browning .50-caliber machinegun in various configurations, but such weapons could engage only low-flying (and relatively slow) aircraft. Like its tanks, the army's mobile field and anti-aircraft artillery had too much *Blitz* and too little *Krieg*.[17]

The champions of strategic bombardment forged a union between service identity and separate mission that fundamentally shaped air policy in

---

[17] Frederick W. Spencer, "A Slow March to Military Effectiveness: The Motorization of the United States Field Artillery, 1917–1941," MA Thesis, The Ohio State University, 1994; Byron E. Greenwald, "The Development of Anti-Aircraft Artillery Organization, Doctrine and Technology in the United States Army, 1919–1941," MA Thesis, The Ohio State University, 1991.

Britain and the United States, came close in Germany, and had its impact in the Soviet Union, Italy, France, and Japan. The Royal Air Force, the first to win independence, pinned its future on mounting bomber attacks; as an independent service it enjoyed an influence unmatched by any other force with the possible exception of the Luftwaffe. The U.S. Army Air Corps, however, almost achieved an equal status, drawing personnel and money from a starved ground army until it earned virtual independence with the creation of General Headquarters Air Force and its assigned bomber forces in 1934.

The Air Corps Tactical School, Maxwell Field, Alabama, formed an alliance with the staff of the office of the chief, air corps and pilot-instructors at the U.S. Army Command and General Staff College to make strategic bombardment the conventional wisdom, despite the beleaguered counterarguments of Claire Chennault, Lewis H. Brereton, and George C. Kenney, who held more eclectic views of air power. The bombardment instructors at the Air Corps Tactical School, who included such influential officers as Kenneth N. Walker, Donald Wilson, Muir Fairchild, and Harold L. George, pioneered bombardment targeting and operational doctrine, but they did not invent the strategic bombardment mission, which had become the favored role of the Air Corps by 1930.

The Royal Air Force Staff College, founded in 1922, became a center for the propagation of the faith in bombing, not a source of original thinking on the strategic uses of air power. Its commandants and curriculum reflected the ideas of the RAF's founding father, Air Marshal Sir Hugh Trenchard. In both the army air corps and the RAF the gospel of the great prophets – Trenchard and Billy Mitchell – retained its power through the work of their well-placed disciples, who then reshaped their air forces as strategic offensive weapons. In Germany, Hermann Göring, no fool, saw no special political advantage in challenging either Hitler or the army general staff, who viewed air power as an instrument of traditional land campaigns. The Luftwaffe dabbled with long-range bombers – which could obviously also perform important reconnaissance roles – but it remained essentially a "balanced" force.[18]

---

[18] Robert F. Furtrell, *Ideas, Concepts, Doctrine: A History of Basic Thinking in the United States Air Force, 1907–1964* (Maxwell AFB, AL, 1971); Maurer Maurer, *Aviation in the U.S. Army, 1919–1939* (Washington, DC, 1987); Williamson Murray, *Luftwaffe* (Baltimore, MD, 1985), pp. 1–27; John Terraine, *The Right of the Line: The Royal Air Force in the European War, 1939–1945* (London, 1985), pp. 3–92; Col. Frank P. Donnini, USAF, "Douhet, Caproni and Early Air Power," *Air Power History,* 37, Summer 1990, pp. 45–52; Robert T. Finney, *History of the Air Corps Tactical School, 1920–1940* (Washington, DC, 1992); Mark Clodfelter, "Pinpointing Devastation: American Air

Aviation politics largely prevented strategic bombardment from becoming the principal mission of the air forces of Italy, France, the Soviet Union, and Japan with the navies and armies (and their ministries) casting their weight against the independent mission. Italy's *Regia Aeronautica*, under the influence of Guilio Douhet and Count Gianni Caproni, began its independent existence in 1923 as a strategic bombardment force, but, under army pressure and without external patrons it shifted back to tactical roles in the 1930s. Its major accomplishment was to block the development of naval aviation. The French Air Force became an independent service in 1933 and preached strategic bombardment as its principal mission; a change of air ministers and military leadership (plus the threat of the Luftwaffe and interceptor obsolescence) pushed it back to air defense and army-support missions in 1937. Even though it doubled its budget in the 1928–1938 decade, it did not double its capabilities because it did not train enough pilots. The Red Air Force also created an independent bombardment force, but abandoned the mission (cowed by the technological challenge of bomber development) in 1939, influenced, no doubt, by the purge of its senior officers in Stalin's decapitation of the professional officer corps. Japanese aviation forces developed upon service lines; the army and navy created their own squadrons within their conceptions of the requirements of a land and naval campaign, neither of which emphasized strategic air war. Ironically, the Japanese Navy built fighters and bombers with greater range than the army's aircraft, fashioned for a China war. Therefore, the navy flew long-range missions in China, since only it could. Japanese bombing had much the same character as German operations – terror bombing of major cities against virtually no air defenses. The ideas that gave the world Guernica, Warsaw, and Rotterdam also contributed the examples of Nanking and Chungking.

Even when air forces did not embrace strategic bombing, they did not necessarily work very hard at making tactical support of the armies effective. Service interest through ministries, general staffs, school systems, and major commands worked to inhibit air-ground operational integration. Characteristic of their whole interwar military experience, the British made major contributions to the theory of tactical air operations,

Campaign Planning before Pearl Harbor," *Journal of Military History*, 58, January 1994, pp. 75–101; Allan D. English, "The RAF Staff College and the Evolution of British Strategic Bombing Policy, 1922–1929," *Journal of Strategic Studies*, 16, September 1993, pp. 408–431; Scot Robertson, *The Development of RAF Strategic Bombing Doctrine, 1919–1939* (Westport, CT, 1995). For a general analysis of ground and air doctrinal formation, see Barry R. Posen, *The Sources of Military Doctrine: France, Britain, and Germany between the World Wars* (Ithaca, NY, 1984).

including close air support, but the RAF and fleet air arm paid no attention to the theory since it never received official endorsement. Although the RAF had useful combat applications of tactical aviation in the Middle East and India, imperial policing shaped neither doctrine nor organization. In truth the British aviation community spent much of its time fighting itself over air defense, strategic bombing, and naval missions, not army support. To the degree that air forces accepted tactical missions, they argued that aircraft and ordnance characteristics made close support too difficult, and urged instead that air forces attack enemy supply lines and reserves. The dramatic exception was the aviation force of the U.S. Marine Corps, which identified the requirements for close support in its study of air roles in amphibious warfare. Even though they had limited operational practice and primitive radios, the marines found that ground controllers could direct air strikes against enemy targets and prevent attacks on friendly forces, provided both the ground forces and attack squadrons practiced together. Having both types of units within a single service with an agreed-upon doctrine made this cooperation much easier.[19]

Innovation within the navies faced diverse institutional barriers, some overcome, some not. The Royal Navy never developed an influential inner circle of senior naval aviators committed to carrier aviation; its leakage started with the temporary disestablishment of the Royal Naval Air Service during and after World War I and the transfer of 60 thousand aviation personnel to the Royal Air Force. Even when the fleet air arm returned in 1924, it answered to both Admiralty and Air Ministry in material matters. Naval flying-officers, in the meantime, were so integrated into the surface navy that they never emerged as a self-conscious corps with special missions, while few reached high rank anyway.

The Americans and the Japanese, on the other hand, provided bureaucratic homes and special opportunities for aviators. Both soon boasted a generation of "air admirals" that numbered William A. Moffett, Joseph Mason Reeves, Ernest J. King, John H. Towers, Kakuji Kakuta, Jinichi Kusaka, and Chuishi Nagumo. The Japanese and American naval air arms developed not just flying schools and bases, but places on naval staffs, war colleges, and operating forces. At the U.S. Naval War College, for example, 127 of 136 strategic war games tested War Plan Orange, with a growing aviation component. Such experiences drew line and air officers to a common strategic vision. By the early 1930s American and Japanese fleet exercises, which tested theoretical doctrine as well as contingency plans, had major air components. Moreover, in the U.S., navy

---

[19] Muller, "Close Air Support."

and army air corps shared a coast defense mission, which stimulated not just competition (e.g., the pursuit of flying records), but also horizontal planning and developmental cooperation. In the other navies, however, aviation officers never reached positions of influence in their own services, let alone competed with the armies and air forces.[20]

The development of submarines escaped the travails of interservice rivalry, although antisubmarine warfare did not, so internal navy politics had even more influence than they did in naval aviation. No navy, not even the *Reichsmarine*, actually stressed submarines, but the British, Japanese, German, and United States navies saw them as useful and relatively cost-effective fleet supplements. Other navies still viewed submarines as principally limited to coastal defense. Submarine forces profited by their very invisibility, which meant far more than submerging below the surface. Despite moral outrage over the "cowardly" conduct of cruiser warfare by the U-boats of 1917–1918, none of the naval treaties of the era did much to constrain submarine development, and the Anglo-German Naval Agreement of 1935 actually encouraged submarine building. The Germans profited from developmental agreements and experiences with the navies of the Scandinavian countries, Russia, Spain, and Japan which allowed otherwise proscribed testing.

Submarines did decline in significance in naval budgets; for the U.S. Navy, submarine investment represented only about 10 percent of the total investment in warships, 1925–1941, and 5 percent of ship-operating costs. Submarine numbers could change dramatically. The U.S. Navy willingly cut its force by two-thirds (ninety-three to thirty-two) from 1930 to 1940 to rid itself of obsolete boats; it wanted more fleet submarines than it obtained, but feasible higher numbers would have made little difference in 1941–1942 since the defects of the Mark XIV torpedo frustrated all American submariners. Even if submariners did not have the same bureaucratic niche as aviators, they included ambitious officers drawn to submarine service by the opportunity for command at junior rank, technological challenges, and hazardous duty pay. In fact, the risks of submarine service – the constant threat of unplanned dives and oxygen poisoning – gave submariners a moral authority and collective self-confidence within their navies that even aviators recognized.[21]

No such aura attached itself to amphibious forces except, perhaps, to

---

[20] Michael Vlahos, *The Blue Sword: The Naval War College and the American Mission, 1919–1941* (Newport, RI, 1980).

[21] Till, "Adopting the Aircraft Carrier"; Holger Herwig, "Innovation Ignored: The Submarine Problem – Germany, Britain, and the United States, 1919–1939," this study, chapter 6.

the landing force – if it handled its public relations as deftly as the U.S. Marine Corps. But the Royal Marines, French *Infanterie de la Marine*, Japanese Special Naval Landing Forces, and the naval/colonial infantry of other nations did not seize the amphibious assault role. Well-crafted and published doctrine does not alone explain the differences between the U.S. Marine Corps and its international counterparts; both the British and the Japanese had conceptual documents as prescient as the *Tentative Manual for Landing Operations* (1934). The British and Japanese (as well as the U.S. Army) taught amphibious operations in their command and staff colleges. The edge developed by the U.S. Marine Corps came in the execution of doctrine. With formation of the fleet marine force in 1933 and the organization of two permanent brigades for amphibious operations, the marine corps gave the navy a valued capability to be tested in the annual fleet exercises. For a time in the 1930s, however, the Japanese Army had the largest and most capable amphibious force, tested in China in 1932 and 1937, but, once ashore, in China the army deemphasized its amphibious forces, then resurrected them in 1941. Only the Imperial Japanese Navy established permanent landing forces and transports, but in insufficient numbers. Again, the British showed an admirable capacity to talk, plan, teach, write, and test theories of amphibious operations (or parts thereof), but in the end provided no significant capability, largely because none of its armed services or an element therein (the Royal Marines) saw much future in the mission. The U.S. Marine Corps, on the other hand, bet its institutional life on amphibious warfare and profited from the help of fellow true-believers in the ranks of the U.S. Navy.[22]

The pattern of military politics in the development of radio communications, cryptology, and radar is generally one of collaboration and bureaucratic deftness. Secrecy prevented any general discussion of doctrine or education on electronic warfare; there is no evidence that the specialty suffered in the developmental stage, although naval officers, at least the Americans, may not have fully appreciated radar's importance at the beginning of World War II. Naval and air communications specialists gave radios, radar, and communications security matters the greatest organizational commitment, but the armies did not ignore the capability. Most nations worked with cryptology and radar within a consortium of intelligence and communications specialists. The combination of technical knowledge and security probably discouraged bureaucratic and political

---

[22] Allan R. Millett, "Assault from the Sea: The Development of Amphibious Warfare between the Wars – the American, British, and Japanese Experiences," this study, chapter 2.

meddling by officers and civilians alike, but such splendid isolation benefitted development, not utilization, of radio communications and electronic intelligence. Even if peacetime intelligence organizations did not receive adequate attention, the communications branches did; the U.S.
Army Signal Corps, a well-funded and respected branch, handled electronic warfare matters in its Signals Intelligence Service. In Germany,
Britain, and Japan the air forces worked with radar with the greatest urgency since they all had air defense and navigation problems. The greatest danger was that a service would tilt too much toward one technical
option; the Japanese Navy, for example, stressed radio direction-finding
intelligence, which gave it excellent data on the U.S. Navy. It became,
however, complacent about the security of its own radio communications
despite the knowledge that the U.S. Navy had broken Japanese codes in
1921–1922 and knew the Japanese negotiating positions at the Washington Conference. In the U.S. Navy, on the other hand, the code-breakers
and radio traffic analysts (with their army allies) created a competitive relationship that eventually benefitted the American armed forces, even
though it ruined careers and reputations. The war between the electronics warfare specialists more than compensated for the mediocre order-of-
battle work of the army's Military Intelligence Division and the Office of
Naval Intelligence.[23]

One cannot attribute the shortcomings of radar development and radio intelligence to interservice competition or faulty organization. In retrospect, all the major military powers might have done better and profited from larger investments, but the barriers differed from country to
country. One common problem was simply the highly sensitive nature of
the research and operations, especially in the United States, where radio
interception was illegal. Even though the Americans targeted Japan as the
focus of their electronics warfare, the services had so few Japanese linguists that they could not translate many of the messages they could de-
code. The Japanese, on the other hand, targeted the United States and the
Soviet Union, but tilted toward decoding Soviet traffic, less difficult to
penetrate than American codes. Moreover, the Japanese intelligence specialists did not do well in their analysis or evaluation of their sources. The
cultural specialists, such as they were in Japan and Germany, worked for
the foreign ministries and offered little help in military evaluation. The
British, however, gloried in their work in electronics warfare and com-

23 Jeffrey M. Dorwart, *Conflict of Duty: The U.S. Navy's Intelligence Dilemma,
1919–1945* (Annapolis, MD, 1983); Col. Bruce W. Bidwell, *History of the Military Intelligence Division, Department of the Army General Staff, 1775–1941* (Frederick, MD,
1986).

munications since interservice conflict, for once, did not condition progress in this area, not the least because it required low investments in people and equipment and did not produce large operational units. Moreover, the British, especially RAF and Royal Navy officers, drew upon a national and organizational conceit that they worked more cleverly than other international officers and could fully exploit the emerging radio technology. In this case, the *amour propre* and the reality almost matched. Britain had a tradition of superior work in military intelligence that drew its inspiration from the fact that the British preferred to dominate their alliances with knowledge and money and to let others do the fighting. World War I had reinforced this bias, and electronic warfare offered an unparalleled way for Britain to reassert its intellectual primacy in the conduct of war at the strategic level and in air and naval operations.

Organizational factors within the armed forces of the major military powers, honed by doctrine and officer education, can offer important insights into the process of innovation, but they cannot be separated from strategic vision and technological developments. There is no historical example of perfect innovation in the interwar period, but those changes that had the greatest impact on World War II had some common characteristics. They involved a commitment that equated organizational survival with the performance of an important wartime mission that might make the difference between victory or defeat. In general, the less the role involved interservice cooperation, the more likely it was to prosper. Moreover, the military function required some bureaucratic (general staff) representation and operational expression (armored division, amphibious brigade, strategic air force, submarine flotilla) that could be exercised and tested. Senior military commanders and their staffs are not won over by manuals and staff college studies. Unless the prophets can point to field successes (even just in training) and a role in important contingency plans, their disciples will be regarded only as *soi-disant* military experts who confuse elegant operational ideas with real combat capability.

CIVIL-MILITARY COLLABORATION:
A CRITICAL INGREDIENT

The history of interwar military intervention demonstrates the importance of civilian participation in the process of change at two levels, political and technological. Both levels of interaction are important, not the least because they compensate for interservice and intraservice friction. Innovators need allies in the civilian political and technological establishments as well as patrons within their service. For the democracies political involvement seemed obvious and structured since politicians in both parliamentary systems (Britain and France) and a presidential-legislative system (the United States) believed in civilian control of the armed forces and exercised it through various mechanisms: the budget process, legislation, investigations, personnel actions, and administrative control at the ministerial level. In a theoretical sense a parliamentary system, especially one supported by an expert career civil service like that of Britain, should have worked best to review innovative proposals, but it did not. On the other hand, the British provided the best linkage between the political and scientific elites and thus insured *avant garde* research and development. The United States actually provided a fertile political environment for the innovators, but had more limited ties between the politicians and the technologists. The French fell somewhere in between, with relative passivity in both political and technological intervention. The institutional posture of the European interwar officer corps rewarded low risk-taking.

The authoritarian regimes did not encourage any collaboration that posed a challenge to the court politics of Hitler, Mussolini, Stalin, and the Japanese military elite. Innovators tended to be sychophants like Guderian or martyrs like Tukhachevsky. Military rebels against the *status quo* could and did appeal directly to their great supreme commanders, but they could just as quickly fall from favor, and they usually alienated many of their superiors and peers in the officer corps. In addition, the sheer demands upon an interested political leader like Hitler (or Winston Churchill, for that matter) meant that innovation was unlikely to be institutionalized or consistently supported through long years of development and testing. Government domination of major corporations, research centers, and the military-technological community of inventors and engineers created a brake upon change; political purges and repression sent a legion of talented European civilian scientists and technologists into angry exile. The refugee nuclear physicists who pushed the United States and Britain toward the atomic bomb were only the most famous examples of the flight from fascism and communism. The Japanese experience followed a different tangent since the leaders of the armed forces commanded obedience in the name of Emperor and mobilized a submissive scientific-engineering establishment to military research through official rewards or the threat of severe sanctions. As in other matters, the Japanese suffered most from bureaucratic compartmentalization and a tendency to avoid confronting major risks and opportunities.[24]

---

[24] Brian Bond and Martin Alexander, "Liddell Hart and DeGaulle: The Doctrines of Limited Liability and Mobile Defense," and Condoleezza Rice, "The Making of Soviet Strategy" in Peter Paret, ed., *Makers of Modern Strategy*, pp. 598–623 and pp. 648–676; Donald Cameron Watt, *Too Serious a Business: European Armed Forces and the Approach of the Second World War* (London, 1985).

The larger the perceived organizational implications of innovation in terms of changes in operational doctrine, personnel reallocation, and cost, the more likely the military bureaucracies would block change in the three democracies. If a reform required some degree of interservice cooperation or even mission modification, the professional judgments (often simple prejudice) of the senior military planners were likely to prevail. In Britain military professionalism countered civilian curiosity; the committee system diffused argument and sharpened pessimism rather than encouraged debate on defense options. When government leaders saw in 1937 that another German war had become likely, they misjudged the timing by three years since the military staffs thought that 1942 was the year of greatest peril. Only gallant politicians like Winston Churchill, Leslie Hore-Belisha, or Lord Swinton inquired into the details of military planning and operational thinking – with career termination the result for the latter two. Armored warfare, amphibious warfare, and close air support had their military champions, but they had to wait to reach senior positions and for a crisis to sway the debates to their favor. Many serving officers, even distinguished soldiers, shrank from finding political champions for their reform ideas. Opposing political-industrial coalitions did not develop to enliven public debate. Although Britain hardly fell into the same political camp as the authoritarian regimes, it, too, put a high priority on consensual government that limited consideration of defense policy outside the quiet offices of Whitehall.[25]

Ground forces reform in France and the United States showed many of the same results as Britain's with some messy differences in method. In France political ideology prevented a serious challenge to the creation of a mobile army, for such a force would have to be manned with career soldiers or long-service regulars; the Left in France found the idea of a professionalized army anathema and hounded Charles de Gaulle and his only political champion, Paul Reynaud, into near silence. The traditional Right – allied as it was with the conservative French officer corps – saw no reason to challenge the prevailing wisdom in the French Army that "firepower killed" and belonged to fixed heavy artillery and protected infantry. Moderate realists who appreciated the German threat believed the allies would provide decisive help. The coalitions of warring ministers, members of the National Assembly, and competing generals grappled with each other throughout the interwar period, but they did not strug-

---

[25] Gaines Post, Jr., *Dilemmas of Appeasement: British Deterrence and Defense, 1934–1937* (Ithaca and London, 1973); John Lee, "Sir Ian Hamilton and the Military Reform Debates of the Early 1920s," paper delivered at the XIX Colloquy, International Commission of Military History, 17–24 July 1993, Istanbul, Turkey.

gle over military reform, but the "soul" of France as the French Army reflected it.

In the United States civilian-military political collaboration arose in the
institutionalized conflict between the Presidency and Congress over domination of military policy. For a decade after World War I the army lived
on war surplus in both equipment and ideas; only an experimental
armored force (1927) championed by Secretary of War Dwight Davis
showed any pulse of radical change. The term of General Douglas
MacArthur as army chief of staff stirred Congressional interest in the
army's mechanization and motorization plans, in part because MacArthur
was a Republican partisan. Leading Democrats in the House of Representatives, especially Ross Collins, suddenly provided a platform for army
critics of the mass artillery – infantry army the War Department planned
to mobilize for the next war. When Malin Craig replaced MacArthur in
1935, he found Congress sympathetic to his plans to focus on the material modernization of designated combat-ready regular and national guard
divisions. In addition, members of Congress always showed great interest
in army contracts. Motorization came faster to the national guard than
the regular army. Although army munitions plants hardly bid against one
another (outside of the War Department, that is), car and truck manufacturers certainly expected their representatives to favor their industrial
constituencies. Corporation officials took great interest in, for example,
the Industrial Mobilization Plan, the army's bible for economic mobilization completed in 1930.[26]

The politics of aviation development showed collaboration at the
political and technological level in every major military power. Only governments could make commercial aviation an initial success through investments in aircraft and landing sites; military research and development
subventions fed new aviation manufacturing companies. With the exception of the United States, airlines became national companies, and even in
the United States such carriers as Pan American World Airways enjoyed

---

[26] For French military politics in the interwar period, see Paul-Marie de la Gorce, *The
French Army* (New York, 1963), pp. 180–282; Alistair Horne, *The French Army and
Politics, 1870–1970* (New York, 1984), pp. 43–65; Ronald Chalmers Hood III, *Royal
Republicans: The French Naval Dynasties between the World Wars* (Baton Rouge,
1985); and Martin S. Alexander, *The Republic in Danger: General Maurice Gamelin and
the Politics of French Defeat, 1933–1940* (Cambridge, 1992). For the German experience, see Michael Geyer, "The Dynamics of Military Revisionism in the Interwar Period," in Diest, ed., *German Military in the Age of Total War*, pp. 100–151. For the U.S.
Army, see Russell F. Weigley, "The Interwar Army, 1919–1941" in Kenneth J. Hagan and
William R. Roberts, eds., *Against All Enemies: Interpretations of American Military History from Colonial Times to the Present* (Westport, CT, 1936), pp. 257–277.

favored relationships with government. It was no accident that Congress provided basic legislation to commercial and military aviation in the same year (1926). By the 1930s, however, American commercial aviation had a momentum of its own, which meant its most dramatic improvements came in the design of large, multi-engine aircraft for trans-continental and trans-oceanic flights. American designers, for example, produced sturdier, yet lighter wings and fuselages, which improved range and payload capacity. Aviation enthusiasts and developers formed an international community in league against all doubters great and small, and they shared technical insights, manufacturing techniques, and aerodynamics research. The great air shows and air races dramatized this collaboration. (Who indeed had tank fairs?)

The air power entrepreneurs even profited in a backhanded way from the moral opponents of population bombardment, whose principled and strident effort to outlaw bombing kept the issue alive in every nation, especially the western democracies. The concept of bombing industrial "choke points" rather than blocks and blocks of apartments engaged academic analysts as well as moralists. (And the United States *did* keep one secret, the invention of the Norden bombsight in 1931.) Aviation pioneering became a commitment in which occupation and nationality counted less than one's commitment to flight and the conquest of all the barriers created by weather and physics; often military developers had a hard time reminding themselves and their colleagues that they were creating aircraft for war, and aircraft ordnance and weapons guidance systems actually lagged behind the development of engines, airframes, and navigation systems.[27]

Whatever the flaws of aviation development, they were not problems of civil-military relations at any level. Civilians, however, could be selective in the *type* of military air missions they embraced. The ethical questions aside, strategic bombardment rallied civilians and military leaders alike; the shared role of prophet in Italy fell to Douhet and Caproni, in

[27] David MacIsaac, "Voices from the Central Blue: The Air Power Theorists," in Paret, ed., *Makers of Modern Strategy*, pp. 624–647; Roger E. Bilstein, *Flight in America* (Baltimore, MD, 1984), pp. 41–123; Benjamin S. Kelsey, *The Dragon's Teeth? The Creation of United States Air Power for World War II* (Washington, DC, 1982); Wayne Biddle, *Barons of the Sky: From Early Flight to Strategic Warfare, The Story of the American Aerospace Industry* (New York, 1991); Edward Homze, *Arming the Luftwaffe: The Reich Air Ministry and the German Aircraft Industry, 1919–1939* (Lincoln, NE, 1976); Faris R. Kirkland, "French Air Strength in May 1940," *Air Power History*, 40, Spring 1993, pp. 22–34; Richard K. Smith, "The Intercontinental Airliner and the Essence of Airplane Performance, 1929–1939," *Technology and Culture*, 24 July 1983, pp. 428–449.

the United States to Billy Mitchell, Donald Douglas, and William E. Boeing. Glenn Curtiss and Glenn Martin worked for naval aviation alongside of Admiral Moffett and Congressman Carl Vinson. But civilians tended to shy away from interservice and intraservice disputes on aviation roles and missions. British political leaders did little to pressure the RAF toward anything other than strategic bombardment; the same pattern occurred in government treatment of the U.S. Army Air Corps, which increased its bombardment squadrons from twelve to twenty-seven in 1926–1934 and its attack squadrons from four to eleven in the same period. The French government spun out in several directions when it allowed the Armée de l'Air to fight understrength in 1940 while excellent combat aircraft languished in storage. The civilian political leadership, moreover, would not overrule a general staff mobilization plan that placed air units under ground army command despite the nominal independence of the air force.

The submarine forces of the interwar period presented no problems in interservice politics, but they did divide politicians and admirals on the issue of wartime roles. That the submarine forces had substantial political support becomes clear through a nonevent: the proposed international agreements on restricting the design, numbers, and tactical missions of submarines had little practical influence. Technical problems and naval doctrinal issues were more important brakes on submarine development, as the German example demonstrates. In the United States, political figures like Charles Evans Hughes and Franklin D. Roosevelt supported the submarine force. As they did in other naval matters, the British and Japanese rebuffed external critics of their plans for submarine and antisubmarine warfare. The German Navy enjoyed political and technological support for its modest Unterseeboot plans even before Hitler took power. The French Navy, which had pioneered in concepts of commerce raiding before World War I, strengthened its political and social alliance with the Right and enjoyed an immunity from meddling with its own internal fleet priorities. The few naval critics in France found themselves condemned as communists and Jews. The navy that suffered most from political activism in roles and missions was the Regia Marina, which lost its aviation component to the Italian Air Force by a fiat from Il Duce and never regained it, but the Italian Navy, including its submarine force, still entered World War II in more modern condition than its sister services.[28]

The more arcane the naval technology, the more likely that civilian technologists could influence development, but the same factor also tended to

[28] Herwig, "Innovation Ignored."

mute political intervention. Submariners worked with civilian marine engineering peers, but neither could upset the priorities set by the naval planning staffs and fleet commanders. The same phenomenon affected mine warfare, an unappreciated naval economy of force option that bedeviled the navies of World War II and still does. The major advance in interwar mine warfare was the influence mine, which could be tethered on the ocean bottom in critical areas and then released and exploded by its response to changes in magnetic field, or screw-noise, or overpressures caused by a passing ship. The Germans (again working with the Russians) took the early lead in developing influence mines and never quite surrendered their advantage to Allied mine-countermeasures programs. But the Germans and Japanese instead lost their aerial, surface, and submarine mine-layers while the United States, in fact, mounted a very successful mining operation against Japanese shipping lanes. These developments, however, occurred because the navies themselves responded to an operational challenge, not because of a great civilian mine warfare constituency.[29]

The history of communications security programs and radar development provides the best example of the merit of close civilian and military cooperation at both the political and technological levels. The two critical developments in radio communications and radar before World War II were increasing transmitter frequency power from 3–30 megahertz (high frequency) to 300–3,000 megahertz (ultra high frequency) while narrowing the wavelengths from shortwaves (100 to 10 meters) to decimetric length (100 to 10 centimeters). Both these accomplishments depended on civilian technical participation in the development process. Electronic warfare, however, would not have profited from publicity or even much political accountability, an advantage that still accrues to electronic intelligence collection and analysis. Technological arcana favors the specialists, as does the conviction that the compromise of electronic warfare systems carries the seeds of national disaster, a holdover from some limited experiences in World War II.

In the interwar period, the British set the pace in electronic warfare after an early lead by Germany with the United States, Japan, and the Soviet Union a step behind. The British development of cryptology benefitted from a Foreign Office–military alliance that created the Government Code and Cipher School under Foreign Office funding and sponsorship, protected by the Official Secrets Act, a legal cover noticeably lacking in the United States. Although British cryptographic activities placed more

---

[29] Gregory K. Hartman, *Weapons That Wait: Mine Warfare in the U.S. Navy* (Annapolis, MD, 1979).

emphasis on political than military affairs, it created a large cadre of experts, a basic understanding of the mathematical and technical problems of the specialty, and a well-placed champion in Winston Churchill, who had a politician's appreciation for the leveraging power of privileged information.

British excellence in intelligence activities attracted a host of European refugees, which enlarged the pool of linguists, mathematicians, and analysts. The British exploitation of radar profited not only from civilian participation in research and development, but from a high-level clash of great politico-scientific leaders, Sir Henry Tizard and Frederick A. Lindemann. In 1935 a scientific committee chaired by Tizard, a chemist of political bent, evaluated radar research by Robert Watson Watt's team and reported to the Air Ministry that radar had great potential as a navigation and target identification tool. Lindemann, an Oxford scientific don and confidant of Winston Churchill, challenged the report and argued that radar had little utility. The Tizard group won its argument, in no small measure through the imagination and foresight of Air Vice Marshal Hugh C.T. Dowding, Fighter Command's future leader in the Battle of Britain. Although unpleasant at the personal level, the Tizard–Lindemann rivalry exposed the technological options and helped direct the British toward systems development rather than technical maximization, a besetting German weakness.[30]

The Germans came closest to replicating the British pattern while the Japanese emphasized the military aspects too much and the Americans the commercial too much in their radar work. The American research on radar had too many ties to commercial exploitation, which took it along developmental lines of marginal military utility. The American effort in both the army and navy had little high-level interest except in aviation; it also showed little appreciation of military research abroad until the British began to share their work after the war began. Although the Japanese, drawing upon their own work and German experience, shared the early advantages of radar, they did not face a situation where radar could make a significant difference in their air and naval operations. For example, Japanese night interceptors in 1944–1945 enjoyed good radar support in the war against the USAAF B-29s, but radar alone could not compensate for numerical disadvantages and lack of pilot experience.

---

[30] Nigel West, *The SIGINT Secrets: The Signal Intelligence War* (New York, 1988); Bruce Norman, *Secret Warfare: The Battle of Codes and Ciphers* (New York, 1973); R.V. Jones, *The Wizard War* (New York, 1978); Tony Devereux, *Messenger Gods of Battle: Radio, Radar, Sonar: The Story of Electronics in War* (Washington, DC, 1991).

Civilian intervention and integration in the process of military innovation in the 1920s and 1930s provides a mixed picture, but on the balance it appears that civilian participation at the political and technological levels produced beneficial results. This participation may have been most impressive in the western democracies, but the German example shows that an authoritarian form of government does not foreclose civil-military collaboration. The French, on the other hand, showed that a democratic form of government does not guarantee responsive military reform. In fact, a popular government run amok can ruin a military establishment as easily as a tyrant like Stalin, although the French armed forces, especially the army, also committed suicide in the interwar period.

CONCLUSION

The dynamics of military innovation in the interwar period reveal a complex pattern of interactions between strategic assumptions, the technological state-of-the-art and future research and development, military organizational politics and operational doctrine, and civil-military interaction. If one took the senior military officers of the period at their word, defense politics turned on budgets, civilian timidity, insufficient intelligence, and a lack of political guidance. This position has merit, but it is incomplete. The civilian political leaders of the period would have blamed their armed services' shortcomings on the reactionary attitudes of senior officers, the bureaucratic sloth of general staffs, an arsenal mentality that refused to accept the innovations of civilian inventors, interservice and intraservice rivalry, regimental clubbiness, and a vision of soldiering (or flying or sailing) that put more emphasis on form than effectiveness. This view is also true, but incomplete. In the face of substantial circumstantial conditions that discouraged innovation, the armed forces of the major military powers, nevertheless, changed themselves after 1918. They did so unevenly, with great uncertainty, at substantial cost, and with different results. Change, however, was the rule, not the exception.

One would be tempted to attribute some sort of western technological superiority for Allied success in World War II, but Japan, on a much shallower industrial and technological base, produced armed forces on a competitive level with the best European forces in the early stages of World War II. The democratic governments had some advantages in the long term since their freedom attracted refugee scientists and technologists. Historians of the period focus upon the force of doctrinal ideas, a habitual weakness of intellectuals. The determinants of relative success in different strategic and operational specialties seem to be a combination of astute political support and guidance usually exercised by a few politicians, attention of civilian and military technologists to the most promising innovations, and creation of staffs and operational organizations that can turn ideas into experimental exercises. The patronage of politicians and senior military leaders is essential, for prophets by their nature tend to end up on the cross of professional prudence. Political intervention is especially crucial in innovations that cross or merge service specialties. Sheer technical innovation, as the Germans proved, does not win wars. Instead, the interaction of technical change and organizational adaptation within a realistic strategic assessment determines whether good ideas turn into real military capabilities.

10

MILITARY INNOVATION
IN PEACETIME

BARRY WATTS AND WILLIAMSON MURRAY

And it is worth noting that nothing is harder to take in hand, more perilous to conduct, or more uncertain in its success, than to take the lead in the introduction of a new order of things.[1]

Military organizations are societies built around and upon the prevailing weapons systems. Intuitively and quite correctly the military man feels that a change in weapon portends a change in the arrangements of his society.[2]

The early Greek imagination envisaged the past and the present as in front of us – we can see them. The future, invisible, is behind us. . . . Paradoxical though it may sound to the modern ear, this image of our journey through time may be truer to reality than the medieval and modern feeling that we face the future as we make our way forward into it.[3]

Looking back over the military history of the twentieth century, what were the fundamental technological, conceptual, operational, and organizational factors that, during times of peace, gave rise to fundamental changes in how military organizations would fight future wars?[4] How long did it take individuals and organizations to move from a vague vi-

---

[1] Nicollo Machiavelli, *The Prince*, trans. W. K. Marriott, Vol. 23, *The Great Books of the Western World* (Chicago, 1952), p. 17.

[2] Elting Morison, *Men, Machines, and Modern Times* (Cambridge, MA, 1966), p. 36.

[3] Bernard Knox, *Backing into the Future: The Classical Tradition and Its Renewal* (New York, 1994), pp. 11–12.

[4] These questions are Andrew Marshall's. See A.W. Marshall, "Historical Innovation: Carrier Aviation Case Study," Memorandum for Distribution, Director of Net Assessment, Office of the Secretary of Defense, 27 June 1994, p. 2.

sion of a new or more effective way of fighting to mature capabilities they could exploit to underwrite success in actual combat? Did individuals or groups matter more in making progress toward such innovations – or is it even sensible to try to separate individual contributions from the organizational contexts in which they inevitably occurred? What barriers to progress were encountered, and how were they overcome? To what extent did successful peacetime innovation depend on having a specific enemy who posed a concrete threat? How important was competition, both between and within the military organizations of a given side, as well as with other nations? Finally, how necessary to successful innovation was conscious awareness of the long-term potential of the new way of waging war? These are the core questions that we will address in this essay.

To furnish the historical "data" for exploring these questions, we have drawn primarily on three instances of peacetime innovation from the years 1918–1939: 1) the development of what came to be known as the *Blitzkrieg;*[5] 2) the rise of land-based or "continental" air power, by which we mean to encompass the various "tactical" uses of air power in support of ground forces as well as the more independent application of air power traditionally termed "strategic bombardment"; and, 3) the maturation of carrier aviation. All three of these cases are relatively well known to historians, and all have been covered to one degree or another in earlier chapters. For these reasons, only one of the three cases that consciously drove the findings of the present chapter – carrier aviation (with emphasis on the American experience) – has been included in its entirety.

---

[5] The term *Blitzkrieg* seems to have been first coined by an American journalist; in recent decades, its meaning has become increasingly ambiguous even among scholars specializing in military affairs (George Raudzens, "Blitzkrieg Ambiguities: Doubtful Usage of a Famous Word," *War & Society,* September 1989, pp. 77–78). Kenneth Macksey has claimed that Adolf Hitler began exploiting the term for its propaganda value in 1936, but Hitler himself had little to do with the complex changes that gave rise to the kind of mobile, armored warfare that the German Army developed during the interwar period (*Guderian: Creator of the Blitzkrieg* [New York, 1976], p. 68). In fact, Dennis Showalter has noted that German officers generally did not begin using the word *Blitzkrieg* until after World War II began (lecture, "Military Revolutions in Modern German History," School for Advanced International Studies, Washington, DC, 16 November 1994). The German term for the kind of mobile, mechanized, combined-arms operations that characterized the German campaigns of 1939–1940 was *Bewegungskrieg,* meaning mobile warfare (Telford Taylor, *The March of Conquest, The German Victories in Western Europe, 1940* [New York, 1958], p. 311). In this chapter, we will use the term *Blitzkrieg* primarily to refer to the particular combination of technological, weapon-system, conceptual, doctrinal, and organizational developments that lay at the core of German operations during World War II. Given this meaning, the term can also be understood as a concrete example of a peacetime military innovation sufficiently fundamental to change the way in which future wars would be fought (at least by competent, well-equipped militaries).

The reader should also be aware of one methodological point. In preparing accounts of all three of these cases for Andrew Marshall in late 1994, we found it necessary to discuss each case's antecedents in World War I, as well as the "report card" on innovation provided by the harsh test of World War II combat experience. While some early reviewers balked at such extensions of the *interwar* cases, our conclusion was that without some attention to antecedents and subsequent effectiveness, it was impossible to draw practical lessons and implications for the early twenty-first century.

Our underlying purpose, then, is not to reexamine these historical episodes for their own sakes, but to use them as a basis for helping decision makers to think creatively about changes in the nature of war that may occur in coming decades. This deeper purpose obviously presumes the possibility of meaningful parallels between past military innovations and changes in the nature of war that may occur in the future. We will begin, therefore, by sketching this broader context.

THE INTERWAR PERIOD AND THE NEXT MILITARY REVOLUTION

The years 1918–1939, which separate World War I from World War II, witnessed profound changes in how technologically advanced military organizations would fight. The Germans restored movement to land warfare through creation of the *Blitzkrieg*; the RAF's Fighter Command capitalized on radio detection and ranging (radar) and development of advanced fighters like the Spitfire to assemble an integrated, air-defense "system of systems"; American and British bomber enthusiasts laid the doctrinal, technical, and organizational foundations for the strategic bombing campaigns of World War II; the Japanese and American navies advanced carrier aviation to the point that, in 1942, carrier-based aircraft supplanted the battleship as the heart of their fleets in the Pacific (during daylight at least); the American and German navies evolved concepts and equipment for submarine warfare on enemy commerce; and, the U.S. Marine Corps devised a viable approach to amphibious assaults on enemy-held territory.

In most of these cases of peacetime military innovation, technological developments played an enabling or facilitating role in precipitating fundamentally new and more effective ways of fighting. In a narrow and specific sense, such innovative developments were revolutionary. Yet the underlying technologies themselves (the internal combustion engine, radio communications, radar, etc.), as well as the new military systems to which they gave birth (airplanes, tanks, amphibious landing craft, aircraft carriers, radar, and so forth), formed only a part of these innovations, if not the smallest part.

Beyond the time required for the fragile airplanes and unreliable tanks of World War I to evolve into more mature weapons by the late 1930s, military services had to develop new organizational structures such as the panzer division or Fighter Command's organizational arrangements for the command and control of Britain's air defense. They had to integrate advanced weapon systems with appropriate tactics, operational concepts, and doctrines in order to realize the full potential of new ways of fighting. There was nothing inevitable about the outcomes; much of the more successful innovation that occurred was the result of *ad hoc* improvisation, bedeviled by changing or uncertain political priorities. These processes were extraordinarily difficult to execute and fraught with profound uncertainties.

To cite the most striking example, the stunning effectiveness of the German campaign against France and the Low Countries in May 1940 was not simply a matter of the Wehrmacht possessing better tanks, or having them in greater numbers. In fact, the Allies possessed a numerical edge of approximately 1.3 to 1 in tanks, and many of their armored fighting vehicles possessed superior protection and armament to German tanks.[6] Moreover, at the outset of the May 1940 campaign, the Allies had force-ratio advantages of around 1.2 to 1 in manpower, a slight edge in divisions (1.03 to 1), and, from Luxembourg to the Swiss border, the French had completed the Maginot Line; only in total aircraft and antiaircraft artillery did the Allies face substantial disadvantages at the theater level.[7]

---

[6] The French Somua 35 and B.I. bis tanks were considered superior to any of the German panzers (Douglas Porch, "Why Did France Fall?," *The Quarterly Journal of Military History*, Spring 1990, p. 33). Similarly, the British Matilda had stronger armor than the German tanks, and the 37-millimeter gun on the German Mark III was inferior to the British two-pounder (F.W. von Mellenthin, *Panzer Battles: A Study of the Employment of Armor in the Second World War*, trans. H. Betzler [Norman, OK, 1956], p. 12). On the other hand, the operational ranges and turret sizes of the panzers used in May 1940 favored the Germans.

[7] Phillip A. Karber, Grant Whitley, Mark Herman, and Douglas Komer, "Assessing the Correlation of Forces: France 1940," BDM Corporation (McLean, VA, June 1979), pp. 2–3; also, Trevor N. Dupuy, *Understanding War: History and Theory of Combat* (New York, 1987), pp. 93–94. While the Germans had a substantial edge in total combat aircraft in May 1940, the opposing sides were nearly equal in single-/twin-engine fighters (ignoring British fighter strength retained in England). The Germans possessed 1,736 fighters; their opponents had some 1,590 (Klaus A. Maier et al., *Das Deutsche Reich und der Zweite Weltkrieg*, vol. 2, *Die Errichtung der Hegemonie auf dem Europäischen Kontinent* [Stuttgart, 1979], p. 282). So the German preponderance in total aircraft (roughly 3,400 to some 1,700 Allied aircraft) was provided by some 1,680 bombers and dive bombers

Consequently, we must look beyond both military hardware and theater-level force ratios to explain the magnitude of the German victory.

What mattered most was that the Germans had evolved sound concepts for mobile, combined-arms warfare and had trained their army to execute those concepts. To begin with, the *Reichswehr*'s post-World-War-I field-service regulations reflected meticulous study of the tactical lessons of 1917–1918. The 1933 *Die Truppenführung* ["Troop Leadership"] – a refinement of the 1921–1923 "Leadership and Battle with Combined Arms" prepared under General Hans von Seeckt[8] – emphasized combined arms, penetration, and exploitation derived from the infiltration tactics the Germans developed at the end of World War I.[9] Furthermore, the German Army retained its tradition of *Auftragstaktik* (mission-oriented orders) throughout the interwar period, an approach that encouraged lower echelon commanders to exploit local opportunities to the maximum extent.[10] In addition, officer training had long emphasized initiative, risk taking, and leading from the front at all levels throughout the army – in other words the basic leadership principles on which modern, mobile war depends.[11] Finally, exercises during the 1920s not only had included multi-divisional maneuvers, but taught German commanders to ignore the continuous front and so encouraged them to pay less attention to their flanks.[12]

In addition to doctrine, concepts, and training, there were also organizational areas where the Germans moved significantly beyond their future opponents. As early as the mid-1920s they had already recognized radio communications as a necessary adjunct to the demands for speed and opportunistic responses to unpredictable situations inherent in their doctrine. Thus, in contrast to their opponents, the Germans took full orga-

(*The Rise and Fall of the German Air Force: 1933–1945*, ed. Cyril March [Poole, 1983], p. 66). It should also be remembered that most of the French fighters available in May 1940 were no match for the Messerschmitt Bf 109E.

[8] James S. Corum, *The Roots of Blitzkrieg: Hans von Seeckt and German Military Reform* (Lawrence, KS, 1992), p. 200.

[9] Robert A. Doughty, *The Breaking Point: Sedan and the Fall of France, 1940* (Hamden, CT, 1990), p. 30.

[10] Ibid., pp. 326–327.

[11] Casualty data for key German formations during the May 1940 campaign such as Heinz Guderian's XIX Panzer Corps indicate that German officers took a disproportionate share of casualties (Doughty, *The Breaking Point*, p. 330). More importantly, the Germans also demonstrated, in contrast to the French, "an ability to continue fighting despite the loss of key leaders. Instead of permitting units to collapse because of unfamiliarity with the mission or with what was happening, officers stepped forward and successfully assumed the responsibility of leadership." (ibid., p. 331).

[12] Corum, *The Roots of Blitzkrieg*, pp. 185–186, 201–202, and 205.

nizational advantage of communications technology not only by in-
stalling up-to-date radios and a radio operator in every tank but also by
adding signal organizations that allowed battalion, regimental, and even
division commanders to command from the front.[13] The panzer division
itself, which represented a mechanized, combined-arms unit in which
tanks were only a part of a larger whole, was a creation that rested on an
intertwining of a realistic reading of the past with considerable intuition
about the future.[14] Last but not least, the Germans worked at developing
air support for ground forces, including armor, in the breakthrough battle (although by the outbreak of the war they had yet to work out effective means for close air support for advancing tank units).[15]

The result in May 1940 of these technological, tactical, doctrinal, and
organizational developments was a bold operational approach that pro-

---

[13] Hermann Balck, trans. Pierre Sprey, "Translation of Taped Conversation with General
Hermann Balck, 12 January 1979," Battelle Institute, Columbus, OH, pp. 20–21. Walter Nehring, who was one of Guderian's staff officers for many years, emphasized after
the war that "from the outset it was realized that, without a comprehensive communication network, the concept of high mobility and deep penetration by panzer divisions
was unthinkable" (Macksey, *Guderian*, p. 51). Corum traces the *Reichwehr's* conviction
that every tank needed a radio back to the views of the German armor expert Ernst
Volckheim in 1924 (*The Roots of Blitzkrieg*, p. 108). Yet, as obvious as the requirement
for a radio in every tank became to the Germans by the late-1920s, in May of 1940 only
a few French tanks would be so equipped (ibid.). But it is also clear that the German view
of the importance of radio communications between and among tanks was heavily influenced by their observation of the British experimental exercises with armored forces
in the late 1920s and early 1930s.

[14] The first panzer battalion was created in 1934; the following year the first three panzer
divisions were formed, although without tanks since equipment was still in short supply
(Macksey, *Guderian*, pp. 63–64). The creation of the first panzer divisions in 1935 occurred at the direction of General Ludwig Beck, then head of the German general staff;
Beck was also head of the committee that produced the 1933 *Truppenführung* under
which the German Army fought to the end of World War II (Corum, *Roots of the
Blitzkrieg*, p. 140).

[15] In Poland the Germans did not yet possess the capability to use air power effectively in
direct support of ground forces when the situation on the ground was relatively fluid or
mobile. In fact, there were a number of incidents during the 1939 campaign in which
German aircraft bombed their own armored spearheads, including one which saw thir-
teen German soldiers killed and twenty-five seriously wounded by the Luftwaffe. In April
1940 the 1st Panzer Division had carried out a test exercise with the Luftwaffe on the
use of close air support with advancing ground forces; the conclusion was that there were
too many intractable problems to be solved, especially given the approaching start date
for the campaign in the west (Williamson Murray, *German Military Effectiveness* [Bal-
timore, MD, 1992], pp. 113 and 115). For the evidence of the first examination of close
air support and tanks in mobile war, see: 1st Panzer Division, Ia Nr. 232/40, 24 April
1940, "Zusammenarbeit Panzer-Stuka," National Archive and Record Service (NARS),
T-314/615/00396.

duced one of the most crushing military victories of the twentieth century. The final German plan resulted in the Wehrmacht's armored forces traversing the Ardennes and relying on well-led, well-coordinated infantry and artillery attacks to force the Meuse River.[16] But once armor was across, the Germans were in a position to exploit the breakthrough at a pace far swifter than the Allies could handle. As a result, Army Group A enveloped the Belgian Army, a substantial portion of the three French armies, and the British Expeditionary Force. In short, the advances in land warfare that defeated France and pushed the British off the continent resulted from the integration of interrelated and complex elements, many of which had long histories. Embedded military hardware such as the tank constituted only one of the components necessary for successful innovations.

Carrier aviation, strategic bombing, the integrated air-defense system that defeated the Luftwaffe in the Battle of Britain, and amphibious warfare all appear to be instances of integrated, combined-systems "military revolution" exemplified by the *Blitzkrieg* of May 1940. Needless to say, innovations of this sort rarely reach fruition over short periods of time. They require military organizations to weave together many disparate elements within a complex tangle of interactions created by the personalities, strivings, values, past experiences, history, visions, and cultures of the individuals and institutions involved. The process of such innovation in peacetime appears to be highly nonlinear. What this observation means is that innovation displays the extreme sensitivity to current and initial conditions that gives rise to the loss of long-term predictability: the most minute differences in initial or current conditions can, over time, give rise to completely different outcomes and can spell the difference between successful innovation and failure.[17] As a result, the combined-systems "rev-

---

[16] The best discussion of the convoluted planning for the 1940 German campaign in the west remains Chapter 5 of Telford Taylor's *The March of Conquest*. Taylor makes clear that while Erich von Manstein and Hitler had been thinking about pushing armored forces through the Ardennes as early as fall 1939, their early conceptions would have used only three to four of the armored divisions, primarily to aid the advance of Army Group B (ibid., pp. 165–167). It was not until German plans fell into Belgian hands from the forced landing of a German aircraft on 10 January 1940 that a major rethinking of the concept for the campaign became possible. Even then, it was General Franz Halder who decided to concentrate the main effort in the Ardennes (ibid., pp. 171–175). In fact, the revised plan that Halder brought to Hitler on 18 February 1940 concentrated the *Schwerpunkt* in the Ardennes to a far greater degree than Manstein had ever proposed (ibid., p. 172). That, of course, never prevented Manstein from arrogating full credit to himself.

[17] Ian Stewart, *Does God Play Dice? The Mathematics of Chaos* (Oxford, England, 1989),

olutions" of the interwar period typically took at least a decade to reach operational maturity, if not two.[18]

Not surprisingly, in the aftermath of May 1940 British and French perceptions of the abruptness with which land warfare had changed tended, understandably, to be quite different. As victims, the natural response of British and French observers was to see themselves as defeated by a sudden, unpredictable advent of a profound change in warfare. Focusing on the most visible aspect of their defeat, French, and British (and American) observers seized on the fast-moving German panzers and images of Stuka support spun by Goebbels' propaganda machine as the essence of an abrupt "revolution" in warfare.

Our principal aim is to examine the combined-systems military "revolutions" of 1918–1939 from the standpoint of peacetime innovation. The *hypothesis* behind this focus is that we are presently in the initial stages of a similar period of change. The advances since the late 1960s in microelectronics, information technologies and software, satellite communications, advanced sensors, and low-observable technologies all suggest extraordinary new capabilities. To cite Marshal Nikolai V. Ogarkov's 1984 characterization of the advances in nonnuclear weaponry – including development of "automated reconnaissance-and-strike complexes,"[19]

---

pp. 66–72, 127–134, and 139–142; also, J.A. Dewar, J.J. Gillogly, and M.L. Juncosa, "Non-Monotonicity, Chaos, and Combat Models," RAND/R-3995-RC, Santa Monica, CA, 1991, pp. 3–6. Having made this basic observation about innovation, however, we would caution the reader *not* to leap to the conclusion that successful innovation, like strategic change in a corporation, is simply a matter of random rolls of the dice. As we will argue in the final section of this chapter, visionary leaders can exercise far more control over the ultimate outcome of the innovation process than the embedded nonlinearities might initially suggest.

18 The integration of radar into a viable air-defense system offers no exception to the proposition that combined-systems innovations from the interwar period typically took a decade or longer to mature. As early as 1904 a young German named Christian Hülsmeyer had patented a "telemobiloscope" which he claimed could transmit radio waves and receive their reflections off a passing object. See the Beyerchen essay in this work, p. 265.

19 The Soviet paradigm for "reconnaissance-strike complexes" was the U.S. Assault Breaker, which began as a program in the 1970s (Paul N. Walker, "Precision-guided Weapons," *Scientific American*, August 1981, pp. 42–43; V. G. Reznichenko, *Taktika* [Moscow, Russia, 1978], p. 9 of Foreign Broadcast Information Service translation, *JPRS [Joint Publications Research Service] Report: Soviet Union, Military Affairs, TACTICS*, JPRS-UMA-88-008-L-1, 29 June 1988; and, S. Davydov and V. Chervonobab, "Conventional, But No Less Dangerous," *Energiya: ekonomika, techniika, ekologiya*, No. 7, 1987, pp. 43–48).

long-range and high-accuracy munitions, and electronic-control systems – "make it possible to sharply increase (by at least an order of magnitude) the destructive potential of conventional weapons, bringing them closer, so to speak, to weapons of mass destruction in terms of effectiveness."[20]

In this context, it is not hard to see certain prospective parallels between our current situation and the combined-systems revolutions of the 1918–1939 period. As Andrew W. Marshall, Director of Net Assessment in the Pentagon for over two decades, began suggesting in the early 1990s,[21] in the analogy between our current situation and the interwar period, we appear today to be somewhere near the beginning, perhaps in the equivalent time frame of 1922 or 1923.[22] If this insight proves correct, then the impressive use of satellites, stealth (low-observable technologies plus appropriate tactics), and precision-guided munitions during the 1991 Persian Gulf War are probably analogous to the November 1917 battle of Cambrai in the development of the *Blitzkrieg*.[23]

Cambrai represented the first large-scale attempt by the British to use a surprise tank attack to break through German entrenchments without

[20] Interview with Marshal of the Soviet Union N.V. Ogarkov, "The Defense of Socialism: Experience of History and the Present Day," *Krasnaya zvezda*, lst ed. in Russian, 9 May 1984, pp. 2–3.

[21] Since the fall of the Berlin Wall in 1989, Andrew Marshall has been perhaps the foremost advocate inside the Pentagon of the possibility that we might be entering a new period of fundamental change in how future wars will be fought. See, for example, Thomas E. Ricks, "Warning Shot: How Wars Are Fought Will Change Radically, Pentagon Planner Says," *The Wall Street Journal*, 15 July 1994, p. A1.

[22] A.W. Marshall, "Some Thoughts on Military Revolutions," Office of Net Assessment (OSD/NA) memorandum, 27 July 1993, p. 2. Andrew F. Krepinevich, who did much of the initial work in OSD/NA on the revolution in military affairs, recalls that Marshall first associated our situation today with 1922 or 1923 (in the analogy to the interwar years) during the summer of 1991. Krepinevich's recollection is vivid because his own initial intuition was to associate Desert Storm with the Spanish Civil War of the 1930s rather than with the 1917 Battle of Cambrai (telephone conversation, 22 November 1994).

[23] Marshall, "Some Thoughts on Military Revolutions," p. 2; see also, Commanders James R. FitzSimonds and Jan M. van Tol, "Revolutions in Military Affairs," *Joint Force Quarterly*, Spring 1994, pp. 24–31. FitzSimonds and van Tol have noted that, perhaps somewhat counterintuitively, the greatest changes in warfare have generally occurred during peacetime rather than wartime ("Revolutions in Military Affairs," p. 26). This insight reflects Steve Rosen's conclusion that peacetime military innovation, being driven by perceptions of structural changes in the national security environment and visions of hypothetical future wars involving new military capabilities, is fundamentally different from wartime innovation (Stephen P. Rosen, *Winning the Next War: Innovation and the Modern Military* [Ithaca, NY, 1991], pp. 52 and 76). During time of war, adaptation rather than innovation may better characterize what occurs.

a massive, extended preliminary artillery bombardment. It offered a tantalizing glimpse of how the tank might eventually change land warfare. But even for postwar tank enthusiasts like J.F.C. Fuller and B.H. Liddell Hart, Cambrai offered only a partial vision of future armored war and yielded little in the way of detailed forecast as to how things would actually work out in the Low Countries and France in May 1940.[24]

The difficulties of foreseeing the future during periods of major military change highlight the motivations and prospective benefits of revisiting the innovations of 1919–1939. On the historical evidence the process of military innovation on this scale does not appear to be linear or precisely predictable.[25] The leaders of the post-1918 German Army did not deliberately set out to create a new way of fighting, but rather aimed to build upon the operational, and particularly the tactical lessons of 1914–1918 in a coherent and effective fashion. By casting their net widely over the experiences of the last war and examining that experience realistically (at least at the tactical and operational levels), they moved, in fits and starts, toward a new conception of fighting. They managed to do so, moreover, despite the constraints of the Treaty of Versailles and, until after Adolf Hitler's rise to power, meager defense budgets.

Indeed, it seems likely that, *even on the eve of the 1940 campaign*, few of even those German officers involved in development of armored warfare during the interwar period had a firm belief that their efforts would transform land warfare. Vagueness and uncertainty in gaming the 1940 campaign as to what the army high command should do if the three panzer corps got across the Meuse before the eighth or ninth day of the offensive confirms the inability of most German generals to predict in detail how a *blitzkrieg* campaign could play out against French and British

---

24 For Fuller's first-hand account of the Battle of Cambrai, see Major General J.F.C. Fuller, *Memoirs of an Unconventional Soldier* (London, 1936), pp. 192–219. Although the Cambrai operation was originally proposed as a rapierlike raid, the British Third Army converted it into an attempt at a decisive battle, prevented the Tank Corps from holding any tanks in reserve, permitted the Cavalry Corps commander to command from the rear, and thereby precluded the cavalry forces from exploiting the breakthrough created by tanks working in coordination with infantry (ibid., pp. 188–191). The German counterattack that followed the Allies initial success at Cambrai was the first large-scale use of infiltration, exploitation tactics on the Western Front and quickly pushed the British back – in some cases beyond their original front (Corum, *The Roots of Blitzkrieg*, p. 9).

25 To deny that future events are precisely predictable is not, however, to deny causality. On this crucial point, see Heinz-Otto Peitgen, Hartmut Jürgens, and Dietmar Saupe, *Chaos and Fractals: New Frontiers of Science* (New York, 1992), pp. 9–14. Nor should the denial of precise or detailed predictability be taken to deny that some individuals can develop surprisingly sound *intuitions* about the future.

opposition.[26] In his memoirs, even Guderian characterized his panzer corps' success in getting its rifle regiments across the Meuse on the fourth day as "almost a miracle."[27] Robert Doughty's meticulous reconstruction of the crossings supports Guderian's characterization. In fact, at numerous junctures, the operations of the three panzer corps that crossed the Meuse during the battles of 13–14 May 1940 succeeded by razor-thin margins and, in some cases, by sheer luck, while, in other cases, the initial attempts to cross failed completely.[28] Nevertheless, the cumulative result of these small differences was that, by the morning of 16 May, "the French Army teetered on the edge of collapse, and in subsequent days the Germans won one of the most decisive operational victories in military history."[29]

---

[26] Guderian was "a brilliant man but an extremist and an egoist, with a personality that bordered on the fanatic" (Corum, *The Roots of Blitzkrieg*, p. 140). As a result, Guderian's memoirs are often unreliable on matters having to do with his personal role in the creation of the *Blitzkrieg* and the conduct of operations during World War II. According to his memoirs, he first proposed that XIXth Panzer Corps could force a crossing of the Meuse on the fifth day of the offensive at a war game that took place at Koblenz on 7 February 1940 (Heinz Guderian, *Panzer Leader*, trans. Constantine Fitzgibbon [New York, 1957], p. 69; for confirmation see, Franz Halder, *The Halder War Diary: 1939–1942*, ed., Charles Burdick and Hans-Adolf Jacobsen [Novato, CA, 1988], p. 95). It appears to have been at this exercise that Hitler questioned Guderian as to whether he would recommend heading for Paris or the Channel coast, if XIXth Panzer Corps forced the Meuse on the fifth day. Guderian replied that while the decision was up to the High Command, in his opinion the main German attack should drive for the English Channel (ibid., pp. 70–71). On 7 February Halder, then head of the German general staff, recorded in his war diary the opinion that there was "no sense" in Guderian's proposal and that a concerted attack across the Meuse would be "impossible before the ninth or tenth day of the offensive" (*The Halder War Diary*, pp. 95 and 96). Further disagreement over the viability of Guderian's proposal that the panzer corps could force an early crossing of the Meuse without waiting for the infantry divisions to catch up ensued at a 14 February war game held at Mayen, headquarters of List's 12th Army (*Panzer Leader*, pp. 69–70). After yet another game in mid-March 1940, also attended by Hitler, Halder's diary entry reflected the following judgment on the issue of what to do after crossing the Meuse: "Decision reserved on further moves after the crossing of the Meuse" (*The Halder War Diary*, p. 106).

[27] Guderian, *Panzer Leader*, p. 84. For Balck's personal recollections of the crossing of the Meuse by 1st Motorized Infantry Regiment, 1st Panzer Division, XIXth Panzer Corps, on 13 May 1940, see Balck, "Translation of a Taped Conversation with General Hermann Balck, 13 April 1979," pp. 2–8. Balck, was the commander of 1st Panzer's infantry regiment at the time. What is clear from the historical accounts of the crossings by the three panzer corps is that Hoth's crossing only succeeded through Rommel's extraordinary leadership, Reinhardt's panzer corps never made a successful crossing until the panzer corps on its flanks had succeeded, and only three of the six crossings initially attempted by Guderian's panzer corps got across.

[28] Doughty, *The Breaking Point*, pp. 83, 135, 139, 164–165, 321, 323–324, and 329–332.

[29] Ibid., p. 331.

From the standpoint of understanding peacetime innovation, there is considerable danger in our retrospective awareness of how specific developments ultimately played out. The very act of historical reconstruction imposes a clarity and coherence on events that was neither present *nor possible* at the time. The riveting example of the May 1940 campaign suggests another troubling insight. If, as hypothesized, we *are* entering a period in which warfare is undergoing fundamental changes, then the dimensions of the Allied defeat in May 1940 underline the potentially catastrophic consequences of failing to keep abreast of such changes.

The interwar period also suggests that an early lead in an area of fundamental change, such as the British enjoyed in tank development in 1918, does not guarantee ultimate success. Arguably, multiple factors contributed to Germany's ultimate collapse in November 1918. Principal among them were approximately one million casualties that the German Army suffered during Ludendorff's offensives of mid-March to mid-July 1918,[30] the arrival in Europe of American troops in substantial numbers, and the considerable (if grudging and sporadic) improvements in Allied defensive and, later, offensive tactics after March 1918.[31] Another major contributor to victory in World War I, even if the tank's role was perhaps not quite as large as advocates like Fuller believed at the time, was the increasing use of armor by Allied forces.[32] By late 1918, the British, especially, had a substantial lead in armored vehicles. Indeed, it is plausible to

---

[30] Timothy Travers, *How the War Was Won, Command and Technology in the British Army on the Western Front 1917–1918* (London, 1992), p. 108. The total of nearly one million German casualties from 21 March through mid-July 1918 is based on figures in the German official history. The monthly figures are 235,544 casualties for March, 257,176 for April, 114,504 for May, 209,435 for June, and 183,300 or so for the first half of July (ibid.). "The reason for these heavy casualties was partly because German offensive tactics often reverted to the traditional mass infantry assaults after initial successes, thus leading to heavy casualties; and partly because German offensive tactics, however ingenious, never really solved the problem in 1918 of continuously attacking defenses that fought back with large amounts of firepower" (ibid.).

[31] To take our own advice, it is well to remember that, as late as August 1918, few, if any, on the Allied side imagined that the Great War would be over before winter (Fuller, *Memoirs of an Unconventional Soldier*, p. 318).

[32] For Fuller's view of the role played by the tank in the Allied victory in World War I, see J.F.C. Fuller, *Tanks in the Great War: 1914–1918* (New York, NY, 1920), pp. 287–288; also, J.F.C. Fuller, *The Conduct of War 1789–1961: A Study of the Impact of the French, Industrial, and Russian Revolutions on War and Its Conduct* (New Brunswick, NJ, 1961), pp. 176–177. More recent research into the tactical revolution that occurred on both sides during 1917–1918 suggests, however, that the importance of the tank in Germany's collapse was exaggerated by advocates like Fuller. In *Tanks in the Great War*, for example, Fuller wrote that the success of British tanks during the battle of Amiens (8–11 August 1918) foreclosed "all hope of [Germany] winning the war by force of arms" (p. 227). While the success at Amiens largely hinged on an effective combined-arms of-

suggest that by 1918 the British Army "was in a position to learn how to use tanks against a reactive enemy and had developed all of the intellectual bases for *blitzkrieg* warfare."[33] How, then, did the British fail and the Germans succeed during the 1920s and 1930s in exploiting the tank to restore movement on the battlefield? Therein lies the deeper mystery that this essay will attempt to probe: understanding better why some military organizations innovate more successfully than others.

Given what we have already suggested about the degree to which military innovation in peacetime is unavoidably nonlinear, contingent, and infected with serendipity, it seems best to avoid theoretical generalizations in probing for answers. Instead, we shall concentrate on the more modest but defensible aim of identifying, based on what history reveals about innovation during the interwar period, some of the specific actions, bureaucratic tactics, and strategies that senior-level officials in the current Pentagon, military as well as civilian, might consider implementing in order to facilitate innovation in coming years. The following observation in Steven Rosen's *Winning the Next War* suggests the kind of insights we believe will be more useful than abstract generalizations: "Peacetime innovation has been possible when senior military officers with traditional credentials, reacting not to intelligence about the enemy but to a structural change in the security environment, have acted to create a new promotion pathway for junior officers practicing a new way of war."[34] The compelling motivation for examining interwar innovation is the hope that the U.S. military can avoid the kinds of failure that enveloped Allied armies in May 1940.

To entertain such a hope suggests that even if precise or detailed theories are not possible, the sharpening of broad intuitions about the nature of war in the early twenty-first century is by no means a futile enterprise. A corollary to the proposition that some military organizations master innovation better than others is that some individuals develop better intuitions about future warfare than others, however inchoate or partial such intuitions might be. Fuller, for instance, recognized during the last year of World

---

fensive in which British tanks played a crucial role, British offensives after Amiens emphasized an infantry-artillery combination with only a few tanks thrown in for good measure (Travers, *How the War Was Won*, pp. 45 and 175). Even though the British failed to exploit the tank in the final months of World War I, Travers' conclusion is that mechanical warfare had become a "genuine alternative in 1918" (ibid., p. 179). Corum argues that the Germans reached much the same judgment during the early 1920s (*The Roots of Blitzkrieg*, p. 22).

[33] Rosen, *Winning the Next War*, p. 127.

[34] Ibid., p. 251. Rosen's observation is particularly appropriate to the Germans rather than the Americans in the interwar period.

War I that the tank was more of a psychological weapon than a material one and that its greatest potential lay in the possibility of unhinging or paralyzing the brain of opposing armies, corps, and divisions rather than in destroying them physically.[35] According to Fuller, he first had this intuition during March 1918 when he saw "tens of thousands of men pulled back by their panic-stricken headquarters" as German units employing infiltration tactics broke Fifth Army's lines by "devouring" entire positions.[36] By May 1918 Fuller had elaborated his insight in a memorandum entitled "Strategic Paralysis as the Object of the Decisive Attack."[37] Jumping ahead to May 1940, the functional shattering of otherwise more or less intact forces that Fuller postulated in 1918 was precisely what the French pilot Antoine de Saint Exupéry observed from overhead during the reconnaissance missions he flew over the collapsing French Army during that campaign. In every region through which the German armor divisions had passed, he wrote, "a French army, even though it seems to be virtually intact, has ceased to be an army. It has been transformed into clotted segments. The armored divisions play the part of a chemical agent precipitating a colloidal solution. Where once an organism existed they leave a mere sum of organs whose unity has been destroyed."[38]

---

35 Fuller, *The Conduct of War 1789–1961*, pp. 241 and 256; also, *Memoirs of an Unconventional Soldier*, pp. 321–322 and 324–325. As Fuller put the basic insight in one of his more succinct formulations: "At Cambrai what was the predominant value of the tank? It was its moral effect. It showed clearly that terror and not destruction was the true aim and end of armed forces. That is to say: To attack the nerves of an army, and through its nerves the will of its commander, is more profitable than batter to pieces the bodies of its men" (J.F.C. Fuller, *Lectures on F.S.R. III (Operations between Mechanized Forces)* [London, 1932], p. 7). Interestingly this passage went on to endorse the even greater potential of air power to execute "brain warfare": "The aeroplane also proved this, and even more dramatically; for not only could it attack the will of an army by avoiding its body, but also the political will and national will behind the army" (ibid.). For other examples of Fuller's appreciation of the moral impact of tanks, see *Tanks in the Great War: 1914–1918*, pp. xvii, 148–149, 152–153, 155–156, 216, 224, 227–229, and 253. Worth adding is that although the German general staff did not push the development of tanks during 1917–1918, from June 1918 to the end of the war the German Army tended to be jittery whenever tanks "were supposed (let alone known) to be present" (Macksey, *Guderian*, pp. 17, 20, and 22). Note, too, that by the spring of 1918 one of the principal concepts of the new German offensive doctrine was the disruption of the enemy's equilibrium.

36 Fuller, *The Conduct of War 1789–1961*, p. 241; also, Timothy T. Lupfer, *The Dynamics of Doctrine: The Changes in German Tactical Doctrine During the First World War* (Fort Leavenworth, KS, July 1981), pp. 41–42 and 48.

37 Fuller, *The Conduct of War 1789–1961*, p. 242. For the full text of Fuller's May 1918 memorandum, later retitled "Plan 1919" – and known to historians by the latter title. See *Memoirs of an Unconventional Soldier*, pp. 322–336.

38 Antoine de Saint Exupéry, *Flight to Arras*, trans. Lewis Galantière (New York, 1942), p. 98.

Notwithstanding Exupéry's testament to Fuller's vision, Doughty has argued persuasively that it is to embrace "myth" to believe that the German high command in May 1940 had purposely adhered to Fuller's concept of "attack by paralyzation," whatever more ambitious hopes panzer enthusiasts like Guderian harbored.[39] The most plausible interpretation of German motives is that the immediate purpose of the thrust through the Ardennes was to achieve the traditional *Kesselschlacht desiderata* of encirclement and annihilation,[40] and the degree of success that the Germans achieved in May 1940 seems to have surprised them as much as their adversaries.[41] Only in retrospect can we see the degree to which Fuller's May 1918 intuition about the prospective operational and strategic effects of armored warfare was closer to the mark than that of the German High Command in May 1940.[42]

CARRIER AVIATION

By employing aspects of the *Blitzkrieg*'s development to illustrate the relevance of interwar innovations to the prospective changes in how wars

---

[39] Doughty, *The Breaking Point*, p. 323. Doughty grants that a few officers may have accepted Fuller's notion of brain warfare. Still, the sharp struggle between Kleist and Guderian that resulted in Guderian's resignation on 17 May 1940 "clearly demonstrates the concerns of the German High Command about the pace and vulnerability of the XIXth Panzer Corps" (ibid.). As for Guderian's personal views, he was intimately familiar with Fuller. In fact, Guderian initially turned to Fuller when he was seeking "guidance with regard to the development of armoured warfare – notwithstanding the implication in a paragraph on page 20 [the English edition alone] of *Panzer Leader* that Liddell Hart provided the principal inspiration" (Macksey, *Guderian*, pp. 40–41, 48, and 65). (As Corum has noted, the paragraph in question was added by Liddell Hart to the English edition and did not appear in Guderian's own German edition – *The Roots of Blitzkrieg*, p. 141.) However, Guderian's conception, as reflected in his pre-World War II writings, seems to have focused less on "brain warfare" than on the notion of exploiting the speed of panzer units "to be able to move faster than hitherto: to keep moving despite the enemy's defensive fire and thus to make it harder for him to build up fresh defensive positions" (Macksey, *Guderian*, p. 72). This emphasis on speed, on getting the "green light to the very end of the road," once panzer units had gotten "out on the loose," reflects the spirit of Guderian's memoirs far more than Fuller's discussions of brain warfare (Guderian, *Panzer Leader*, p. 75).

[40] Doughty, *The Breaking Point*, p. 323.

[41] R.V. Jones' view in June 1940 was that the Germans had been so surprised by their own success that, by then, they "had no coherent plan for the immediate future." R.V. Jones, *The Wizard War: British Scientific Intelligence 1939–1945* (New York, 1978), p. 92.

[42] The principal shortcoming in Fuller's thinking about armored warfare during the 1920s and 1930s was that he envisioned fast-moving tanks driving directly at command elements and headquarters without the accompanying combined-arms support that was so central to German success in May 1940.

might be fought in the decades ahead, the preceding discussion provided some historical insight into this particular example of military innovation in peacetime. This section offers a more complete account of another interwar case study in innovation: the development of carrier aviation by the American and British navies, with primary emphasis on the American experience.[43] We have chosen carrier aviation over land-based air power for extended treatment for two reasons. First, carrier aviation is a somewhat cleaner case. Admittedly, tension in American and British navies between the wars over whether to use aircraft to support or replace the battleship line certainly arose from time to time. But it never appears to have resulted in as divisive or bitter a quarrel as the interwar struggles between Anglo-American strategic bombing enthusiasts and their nonflying army brethren who saw aircraft primarily as fire support for ground forces. Second, the interwar development of naval aviation, especially in the American and Japanese cases, provides instances of relatively successful peacetime innovation. By comparison, the efficacy of strategic bombing remains to this day less obvious and more controversial – at least in the views of some observers. These considerations led us, therefore, to the judgment that naval aviation would be an easier case study from which to draw practical implications for the future.

With the aid of historical hindsight, the American interwar development of naval aviation appears to have been one of incremental-but-successful evolution from battleship gunfire to carrier aircraft as the primary offensive means of defeating opposing fleets. As we will see, American success in particular was contingent on organizational interactions and processes through which the naval aviation community reached technological decisions about the kinds of carriers and carrier aircraft it would need. In contrast, the Royal Navy's failure to develop carrier capabilities comparable to those of the U.S. and Japanese navies represents an exam-

---

[43] Our decision to concentrate mainly on the Americans and British obviously neglects the Japanese. We chose not to explore the Japanese experience in any depth for two reasons. First, compared to the American and British cases, the primary source material on the development of carrier aviation by the Japanese between the world wars is still rather thin – especially in English. Second, the Japanese acquired most of their carrier-aviation technology from foreign sources, and were so obsessed during the interwar period with the tactics of "out-ranging" the U.S. Navy's battleships in a decisive fleet engagement, that they never developed a clear vision of fast carriers as a substitute for battleships (Commander James R. FitzSimonds, "The Development of Japanese Naval Air Power," internal memorandum, OSD/Net Assessment, 30 June 1994, pp. 2–3). For a discussion of out-ranging, see Rear Admiral Yôichi Hirama, "Japanese Preparations for World War II," *Naval War College Review*, Spring 1991, pp. 72–74.

ple of flawed, if not failed, innovation, especially in light of the British lead in 1918. At the close of World War I, the Royal Navy possessed "a fleet of nearly a dozen carriers of one sort or another at a time when no other naval power had even one."[44] Indeed, by the end of 1918 the British fleet even included the newly commissioned (but not yet operational) HMS *Argus*, whose fully flush deck made her the first true aircraft carrier.[45] Additionally, during World War I the British had accumulated considerable operational experience with carriers and naval aviation.

This experience encompassed the operation of carriers as part of the main battle fleet as well as employment in independent attacks against targets ashore. The first carrier air raid in history, an attack on the Cuxhaven Zeppelin base near Wilhelmshaven by seven British seaplanes from three improvised "carriers" in the Heligoland Bight, took place on Christmas Day 1914.[46] But, despite the lead conferred by nearly four years of wartime carrier operations and development, the British entered World War II with only four first-line carriers (plus three obsolescent ones); their carrier aircraft were markedly inferior in numbers and quality to those of the Americans and Japanese; and, most crucially, even first-line British carriers of 1939 could not generate pulses of aircraft striking power comparable to those attainable from Japanese and, especially, American front-line carriers.[47] Explaining this curious turn of events will be an important part of this case study. As will become evident, one cannot satisfactorily explain the poor technical choices made by the British concerning carriers and carrier aircraft during the interwar period either by the more com-

---

[44] See the Till chapter in this work, p. 194. Given the loss of HMS *Campania* on 5 November 1918, the Royal Navy appears to have had the following carriers at the close of World War I: *Furious, Argus, Vindictive, Nairana, Pegasus, Vindex, Ark Royal, Manxman, Engadine, Empress,* and *Riveria* (Norman Friedman, *British Carrier Aviation: The Evolution of the Ships and their Aircraft* [Annapolis, MD, 1988], pp. 47 and 90). These vessels ranged from early seaplane carriers, such as *Empress* and *Riveria*, to *Ark Royal*, the first ship largely designed and built as an aircraft carrier, and *Argus*, the first flat-deck carrier (ibid., pp. 29, 30, and 65).

[45] Friedman, *British Carrier Aviation*, pp. 67–68. In September 1916, the British achieved success in making arrested landings with an Avro 504 biplane equipped with a hook; it appears to have been these experiments that encouraged the decision to complete the liner *Conte Rosso* as the flat-deck carrier *Argus* (ibid., p. 61).

[46] Ibid., p. 32. The improvised carriers that conducted the Christmas 1914 raid were *Engadine, Riveria,* and *Empress*. All three were troop carriers modified to carry small numbers of seaplanes. For flight operations, the seaplanes were lowered into the water using cranes.

[47] Till, "Adopting the Aircraft Carrier," pp. 198, 202, 208–9, and 224–225; Friedman, *British Carrier Aviation*, pp. 18–19.

plex strategic and economic circumstances that confronted Britain, or as a particularly unfortunate run of bad luck.[48]

American naval interest in aviation predated World War I. In 1900, realistic ranges for gunfire between capital ships of the line were in the vicinity of 1,500 yards and, even at those ranges, accuracy under combat conditions remained well under 10 percent.[49] By the end of World War I, practical battle ranges for battleships extended tenfold to some 15,000 yards.[50] During this period gunfire accuracy also improved by two, if not three, orders of magnitude due to the innovation of continuous-aim fire by William S. Sims.[51] This rapid expansion in the range and accuracy of large naval guns against ships capable of high sustained speeds hinged on developing reliable and accurate means of fire control.

---

[48] The fact that many of the technical decisions about carriers and carrier aircraft made by the British between the two world wars appear quite sensible when placed in historical context should not obscure the fact that they were later shown to have been wrong. Till's recent argument that the British, Americans, and Japanese were all three "reasonably successful" in developing carrier aviation during the interwar years, and that "one can trace specific failures to apparently minor technical errors and problems," obscures this fundamental point (Till, "Adopting the Aircraft Carrier," p. 225). For example, the tacit British decision *not* to adopt deck parks can be seen as having been altogether rational given the Royal Navy's circumstances and the state of technical knowledge at the time. Nonetheless, it imposed "very severe limits" on British carrier aircraft and, arguably, was "the single most important British interwar naval air decision" (Friedman, *British Carrier Aviation*, p. 11).

[49] During the war with Spain in 1898, for example, Admiral George Dewey's ships had fired 5,859 shells and achieved only 142 hits, for a hit rate of 2.4 percent (David F. Trask, "William Sowden Sims: The Victory Ashore," *Admirals of the New Steel Navy*, ed. James C. Bradford [Annapolis, MD, 1990], p. 285).

[50] Norman Friedman, Thomas C. Hone, and Mark D. Mandeles, "The Introduction of Carrier Aviation into the U.S. Navy and the Royal Navy: Military-Technical Revolutions, Organizations, and the Problem of Decision," draft report for the Director of Net Assessment, Washington, DC, 14 July 1994, p. 40.

[51] "In 1899 five ships of the [U.S.] Atlantic Squadron fired five minutes each at a lightship hulk at the conventional range of 1,600 yards. After twenty-five minutes of banging away, two hits had been made on the sails of the elderly vessel. Six years later one naval gunner made fifteen hits in one minute at a target 75 by 25 feet at the same range – 1,600 yards; half of them hit in a bull's-eye 50 inches square." (Morison, *Men, Machines, and Modern Times*, p. 22). Sims learned all there was to know about continuous-aim fire from Admiral Sir Percy Scott while both were serving on the China station, Sims as a junior officer and Percy as the commanding officer of HMS *Terrible* (ibid., pp. 27–28). Sims tried to interest the Bureaus of Ordnance and Navigation in this innovation to no avail. In 1902, after he wrote directly to President Theodore Roosevelt, Sims was brought back to Washington and installed as Inspector of Target Practice, a post he held through the remaining six years of the administration (ibid., p. 31). When Sims departed that post in 1908, he was universally acclaimed as the man who taught the navy how to shoot.

It was this immediate tactical problem of fire control that convinced then Lieutenant (jg) (later Admiral) John H. Towers even before World War I of "the need to get into the air"; his experience in 1910 as a gunfire spotter on the USS *Michigan*, whose 12-inch guns could reach over the visual horizon, led him to see aircraft as a means of providing accurate long-range gunfire.[52] By early 1911, Captain Washington Chambers and Curtiss pilot Eugene Ely had demonstrated that aircraft could take off and land on a ship.[53] In February 1913, Rear Admiral David W. Taylor, the new chief of the Bureau of Construction and Repair, approved construction of a wind tunnel in the Washington Navy Yard so that his staff "could begin systematic experiments in aerodynamics" resembling the work in hydrodynamics already being done in the Yard's towing tank.[54] During August 1913, the navy's General Board reported to Secretary of the Navy Josephus Daniels that a "complete and *trained* air fleet" had become "a necessary adjunct" to the navy.[55] And, in 1914, the navy detailed J.C. Hunsaker to participate in establishing the first graduate degree program in aeronautical engineering in the United States at the Massachusetts Institute of Technology.[56] Thus, by 1914 senior naval leaders had come to see aviation as a "required adjunct to fleet operations" and had taken steps to insure that their service would be at the forefront of aviation progress.[57]

---

[52] Friedman, Hone, and Mandeles, "The Introduction of Carrier Aviation into the U.S. Navy and the Royal Navy," pp. 39–40. The USS *Michigan* was the first American dreadnought battleship.

[53] Charles M. Melhorn, *Two-Block Fox: The Rise of the Aircraft Carrier 1911–1929* (Annapolis, MD, 1974), p. 8. The makeshift system Ely used to land his aircraft on the *Pennsylvania* on 18 January 1911 "incorporated the basic elements of a modern arrested landing system: hook, wire, deaccelerating assist, and barrier" (ibid.). Shortly after this successful experiment, however, Chambers became enamored with seaplanes and was still championing them over aircraft carriers for wheeled naval aircraft as late as 1930 (ibid., p. 9).

[54] Friedman, Hone, and Mandeles, "The Introduction of Carrier Aviation into the U.S. Navy and the Royal Navy," p. 46.

[55] William R. Braisted, "Mark Lambert Bristol: Naval Diplomat Extraordinary of the Battleship Age," *Admirals of the New Steel Navy*, ed. James C. Bradford (Annapolis, MD, 1990), p. 334.

[56] Clarke van Vleet and William J. Armstrong, *United States Naval Aviation: 1910–1980* (Washington, DC, 1981), NAVAIR oo–80P-1, p. 9.

[57] Friedman, Hone, and Mandeles, "The Introduction of Carrier Aviation into the U.S. Navy and the Royal Navy," p. 52. In the years between the war with Spain and World War I, two communities within the USN – regular line officers, on the one hand, and technical officers from the bureaus of Ordnance, Steam Engineering, and Construction and Repair, on the other – promoted and managed the transformation of the service's technology. Innovations during this period included steam turbines, long-range guns,

This sense of naval aviation's growing importance for future war at sea covered a range of possible aircraft uses. By 1912 Lieutenant Commander Henry C. Mustin, an expert in naval gunnery and early naval aviator, realized that battleship and battle-cruiser gunfire, effectively controlled by aircraft, could potentially win a fleet engagement in a matter of minutes.[58] He not only began to advocate aircraft-carrying ships, but also seems to have been "the first U.S. Navy officer to see clearly that the striking power of aircraft at sea *could and would* equal that of the battleship."[59] Mustin's early leap from employing aircraft to direct battleship gunfire against fast-moving ships to using them to deliver ordnance – which he may have made as early as 1914 – llustrates how hard it is to pinpoint the precise origins of carrier concepts. What does seem clear is that, by 1914–1915, three distinct concepts had arisen within both the British and American navies:

> One concept was to use carrier-based aircraft to attack land targets. Under this concept, the carrier was just a forward airfield. It did not work in concert with, or form an integral part of, a whole fleet. Another concept was that the carrier, though part of a larger fleet, was nonetheless an auxiliary to existing weapons such as the battleship. Under this concept, carrier aircraft scouted, drove off enemy scouts, and directed the gunfire of heavy ships. Yet another, different (and revolutionary) concept was that of the carrier and its aircraft as the centerpiece of a long-range striking force – the new battle line, so to speak.[60]

During World War I, British carrier operations emphasized the first two roles: land attack and fleet support. The reason lay in interlocking technical problems that required solutions to develop effective carriers, aircraft, and weapons capable of attacking battleships. The numbers of aircraft that World War I carriers could generate remained limited, and the aircraft themselves were not capable of delivering ordnance with much potential to sink capital ships.

---

fire-control calculators, submarines, signals intelligence, and the shift from coal- to oil-fired ship propulsion (ibid., p. 43). In the context of all these technological changes, it is hardly surprising that the U.S. Navy embraced aviation as well.

[58] Aerial spotting allowed accurate battleship gunfire to extend well beyond the visual horizon. By 1922, for example, effective battleship gunfire had extended from 20,000 to 24,000 yards with the aid of spotter aircraft (Norman Friedman, *U.S. Aircraft Carriers: An Illustrated Design History* [Annapolis, MD, 1983], p. 33). When the Japanese battleship *Yamato* was launched in 1940, its 18.1-inch guns could reach out to 40,000 meters (Yôichi Hirama, "Japanese Preparations for World War II," p. 73).

[59] Friedman, Hone, and Mandeles, "The Introduction of Carrier Aviation into the U.S. Navy and the Royal Navy," pp. 49–50 and 203.

[60] Ibid., pp. 174–175.

The most promising aircraft weapon for such strikes at the beginning of war was the high-speed torpedo.[61] Its promise led the Royal Navy to concentrate the initial "carrier" efforts on torpedo-carrying floatplanes that specially modified ships could carry to within striking distance of the target.[62] While the British achieved some success during 1915 against Turkish shipping in the Sea of Marmara by using 14-inch (810-pound) torpedoes delivered from floatplanes, they needed heavier weapons in order to sink, rather than damage capital ships.[63] But the drag and weight associated with seaplane floats so limited aircraft performance that the British could not lift larger torpedoes until they had shifted to wheeled aircraft. Not until 1918 did the Royal Navy field sufficient quantities of wheeled aircraft, the Sopwith Cuckoo, with the performance to carry the 18-inch (1,000-pound) torpedo needed for a mass strike on the German fleet in its base.[64]

The use of sea-based aircraft to attain air superiority over the fleet also led the Royal Navy from floatplanes to wheeled aircraft over the war's course. While the British used floatplanes initially, once the Germans began employing relatively high-flying Zeppelins to shadow the Grand Fleet, the 1917 aircraft committee quickly made the anti-Zeppelin mission a first priority of shipborne aircraft operating with the fleet.[65] Since only wheeled fighters like the Sopwith Pup had the performance to intercept Zeppelins, this mission, too, led to the conclusion that the Royal Navy needed the capability to recover high-performance, wheeled aircraft at sea.[66] In turn, employment of wheeled aircraft at sea required more than floatplane carriers. The British required ships with flight decks for the launch and recovery of aircraft. HMS *Furious*, initially designed with a "flying-off" deck only, was modified in 1918 to include a "flying-on" deck to permit recovery of wheeled aircraft.[67] Unfortunately, the ship's

---

[61] The Whitehead torpedo is perhaps best characterized as a "ram with reach" (Wayne P. Hughes, Jr., *Fleet Tactics: Theory and Practice* [Annapolis, MD, 1986], p. 62). The possibility of marrying the airplane and the torpedo was recognized early in the development of naval aviation. In the United States, Rear Admiral Bradley Fiske, USN, was granted a patent on his proposal to use the airplane as a delivery vehicle for the torpedo in 1912 (Melhorn, *Two-Block Fox*, pp. 10–11).

[62] Friedman, *British Carrier Aviation*, p. 26.   [63] Ibid., pp. 37, 41, and 49.

[64] Ibid., p. 41. Even then, the "lightweight" 18-inch torpedo only contained 170 pounds of explosives and would have been questionable against capital ships.

[65] Ibid., p. 48.   [66] Melhorn, *Two-Block Fox*, pp. 15–16.

[67] The air turbulence from the bridge structure of *Furious* made landing on her rear deck "almost as hazardous as ditching in the sea," and *Furious* remained a one-shot carrier (Geoffrey Till, *Air Power and the Royal Navy: 1914–1945, A Historical Survey* [London, England, 1979], p. 62). In its justly famous Tondern raid of 19 July 1918, for in-

split-deck configuration with a truncated central superstructure did not permit safe recovery of aircraft, whereas the flush-deck *Argus* was still undergoing trials as the war ended.[68]

Not surprisingly, British wartime experience prompted the Americans to give serious thought to the future of aviation. In 1919, how aggressively the U.S. Navy should pursue such development became a major issue for the General Board.[69] As early as 1916, American appreciation for the growing importance and utility of aircraft in the war prompted Congress to authorize creation of separate navy and navy reserve flying corps.[70] That same year, the navy also negotiated its first aircraft production contract with Glenn Curtiss.[71] By the end of the war, British experience with naval aircraft in such roles as land attack and fleet support, as well as the Royal Navy's development of specialized ships, provided further motivation for Americans to press ahead. However, at the end of 1918 U.S. naval opinion was not unanimous as to how best to proceed. Conservatives, including the chief of naval operations, saw the capital ship as the decisive element of naval warfare; aviation could provide reconnaissance and over-the-horizon spotting, but was only a support element for the battle line – a not unreasonable view at the time.[72] Opponents of the conservative position ranged from outspoken aviation advocates such as Admirals Sims and William F. Fullam, who foresaw the carrier as an agent of revolutionary change in naval warfare, to less radical proponents who, while convinced that the aircraft carrier was indispensable to future naval operations, stressed the carrier's importance as a support to the naval gun.[73]

---

stance, *Furious* launched seven Sopwith Camels which flew 80 miles to find, bomb, and destroy Zeppelins L54 and L60 in their shed; but five of the aircraft were forced to fly on to neutral Denmark while the two that returned to the ship ditched alongside rather than attempt to land on the carrier (H.A. Jones, *The War in the Air: Being the Story of the Part Played in the Great War By the Royal Air Force*, vol. 6 [London, 1937], pp. 365–367).

[68] Friedman, *British Carrier Aviation*, pp. 27, 55–56, 61–62, and 67.

[69] The General Board was a group of senior flag officers who, during the interwar period, advised the Secretary of the Navy on "naval building programs, naval arms limitations, and such other policy matters as the secretary might refer to them" (Braisted, "Mark Lambert Bristol: Naval Diplomat Extraordinary of the Battleship Age," *Admirals of the New Steel Navy*, p. 356).

[70] Friedman, Hone, and Mandeles, "The Introduction of Carrier Aviation into the U.S. Navy and the Royal Navy," p. 51.

[71] Van Vleet and Armstrong, *United States Naval Aviation: 1910–1980*, p. 15.

[72] Melhorn, *Two-Block Fox*, pp. 30–31.

[73] After his retirement, Fullam began to argue in forums such as the *New York Herald* that naval air power would eventually drive dreadnoughts and battle cruisers from the sea (Rear Admiral W. F. Fullam, "Battleships and Air Power: Provisions Must Be Made in

Thus, when the General Board convened in January 1919, the fundamental question was not whether to adapt aviation to naval purposes, but how best to do so and at what pace.[74]

The uncertainties were immense. What, for example, was the proper air vehicle? Was it the dirigible, the floatplane (whether launched from battleships or carried by a seaplane tender), or wheeled aircraft operating from a flush-deck carrier like HMS *Argus*? If the answer were the latter, were the frail planes then available really capable of supplanting light cruisers as the "eyes of the fleet?" Looking further ahead, could carrier aircraft acquire, as some foresaw, the capability to sink capital ships? The General Board, lacking aviation expertise, heard testimony from a wide range of witnesses with experience and knowledge and its hearings lasted into autumn 1919. While most witnesses were naval officers, the board also heard from army Brigadier General William Mitchell, who had just returned from France where he had coordinated the employment of 1,480 aircraft in support of U.S. ground forces in the battle of St. Mihiel.[75]

The board's conclusion in 1919 was that aircraft had become an essential arm of the fleet. Hence, it recommended to Secretary of the Navy Josephus Daniels that the navy undertake a broad program of peacetime development to establish a naval air service "capable of accompanying and operating with the fleet in all waters of the globe."[76] On that basis, Daniels dropped his earlier opposition to construction of an aircraft car-

Future for Protection from Above," *Sea Power*, December 1919, p. 274). Melhorn, *Two-Block Fox*, pp. 31–32.

[74] Melhorn, *Two-Block Fox*, pp. 32–33. Melhorn notes that the General Board "had already made a tentative commitment to the aircraft carrier when it convened" (ibid., p. 32).

[75] Alfred H. Hurley, *Billy Mitchell: Crusader for Air Power* (Bloomington, IN, 1975), pp. 35–36. Mitchell's initial testimony before the Navy General Board in spring 1919 emphasized the need for wheeled fighters and carriers that could operate independently of the rest of the fleet, and was relatively compatible with the views of naval aviators (Melhorn, *Two-Block Fox*, pp. 40–41). Later testimony from Mitchell before congressional committees, however, emphasized increasingly the need to unify all aviation in a separate air force on the British model, a position that clearly threatened the desires of naval officers to retain control over their own aviation.

[76] Van Vleet and Armstrong, *United States Naval Aviation: 1910–1980*, p. 38. One event that seems to have helped the development of a consensus within the USN favoring carriers was the success achieved by a novice airborne spotter in coaching the gunfire of the *Texas* to an average error of only sixty-four yards, many times better than was done by ship's spotters (Melhorn, *Two-Block Fox*, p. 37). Commander van Tol has speculated, however, that the public testimony of senior champions like Mitchell, Fullam, and Sims may have been even more decisive in pushing the USN to make a strong institutional commitment to develop carrier aviation (Commander Jan van Tol, "Historical Innovation: Carrier Aviation Case Study," OSD/NA memorandum, 27 June 1994, p. 4).

rier. When several influential military figures, including Mitchell, also testified in favor of the aircraft carrier, Congress followed suit. The Naval Appropriations Act passed in 1919 contained a number of provisions important to the development of naval aviation, the most important of which was conversion of the collier *Jupiter* into an aircraft carrier, later commissioned as the USS *Langley*.[77]

Over the next six years, a series of interwoven events and developments set the course for American naval aviation. These developments occurred in four basic areas: 1) key individuals such as Sims and William Moffett attained bureaucratic positions from which they promoted and influenced development of naval aviation; 2) aviation emerged as a recognized and separately funded enterprise within the Department of the Navy; 3) changes in the external environment modified development of U.S. naval aviation; and, 4) an *ad hoc* institutional process emerged for answering both conceptual and technical questions about how best to proceed in developing carriers and carrier aircraft. Sims had received appointment as president of the Naval War College in January 1917. After President Woodrow Wilson declared war on the Central Powers, Sims was ordered to represent the navy in Britain, which he did during 1917–1918 with the same energy, skill, and intolerance for those who disagreed with him that he had displayed during his efforts to reform naval gunnery.[78] At the war's conclusion, he successfully managed to return to the Naval War College, where he remained until retirement in 1922. When Sims returned to Newport in December 1918, he was an outspoken proponent of the notion that carriers would revolutionize naval warfare. He immediately began adapting war gaming at the war college into a tool to educate naval officers by providing surrogate decision-making experience in naval warfare; his tactical games "contributed substantially to the development of ideas about how to employ the aircraft carrier."[79] Ultimately,

---

77 Friedman, Hone, and Mandeles, "The Introduction of Carrier Aviation into the U.S. Navy and the Royal Navy," p. 58.
78 Trask, "William Sowden Sims: The Victory Ashore," *Admirals of the New Steel Navy*, pp. 282, 286, and 289.
79 Peter P. Perla, *The Art of Wargaming: A Guide for Professionals and Hobbyists* (Annapolis, MD, 1990), p. 71. During 1919–1922 Sims oversaw the further development of two kinds of games at the Naval War College: strategic games, based on naval charts containing symbols for forces, which focused overwhelmingly during the interwar period on exploring the major problems posed by a Pacific war between the U.S. and Japan; and, board maneuvers, which were tactical games, played on a specially inscribed wooden table with three-minute time steps, designed to compare the value of different tactical formations, offensive and defensive techniques, and force mixes (ibid., pp. 71–75). While the tactical gaming materially aided naval aviation, its dominance by the Fire Effects Sys-

Newport's war gaming became a key element in the institutional process by which the U.S. Navy worked out answers to fundamental issues that confronted all navies in developing carrier aviation beyond the Royal Navy's achievements in World War I.

The other officer who achieved a key bureaucratic position from which to shape naval aviation was, of course, Moffett. A line officer who endorsed the philosophy of worldwide U.S. maritime expansion, Moffett lived through the bitter resistance of line officers to absorbing engineers into their own ranks – a traumatic experience that lasted from the war with Spain through the beginning of World War I.[80] Moffett did not take aviation seriously until the war in Europe demonstrated the usefulness of airplanes and he himself became involved, as commander of the Great Lakes Naval Training Station near Chicago, in satisfying the navy's pressing need for aviation mechanics in 1918.[81] When Moffett received command of the battleship *Mississippi* in December 1918, he had a turret ramp installed to launch scout planes to spot the fall of shells. In May and June of 1920, spotter planes under Captain Henry Mustin and Commander John Towers proved so effective in gunnery exercises that the *Mississippi* achieved "scores so high that they almost equaled those of all the other battleships combined."[82]

Following his tour in command of the *Mississippi*, Moffett left the West Coast on 9 December 1920 with orders to report to the Navy Department. En route to Washington he stopped in Illinois where he renewed links with industrialists such as William K. Wrigley and J. Ogden Armour.[83] Wrigley had by then become prominent in the Republican party, and Moffett, at Mustin's urging, asked Wrigley to recommend himself to

---

tem, which assessed damage from naval shellfire, tended to give many naval officers "a false sense of comfortably paced, discretely phased combat" which did not prepare them well for the "mayhem of the close-in clashes of the Solomons where forces rapidly approached each other at point-blank ranges, and ships' combat lives were measured in minutes" (ibid., p. 76).

80 Clark G. Reynolds, "William A. Moffett: Steward of the Air Revolution," *Admirals of the New Steel Navy*, ed. James C. Bradford (Annapolis, MD, 1990), pp. 374 and 376; also, Friedman, Hone, and Mandeles, "The Introduction of Carrier Aviation into the U.S. Navy and the Royal Navy," p. 43. Moffett served under Captain Alfred Thayer Mahan's command aboard the protected cruiser *Chicago* in 1893–1895, and went to the Naval War College in summer 1896 during Mahan's last tour.

81 Reynolds, "William A. Moffett: Steward of the Air Revolution," p. 377. It was during his tour as commander of the Great Lakes Naval Training Station that Moffett developed an especially close relationship with the chewing-gum magnate William K. Wrigley.

82 Ibid., p. 378.

83 William F. Trimble, *Admiral William A. Moffett: Architect of Naval Aviation* (Washington, DC, 1994), p. 65.

lead naval aviation. Wrigley delightedly agreed as did President Warren Harding, and in March 1921 Moffett relieved Captain Thomas Craven as director of naval aviation.[84] Congress passed legislation establishing a Bureau of Aeronautics (BuAer) as a flag-rank billet on 12 July 1921, and Harding nominated Moffett to be the first BuAer chief within a week of signing the legislation.[85] Moffett became BuAer on 26 July 1921. While the job was nominally a four-year assignment, he exploited his political connections to receive two additional four-year terms as BuAer and successfully rebuffed attempts to send him to sea or to shunt him off to the General Board.[86] As a result, Moffett remained chief of BuAer from his original appointment in July 1921 until his death in the crash of the dirigible *Akron* in April 1933.

During Moffett's long tenure at BuAer, his contributions to American naval aviation were many. Four bear mentioning. First, his bureaucratic skill and political connections were crucial in heading off Mitchell's efforts to create a unified air force on the model of the Royal Air Force, as well as in securing adequate funding for naval aviation. Moffett set out to bind naval aviators to the navy, rather than see them subsumed into a unified air force as had happened to the Royal Naval Air Service with creation of the RAF in 1918; he had ultimately prevailed by early 1926.[87]

Second, while Moffett's technological choices were not always correct (as his ill-fated enthusiasm for airships proved), his long-term commit-

---

[84] Reynolds, "William A. Moffett: Steward of the Air Revolution," p. 378. While Reynolds certainly suggests that Moffett energized his political connections in order to replace Craven and become the first head of the Bureau of Aeronautics, Steve Rosen has cited documentary evidence that at the time senior naval leaders genuinely believed that Moffett was the best man for the job (*Winning the Next War*, p. 77).

[85] Trimble, *Admiral William A. Moffett*, p. 80.

[86] Reynolds, "William A. Moffett: Steward of the Air Revolution," p. 380; also Trimble, *Admiral William A. Moffett*, pp. 150–151 and 193–195. Moffett's successor as chief of BuAer was Ernest J. King, who would serve as chief of naval operations during World War II (*Admiral William A. Moffett*, p. 273).

[87] Melhorn, *Two-Block Fox*, p. 73; Trimble, *Admiral William A. Moffett*, pp. 156–166; Reynolds, "William A. Moffett: Steward of the Air Revolution," p. 384; and, Friedman, Hone, and Mandeles, "The Introduction of Carrier Aviation into the U.S. Navy and the Royal Navy," p. 208. 1925 closed with the aviation board headed by Dwight W. Morrow coming out strongly in favor of the continued independence of naval aviation, as well as Mitchell being found guilty by an army court-martial of the charges that resulted from his public accusation of "incompetency, criminal negligence and almost treasonable administration of our national defense by the Navy and War Departments" in the wake of the 30 August 1925 loss at sea of a PC-9 attempting to fly from San Francisco to Hawaii and the 3 September 1925 crash of the navy airship *Shenandoah* (Trimble, *Admiral William A. Moffett*, pp. 160 and 165–166).

ment to air-cooled, radial engines proved crucial to the fielding by the eve of World War II of rugged, reliable, high-wing-loaded aircraft with the performance to engage land-based fighters on relatively equal terms and the structural strength to withstand the G-forces involved in dive bombing.[88] Third, under Moffett's leadership naval aviation took the risks necessary to experiment as to what would work and what would not.[89] Finally, like Sims at the Naval War College, Moffett involved BuAer in the institutional process by which the naval aviation community thought through the fundamental problems of realizing the potential of naval air power.

The bureaucratic emergence of naval aviation began with establishment of BuAer in July 1921. When Moffett relieved Craven as the director of naval aviation in March 1921, aviation matters were "split up into ten divisions or bureaus of the Navy Department," each of which was "a law unto itself."[90] Creation of a separate bureau provided Moffett with the opportunity to bring order to what had arguably been chaos. Furthermore, following BuAer's establishment, the navy began to make disproportionately large investments in naval aviation. From 1922 to 1928, naval appropriations dropped from $537 million to $358 million, a decline of 33 percent; nevertheless, from 1923 to 1928, appropriations for naval air grew from $14.7 million to $25 million, an increase of more than 70 percent.[91] Finally, Moffett and his allies made naval aviation an at-

---

[88] Friedman, Hone, and Mandeles, "The Introduction of Carrier Aviation into the U.S. Navy and the Royal Navy," pp. 108, 177, and 192–193; also, Melhorn, *Two-Block Fox*, p. 99. The American development of air-cooled, radial engines permitted aircraft like the Grumman F4F-4 to have substantially higher wing loadings than, for example, the Japanese Zero, which achieved biplane maneuverability in a comparatively fragile airframe that even lacked armor for the pilot (Thomas C. Hone and Mark D. Mandeles, "Interwar Innovation in Three Navies: U.S. Navy, Royal Navy, Imperial Japanese Navy," *Naval War College Review*, Spring 1987, pp. 77–78). Melhorn has emphasized the importance of air-cooled radial engines in the emergence of dive bombers as genuine ship killers due to the greater accuracy of dive versus level bombing (*Two-Block Fox*, pp. 110–111).

[89] An example is Moffett's decision not to penalize naval aviators who lost carrier aircraft through honest mistakes, thereby giving them leeway to learn through trial and error (Melhorn, *Two-Block Fox*, p. 146, note 21). As a 1927 BuAer manual on aircraft tactics put it, "Risk is never to be avoided" (Friedman, Hone, and Mandeles, "The Introduction of Carrier Aviation into the U.S. Navy and the Royal Navy," p. 107).

[90] Trimble, *Admiral William A. Moffett*, p. 76.

[91] Ibid., pp. 278–279. The budget category "Aviation, Naval" included "all direct appropriations for aircraft and engine procurement and maintenance, operations, and the miscellaneous services under the Bureau of Aeronautics' authority" (ibid., p. 168). A similar pattern can be seen in personnel. By 1929 the U.S. Navy had 1,500 fewer men than it had had in 1923, but naval aviation had grown by some 6,750 men over the same period (Melhorn, *Two-Block Fox*, p. 95).

tractive career for the service's "best and brightest." By the mid-1920s Moffett had secured the authority "to draw the best graduates of the Naval Academy into aviation."[92] More importantly, building on the Morrow Board's recommendation in 1925 that the command of flying activities, including command of aircraft carriers and naval air stations, be the preserve of qualified aviators, Moffett provided career paths for flying officers that led to command at sea.[93] Thus, he integrated aviators into the line navy, much as had occurred at the beginning of the century with engineers.

The importance of the latter achievement is perhaps best appreciated in comparison with what occurred with the British and Japanese navies during the interwar years. The Japanese chose to rely primarily on enlisted aviators rather than officers. By 1941, approximately 90 percent of Imperial Japanese naval pilots were enlisted compared to only 20–30 percent in the United States Navy.[94] This choice limited upper mobility for Japanese aviators who best understood the growing potential of naval aviation during the 1930s. The British ran into a similar problem, but for different reasons. In April 1918, their army and naval air arms separated from their services and combined to form the RAF. As a result, the Royal Navy lost nearly 55,000 officer and enlisted personnel of the Royal Naval Air Service and their accumulated experience.[95] Although the navy formed a Fleet Air Arm in 1924 (albeit under dual control with the RAF) and thinking about naval aviation did not altogether cease in the Admiralty during the interwar years, the transfer of personnel in 1918 effectively denuded the Royal Navy of those needed to formulate and execute

92 Hone and Mandeles, "Interwar Innovation in Three Navies," p. 74; Trimble, *Admiral William A. Moffett*, pp. 198–199. Admiral R. E. Coontz recommended that all Naval Academy graduates be given a course in aeronautics, and that the majority of graduates who qualified for aviation training be assigned to flight school upon graduation (Friedman, Hone, and Mandeles, "The Introduction of Carrier Aviation into the U.S. Navy and the Royal Navy," p. 109).

93 Friedman, Hone, and Mandeles, "The Introduction of Carrier Aviation into the U.S. Navy and the Royal Navy," p. 82; Melhorn, *Two-Block Fox*, p. 98. Congress ordained in June 1927 that only aviators could command carriers and naval air stations in the U.S. Navy (Till, *Air Power and the Royal Navy* p. 45). A subtle problem regarding the integration of naval aviators into the USN during the interwar years was that, due to the high attrition they were expected to suffer in wartime, the navy needed to train so many of them that they threatened to crowd out surface line officers for promotion. As Steven Rosen has noted, the problem "was ameliorated by a system in which aviators could go into the navy reserve in peacetime, to be activated in time of war" (Rosen, *Winning the Next War*, p. 77).

94 FitzSimonds, "The Development of Japanese Naval Air Power," p. 8.

95 Till, *Air Power and the Royal Navy*, pp. 30 and 116–117.

sound policies, secure the best equipment, develop an institutional process for thinking through conceptual and technical problems in carrier aviation, or even win friends in high places.[96]

Changes in the external environment also affected development in the U.S. In contrast to the fate that befell the British, American naval aviators ultimately succeeded in fending off the efforts to unify American air power under an independent service. The U.S. Navy was less successful, however, in avoiding the impact of naval arms control. Starting in late 1920, Senator William E. Borah began pushing for arms limitations to stem what he saw as a fast-developing naval arms race in capital ships among the major powers; Borah believed that limiting naval construction would reduce tensions between the powers and lessen tax burdens on all.[97] Harding called for an international conference on naval construction on 12 July 1921.[98] The result was the Washington Naval Treaty, signed in Washington on 6 February 1922 by representatives of the United States, the British Empire, France, Italy, and Japan. The treaty established tonnage ratio of 5-5-3 for the capital ships of Britain, the United States, and Japan, respectively, and a lesser figure for France and Italy; the same 5-5-3 ratio set overall tonnage limits for aircraft carriers of 135,000-135,000-81,000 tons, while limiting new carriers to 27,000 tons with the exception that each nation could build two carriers of not more than 33,000 tons by converting existing or partially constructed ships.[99]

How did the Washington Naval Treaty affect the carrier programs of the British, American, and Japanese navies? As of early 1919, the Royal Navy's plan was to retain four old seaplane carriers – Vindex, Nairara, Pegasus, and Vindictive (totaling some 17,000 tons displacement) – and modify or complete four others – Furious, Argus, Eagle, and Hermes

---

[96] Ibid., pp. 40 and 111. "More to the point, the transfer of the [Royal] Navy's first generation of flyers into the RAF meant that there were hardly any aviators senior enough to command the big carriers. Whereas in 1926 the US Navy had one vice-admiral, three rear-admirals, two captains and sixty-three commanders who received flying pay, the Royal Navy had only one rear-admiral and a few commanders and junior captains by the start of the Second World War." (ibid., p. 45).

[97] Melhorn, Two-Block Fox, p. 63.

[98] Harding's call for a naval arms-control conference came as the highly publicized bombing tests off the Virginia capes by navy and army aircraft to explore the vulnerability of naval combatants to aerial bombing began. These tests began on 21 June 1919 when twelve bombs from Navy F5Ls sank German U-boat U-117; they culminated on 21 July with the sinking of the battleship Ostfriesland following the delivery of eleven 1,000- and 2,000-pound bombs from army aircraft (Van Vleet and Armstrong, United States Naval Aviation: 1910–1980, pp. 49–50).

[99] Ibid., p. 51.

(more than 71,000 tons).[100] By the time of the Washington naval conference, *Argus* had joined the fleet, but *Eagle* was not yet completed, while *Hermes* and *Furious* were delayed into 1924 and 1925 respectively.[101] In subsequent years, the British proved reluctant to abandon their investment in *Argus*, *Hermes*, and *Eagle*. Thus, the Royal Navy failed to build replacement carriers even though *Argus*, *Hermes*, and *Eagle* were too slow to keep up with the fleet and, collectively, could only operate a total of forty-eight aircraft.[102] Furthermore, since the British had no large-hull, fast battle cruisers under construction for conversion to carriers, the best they could do was to convert two World War I light cruisers, *Courageous* and *Glorious*, which offered hulls in the vicinity of 23,000 tons.[103]

In contrast, the Americans and Japanese had large-hulled battle cruisers under construction at the time of the treaty, and each converted two hulls into aircraft carriers: the U.S. Navy's *Lexington* and *Saratoga*, and the Japanese *Kaga* and *Akagi*.[104] Looking back, we know today that carriers of this size were crucial to World War II carrier operations. Ships like *Lexington* and *Saratoga* had difficulty from the outset staying under the treaty limit of 36,000 tons.[105] Consequently, they were not only large enough to accommodate the heavier, higher performance naval aircraft of World War II, but could operate those aircraft in far greater numbers than Royal Navy's carriers.[106] Indirectly, therefore, the Washington Naval Treaty constrained the British more than the Americans and Japanese, at least partially as a result of the early lead in naval aviation that the Royal Navy had achieved in the last war.[107]

---

[100] Till, *Air Power and the Royal Navy*, p. 64.   [101] Ibid., p. 65.

[102] Ibid., pp. 66 and 71.

[103] Friedman, *British Carrier Aviation*, p. 106; Till, *Air Power and the Royal Navy*, p. 65. Till's view is that the British decision in the early 1920s to opt for larger numbers of smaller carriers in the 20–23,000 ton range was "probably correct" seems hard to sustain in light of World War II carrier operations, especially in the Pacific. Regardless, it does seem clear that this sort of tradeoff was terribly difficult to make fifteen to twenty years before the four major carrier engagements of 1942 – the battles of the Coral Sea, Midway, the Eastern Solomons, and the Santa Cruz islands – provided empirical evidence on the issue.

[104] A modernization clause in the Washington Naval Treaty in fact permitted *Lexington* and *Saratoga* to approach 36,000 tons displacement (Friedman, *U.S. Aircraft Carriers*, p. 43). Inevitably, however, the displacement of these ships grew over time.

[105] The Washington Naval Treaty also permitted the 33,000-ton limit for the one-time conversion of two cruisers to be exceeded by another 3,000 tons for additional protection such as deck armor and underwater blisters (Friedman, *U.S. Aircraft Carriers*, p. 43).

[106] Ibid., pp. 48–49.

[107] To say that the Washington Naval Treaty ultimately constrained the British more than the Americans or Japanese in development of effective carrier aviation between the two

To appreciate the often-subtle but cascading effects of the fact that the British were stuck with carriers displacing 23,000 tons or less, one needs to consider the institutional process that emerged in the United States for answering fundamental questions about how best to proceed in carrier and aircraft development. The first American carrier, the converted collier *Langley*, was not commissioned until March 1922 and did not join Aircraft Squadron Battle Fleet as a second-line combat ship until December 1924.[108] Initially, therefore, American thinking about carrier operations and requirements rested on the experience of others. The alternative that Sims provided at this stage consisted of simulations at the Naval War College in Newport, Rhode Island. What Sims and his war gamers developed, particularly in tactical simulations, was a systematic analytic device for exploring naval aviation's potential.[109]

One of the insights emerging from Newport's games in the early 1920s was the realization that the tactical dynamics of offensive carrier operations differed fundamentally from battleship engagements. When battle lines of dreadnoughts engaged, the fires of the two sides came more or less in steady streams, and each side could redirect or concentrate its "stream" of fire on the enemy's surviving ships as the engagement progressed. Anticipating the attrition equations of Frederick W. Lanchester for warfare under "modern" conditions (that is, with long-range fire weapons),[110]

---

world wars is not to say that the treaty significantly constrained the development of carrier aviation overall. Jan van Tol has noted that the Washington Naval Treaty reduced the investments that the American, British, and Japanese navies might otherwise have made in battleships (Commander van Tol, "Historical Innovation: Carrier Aviation Case Study," p. 6). So if battleships had not been constrained, all three navies might have ended up with less to invest in naval aviation than they ended up with under the treaty.

[108] Garth L. Pawlowski, *Flat-Tops and Fledglings: A History of American Aircraft Carriers* (New York, NY, 1971), pp. 19 and 20. *Langley*, not even able to keep up with battleships, was never more than an experimental carrier and did not figure in tonnage limitations of the Washington Naval Treaty because it was in existence in 1921 (Friedman, *U.S. Aircraft Carriers*, p. 37). Only with the passage of the Vinson-Trammell Act in 1934 did *Langley* count in the total carrier tonnage the U.S. was allowed under law (ibid.).

[109] Friedman, Hone, and Mandeles, "The Introduction of Carrier Aviation into the U.S. Navy and the Royal Navy," pp. 64–65 and 73.

[110] Lanchester's so-called "laws of war" postulated two distinct relationships – his "linear" and "square" laws – between casualties, force ratios, and defeat in tactical engagements, depending on whether the opposing sides are armed with "ancient" weapons such as swords, or with "modern long-range" weapons like rifles (F. W. Lanchester, *Aircraft in Warfare: The Dawn of the Fourth Arm* [London, 1916], pp. 40–41). Lanchester was a British automotive engineer, and he used differential equations to describe the differing effects of superior numbers on attrition depending on whether ancient hand weapons or modern fire weapons were being used by the opposing sides.

three naval officers – J. V. Chase, Bradley A. Fiske, and Ambrose Baudry – depicted the cumulative results of a dreadnought battle between fleets of differing strengths in terms of a "square-law" relationship.[111] Simply stated, this relationship suggested that, all other things being equal, the effectiveness of opposing forces in inflicting attrition on each other equaled the square of their respective fighting power at any point in the engagement.[112] If a battle of annihilation occurred between two battle lines in which one possesses twice the fighting power (for example, twenty battleships versus ten), the numerically superior side would annihilate its opponent at a cost of approximately 17 percent of its initial force (roughly three battleships lost and one damaged in return for destroying all ten of its opponent's ships).[113] By comparison, if the attrition in such a contest instead followed Lanchester's "linear" relationship then the superior side could still defeat the smaller, but would lose ten battleships in sinking ten of the enemy's.

Tactical gaming at Newport in the early 1920s indicated that carrier strikes came in discrete pulses of combat power rather than in continuous streams, and that the effectiveness of such pulses on the enemy was a linear function of the number of aircraft that attacking carriers could launch in a given pulse or strike.[114] Hence, the fundamental measure of offensive carrier effectiveness was the number of aircraft it could launch for a given mission. This insight had far-reaching implications for the development of American naval aviation. It suggested that in fleet engagements, striking first with carrier aircraft conferred enormous advantages; in addition, it indicated that to gain air superiority over the enemy fleet, the initial object of carrier strikes should be the enemy's carriers.[115] On the

---

111 Hughes, *Fleet Tactics*, p. 66.
112 For a relatively simple derivation of the square-law relationship, see Hughes, *Fleet Tactics*, pp. 35–36. The attrition model assumes that both sides choose to fight to the death, which is not what occurred in the 1916 Battle of Jutland.
113 Ibid., p. 69. This result assumes a salvo model of the engagement. If a "continuous fire" Lanchester model is used, the numerically superior side will annihilate the other after only losing about 13 percent of its initial fighting power rather than 17 percent.
114 Friedman, Hone, and Mandeles, "The Introduction of Carrier Aviation into the U.S. Navy and the Royal Navy," p. 73. As Hughes has demonstrated, the carrier battles that occurred in the Pacific during World War II confirm that the dynamics of offensive carrier operations were driven by a pulse of air power. See *Fleet Tactics*, pp. 94–101. The one notable difference between war-college theory and operational practice in 1942 was that both American and Japanese naval aviators assumed, before as well as during much of World War II, that a strike by a single carrier air wing could sink two or three enemy carriers (ibid., pp. 94–95, 104, and 109). In reality, a strike by an American or Japanese carrier air wing in 1942 seems at best to have only been capable of sinking or disabling a single carrier.
115 Commander van Tol, "Historical Innovation: Carrier Aviation Case Study," p. 7.

size of carrier air wings and their operation, Newport's tactical gaming underlined not only that the more planes a carrier could take to sea, the better, but that lowering aircraft launch, recovery, and on-board handling times was the crux of effective carrier air operations.[116]

Perhaps the most important ramification of such ideas was that they fed into an interactive, evidence-driven institutional process that involved the General Board, the Bureau of Aeronautics, war planners in the office of the chief of naval operations, active aviators in the fleet, and, through annual exercises, the fleet itself. One can see a striking example of this interplay and feedback in the experiments with the USS *Langley* conducted by Captain (later Admiral) Joseph M. Reeves. Reeves began the senior officers' course at Newport in fall 1923; after graduation he became head of the Tactics Department where he supervised the 1924–25 games; in summer 1925 he completed the aviation observer course at Pensacola and then received appointment as Commander, Aircraft Squadrons, Battle Force.[117] At that time, the *Langley* was the navy's only carrier, and it was to improving the aircraft operations on the *Langley* that Reeves turned his attention in late 1925.[118]

Having learned from Newport simulations that the number of aircraft a carrier could get into the air was crucial, Reeves shortened aircraft launch and recovery times, and increased the number of planes operating off the *Langley*. By August 1926, *Langley* had acquired a crash barrier to prevent recovering aircraft (whose tail hook had missed the arresting cables) from crashing into aircraft that had already landed, the "deck-park" practice of moving aircraft from one end of the flight deck to the other for launch or recovery was in operation, and *Langley*'s crew could launch

---

[116] Operational experience would eventually show that, for the U.S. carrier operations, "the central issue was the size of a single deckload strike the flight deck could accommodate" (Friedman, *U.S. Aircraft Carriers*, p. 47). This formulation presumes the practice of deck parks, meaning that carrier operations were conducted using the aircraft on the flight deck. This practice required effective crash barriers during recovery and moving recovered aircraft to the rear of the flight deck in order to initiate another launch. "Arguably the single most important British interwar naval air decision was a tacit one: not to adopt deck parks." (Friedman, *British Carrier Aviation*, p. 11).

[117] Friedman, Hone, and Mandeles, "The Introduction of Carrier Aviation into the U.S. Navy and the Royal Navy," pp. 80–81. The observers course constituted catch-up training on aviation and flying for senior officers who were not pilots. Moffett originally established it when Mitchell tried to block Moffett's appointment to head BuAer on the grounds that Moffett was not a qualified aviator. "Moffett simply set up an 'aviation observer' course at Pensacola, completion of which would qualify him as an 'aviator,' and had himself ordered to the course" (Melhorn, *Two-Block Fox*, p. 69).

[118] *Lexington* and *Saratoga* did not join the U.S. fleet until 1928.

aircraft every fifteen seconds and recover them every ninety seconds.[119] In one year, Reeves increased *Langley's* aircraft complement from four-teen aircraft to four combat squadrons capable of operating forty-eight aircraft in combat.[120] In addition, *Langley's* pilots had begun experi-menting with dive-bombing, which quickly demonstrated its accuracy for attacking ships that were under way.[121]

Many things flowed from these innovations. Reeves' success in in-creasing the numbers of aircraft the *Langley* could launch at a time gen-erated empirical data that fed back into the rules governing subsequent war games at Newport. Moffett then capitalized on Reeves' early success to generate support for a five-year, 1,000-airplane program that allowed BuAer to pursue specialized carrier aircraft – fighters, spotters, dive-bombers, etc. – rather than multipurpose aircraft.[122] Finally, by the time *Lexington* and *Saratoga* went to sea for Fleet Problem IX (held off Pana-ma during 23–27 January 1929), the fleet had extended the deck-park concept to include refueling and rearming on the flight deck itself, and *Saratoga* boasted an air wing of 110 airplanes with 100 pilots.[123]

These complex interactions exemplify the institutional process for de-veloping carrier aviation within the navy during the immediate postwar years. While the process may not have been entirely objective – few, if any, institutional processes ever are – it rested primarily on experimentation and evidence, as well as acceptance by those involved in naval aviation. Initially, simulations explored concepts and new ways of fighting that the

---

119 Friedman, Hone, and Mandeles, "The Introduction of Carrier Aviation into the U.S. Navy and the Royal Navy," p. 85.

120 Ibid., pp. 90–91.

121 Ibid., p. 92. While the accuracy of dive bombing emerged during the 1920s, it was not until the 1930s that dive bombers could carry bombs large enough to be true ship killers. "In 1929, the F8C-4, used as a bomber, was powered by a 450 hp. engine, had a gross weight of 4,020 lbs., flew at 146 mph, and had a range of 720 statute miles carrying two 100-lb. bombs. The SBC-4 of 1937 had, in comparison, a 950 hp. engine, a gross weight of 7,632 lbs., a maximum speed of 237 mph, and a range of 590 statute miles carrying a 1,000 lb. bomb. The monoplane SBD-2 of 1941 could carry 1,200 lbs. of ordnance 1,100 miles" (ibid., p. 105).

122 Ibid., pp. 91–93.

123 Ibid., p. 95. Fleet Problem IX is generally viewed as the first real test by the U.S. Navy of multiple carriers launching significant numbers of aircraft (ibid.). The climax of this exercise came when *Saratoga*, commanded by Reeves, left the main force of battleships and, accompanied by one light cruiser, made a high-speed run from the west and launched a seventy-plane strike against the locks of the Panama Canal from a range of 140 miles (ibid., p. 96). This strike, however, was not part of the original plan for the exercise and seems to have come about simply because the destroyer screen for *Sarato-ga's* battleship escort did not have the fuel to stay with it (ibid.). Nevertheless, after Fleet Exercise IX, carriers "were accepted as fleet units" (ibid., p. 97).

navy could not test with its existing equipment. As the first carriers and carrier aircraft became available, broad insights such as the realization of how important it was to maximize the number of aircraft that carriers could launch became linked to specific technical problems such as finding ways to recover aircraft without the need to move each aircraft below deck before the next could land.

From these linkages flowed a series of technical choices in carriers, their aircraft, and related equipment. In hindsight, the four carrier battles of 1942 in the Pacific largely validated these choices. Granted, the processes by which the navy made its choices were seldom smooth or straightforward. Linking conceptual thinking about the future to specific technical choices in the present is seldom easy. Nor would it be accurate to imply that these processes enabled American naval aviators during the 1920s to see the future more clearly or in greater detail than their British or Japanese counterparts. Even on the eve of the Japanese attack on Pearl Harbor, it seems doubtful that U.S. naval leaders foresaw the aircraft carrier's ascendance any more clearly than *Wehrmacht* leaders recognized the full implications of the *Blitzkrieg* on the eve of the May 1940 campaign. Nevertheless, the U.S. Navy did develop institutional arrangements for testing visions of the future, and those arrangements were, for the most part, based on using experimentation and evidence that connected conceptual thinking with specific technical choices.

The fundamental point about the Royal Navy's efforts to develop carrier aviation during the period is that the British had no such institutional processes.[124] During the interwar period it was not so much that the British failed to recognize that carrier effectiveness depended on the number of airplanes that they could launch. Rather, as the only navy with carrier combat experience, British naval officers presumed that the relatively small numbers of aircraft they could generate any one time represented the best that anyone could do. Admiralty assessments of U.S. carrier developments during the 1930s indicate that the British discounted, if not disbelieved, American claims about the numbers of aircraft operating at one time from U.S. carriers.[125]

---

[124] Ibid., p. 121.

[125] Ibid., p. 143. In a 4 October 1928 letter, Reeves noted that he had hidden the *Langley's* true aircraft complement from Vice Admiral Fuller of the Royal Navy (ibid., p. 94, note 133). In 1928 Reeves ordered the *Langley's* captain to board forty-two aircraft – three times its normal complement – for fleet exercises off Hawaii; since the British could not operate more than twelve aircraft from a carrier at that time, it was not surprising that they refused to believe that *Langley* could operate with even twenty-four (Rosen, *Winning the Next War*, p. 71). Documentary evidence that the British did not appreciate as

Consequently, the British were content to stay with smaller carriers in the 20–23,000-ton range, which constrained storage for aviation fuel regardless of the number of aircraft carried. They failed to adopt deck parks and preferred to stow aircraft not engaged in flight operations in the hangar deck, and without effective catapults or arresting gear throughout most of the period (even though the Royal Navy had originally invented both), the launch and recovery of their carrier aircraft required more space than American operations and further constrained the number of aircraft on deck. The Royal Navy chose armored decks and antiaircraft guns rather than defensive fighters as the best way to defend their carriers from air attack and maximize the number of strike aircraft they could take to sea.[126] They ignored dive-bombing until late in World War II, and they stayed with low-performance, multipurpose, multiseat planes to keep landing and takeoff speeds below sixty knots.[127] Virtually all these technical choices were to one degree or another flawed – individually, and even more so in their cascading, cumulative effects. With each step down a path that started with trying to keep planes "hangared" below flight deck except during launch and recovery, the British found it harder and harder to strike out in new directions.

Again, one can argue that, starting with the formation of the RAF, the Royal Navy was virtually prohibited from maintaining the institutional processes that might have produced better decisions and technical choices. Still, given the absence of such arrangements, as well as the bureaucratic and technical trappings that they would have entailed, it seems dif-

---

late as 1935 how large the operating complements of U.S. carriers had become can be found in the foreign developments section of the Royal Navy's annual "Progress in Tactics" during the 1930s, especially in the 1935 edition. Friedman still recalls a conversation he had with a British naval constructor that revealed this individual had still not grasped, some decades after its combat use during World War II, the American notion of a deck park (Friedman, telephone conversation with Barry D. Watts, 4 January 1995).

[126] The issue of armored carriers is particularly difficult to assess. Japanese and American carriers proved, as the British had foreseen, so vulnerable in 1942 that, after the battle of the Santa Cruz Islands in October, both sides were reduced to a single surviving carrier and their naval air wings had suffered grievous attrition (Hughes, *Fleet Tactics*, p. 100). The situation in the Pacific by 1944 was considerably different – at least on the American side. By sorting out air tactics and adding anti-air warfare ships and weapons (including the variable-time or proximity fuse), the defensive capabilities of U.S. fast carrier task forces became quite formidable (Hirama, "Japanese Naval Preparations for World War II," pp. 78–79). Consequently, what "had been a battle to sink carriers in 1942 had become a battle to destroy aircraft" by 1944 (Hughes, *Fleet Tactics*, p. 102).

[127] Commander van Tol, "Historical Innovation: Carrier Aviation Case Study," pp. 14–15; Friedman, Hone, and Mandeles, "The Introduction of Carrier Aviation into the U.S. Navy and the Royal Navy," pp. 147, 149–152, and 154–155.

ficult to maintain that so many bad choices were simply a matter of bad luck. The British failure to push carrier aviation as far as the Americans and Japanese in spite of their early lead resulted primarily from the absence of systematic, rigorous, evidentiary-based processes for testing visions and linking them to sound technical decisions about carriers and their aircraft. Even more fundamentally, they lacked the kind of visionary drive that Moffett, Sims, and others provided the U.S. Navy.

IMPLICATIONS FOR THE NEXT MILITARY REVOLUTION

The question with which we began was: What kinds of technological, operational, and organizational factors have, during times of peace, given rise to fundamental changes in how wars are fought? Our underlying aim was to identify, based on an examination of certain aspects of the historical record on innovation during the years 1918–1939, some of the specific actions, bureaucratic tactics, and strategies that senior current American defense officials today, whether military or civilian, might consider to facilitate and foster innovation in the years ahead. The motivation for both question and purpose was the *hypothesis* that we are now in the early stages of a period in which advances in precision weaponry, sensing and surveillance, computational and information-processing capabilities, and related systems will trigger substantial changes in future wars, changes at least as profound and far reaching as the combined-systems "revolutions" of the interwar period. Given this context, what implications can one draw from cases such as the *Blitzkrieg* or the development of carrier aviation?

To begin with, one must never lose sight of the impact contingency often has on successful innovation. Sometimes the historical cards that military institutions are dealt increase the British Army's deep-rooted regimental system and disdain for the serious examination of the last war certainly stacked the odds against development and adoption of mobile armored warfare. Yet despite these odds, the British Army managed to conduct enough experimentation with tanks by 1926 to assist substantially the German development of armored formations in the mid-1930s. But having moved towards defying the odds, contingency then intervened. When the chief of the imperial general staff, Lord Milne, offered command of the experimental armored unit to Britain's most experienced tank advocate, J.F.C. Fuller turned the assignment down because he, as a lieutenant colonel, had not received everything that he wished. Thereafter, Fuller became more and more of an ideologue whose strident advocacy of an all-tank army served to keep British innovation with armor outside of the mainstream of the army.

A similarly contingent turning point came for the British with the signing of the Washington Naval Treaty in early 1922. Because the Japanese and the Americans were struggling to catch up with the Royal Navy in capital ships, each had two large battle-cruisers under construction at the time, whereas the British, for reasons that reached back to their pre-World War I dreadnought competition with the Germans, did not.[128] For the Americans and Japanese, the conversion of these hulls to aircraft carriers, as permitted by the treaty, not only made economic sense, but gave their navies large-deck carriers whose displacements exceeded those the British elected to stay with during most of the interwar period by approximately 10,000 tons.

At the same time, the British decision in 1922 to stick with converting cruisers in the vicinity of 23,000 tons made economic sense but started the Royal Navy down a long path that, by 1939, would leave it far behind the Americans and Japanese. As a reviewer of this essay has suggested, the contingent nature of these decisions should not be taken to imply that the British Admiralty after 1922 downplayed carriers, failed to see their value, or "made systematic errors" regarding their development. What it does suggest though, is that even when a military service does a "lot of things right most of the time, just a temporary failure in finance or policy or thought for the space of a few years can create major difficulties for innovation."[129]

The evidence points, first of all, to the importance of developing visions of the future. Military institutions not only need to make the initial intellectual investments to develop visions of future war, but they must continue agonizing over such visions to discern how those wars might differ from previous conflicts due to changes in military technology and weaponry, national purposes, and the international security environment. As both the *Blitzkrieg* and interwar carrier aviation attest, any vision of future war is almost certain to be vague and incomplete rather than detailed and precise, much less predictive in any scientific sense. Nevertheless, without the intellectual effort and institutional commitment to evolve a vision of future war, military institutions will almost certainly fail to take the first halting steps toward peacetime innovation.

---

[128] In 1905 the Royal Navy determined to trump the German naval build up by constructing the *Dreadnought*. By so doing they provided themselves with a cushion of quantitative superiority that played a significant role in their domination of the High Sea Fleet throughout World War I. But that naval superiority (over all the world's navies) also led the British into a situation where they had no battle-cruisers under construction in 1922. Hence, British innovation was substantially constrained by the available platforms – a situation very much the result of their intelligent response to an earlier threat.

[129] Comment by Cambridge University Press' reader for this manuscript.

A related observation is that a commitment to any particular institutional vision by senior leaders tends to have long-lasting consequences, whether for good or ill. Hans von Seeckt's post–World-War-I vision of mobile warfare executed by a highly professional, well-trained, well-led army illustrates how long-lasting and powerful the choice of a sound vision of future warfare can be. Based on his experiences from the Eastern Front in World War I, where well-trained, well-led, and well-equipped German forces had consistently defeated larger enemy forces, Seeckt concluded that, in future wars, numbers would "no longer be the key to victory."[130] As a result, he "broke dramatically with German military tradition by advocating the creation of a small, elite professional army based on voluntary recruitment rather than conscription."[131] When, by 1920, Seeckt had been appointed both chief of the *Truppenamt* (the successor to the imperial general staff) and commander of the postwar German Army, he was handed a unique opportunity to impose this vision on the German officer corps, and that is precisely what he proceeded to do over the next six years. Not only did he place the general staff and its value system at the heart of the army, but he was able to mold the *Reichswehr* into a small, highly trained professional army, inculcated with his vision of future war.

Vision, however, is not enough to produce successful innovation. One's view of future conflict must also be balanced and well connected to operational realities. The Germans' development of *Blitzkrieg* during the interwar period was structured by their broader aim of developing mobile warfare within the strategic context of a continental power potentially facing adversaries on two fronts. The overriding goal of sea control played much the same leavening role in the U.S. Navy's development of carrier aviation between 1918 and 1941.[132]

By comparison the strategic bombing theories developed in the United States and Britain between the wars somehow lost the balance, or intimate connection with operational reality, that the overarching goals of mobile warfare and sea control provided soldiers and sailors in developing *Blitzkrieg* and carrier aviation. In the British case, Bomber Command exhibited an astonishing lack of realism during the late 1930s and the first two years of World War II about the practical problems of locating tar-

---

[130] Corum, *The Roots of Blitzkrieg*, p. 30.   [131] Ibid., p. 29.

[132] As David Mets has noted, command of the sea was the first goal for the U.S. Navy throughout the interwar period; it was the means of achieving that goal – battleship guns or carrier aircraft – that underwent transformation, and the USN was prudent not to bet on aircraft prematurely (David R. Mets, "The Influence of Aviation on the Evolution of Naval Thought," unpublished draft, School of Advanced Airpower Studies, Maxwell AFB, Alabama, 22 November 1994, p. 20).

gets in poor weather or at night, to say nothing of accurately bombing them once located. While most of these problems were identified by 1938, analysis of Bomber Command's operations in late 1941 revealed that, over Germany as a whole, less than 17 percent of the bomber crews dispatched claimed to have dropped within five miles of their aiming points and, over the Ruhr, this percentage was under 7 percent.[133] Equally disconnected from reality was the U.S. Eighth Air Force's stubborn commitment, as late as the beginning of October 1943, to the doctrine of bomber invincibility, which Air Corps Tactical School bomber advocates created after 1926. By late 1942, this doctrine had been interpreted to mean that 300 heavy bombers operating with escort fighters could "attack any target in German by daylight with less than 4 percent losses."[134] Despite the contrary evidence provided by the 1940 Battle of Britain and Eighth Air Force's own early missions over occupied Europe,[135] Eighth Air Force leaders clung to this doctrine until, during the six disastrous deep-penetration missions of 2–14 October 1943, German fighters decisively proved otherwise by inflicting unsustainable attrition on American bomber formations.[136] In both of these instances, the reasons why bomber advocates lost touch with reality are hard to pinpoint even today.

[133] PRO AIR 2/2598, Air Ministry File #541137 (1938); D. M. Butt, "Report by Mr. Butt to Bomber Command on his Examination of Night Photographs, 18th August 1941," Sir Charles Webster and Noble Frankland, *The Strategic Air Offensive against Germany* (London, 1961), vol. 4, *Annexes and Appendices*, Appendix 13, p. 205.

[134] Major General Ira C. Eaker, 20 October 1942 letter to General H. H. Arnold, quoted in Bernard Boylan, "Development of the Long-Range Escort Fighter," USAF Historical Study No. 136, Air University, Maxwell AFB, Alabama, September 1955, pp. 68 and 265.

[135] Regarding the conclusions that American bomber advocates drew from the failure of German daylight bombers to achieve air superiority over southern England during the Battle of Britain, see Major General Haywood S. Hansell, Jr., *The Air Plan That Defeated Hitler* (Atlanta, 1972), pp. 53–54; for an analysis of the "lessons" that Eighth Air Force drew from early American missions over the Continent, see Lieutenant Colonel Thomas A. Fabyanic, "Strategic Air Attack in the United States Air Force: A Case Study," Air War College Report No. 5899, April 1976, pp. 122–130.

[136] Wesley F. Craven and James Lea Cate, *The Army Air Forces in World War II*, vol. 2, *Europe: Torch to Pointblank, August 1942 to December 1943* (Chicago, 1949), pp. 699 and 849–850; Roger A. Freeman, with Alan Crouchman and Vic Maslen, *Mighty Eighth War Diary* (New York, 1981), pp. 120–133; and, Major General Orvil A. Anderson, "Eighth Air Force Tactical Development: August 1942-May 1945," Eighth Air Force and Army Air Forces Evaluation Board (European Theater of Operations), London, 9 July 1945, p. 116. Eighth's overall loss *rate* as a percentage of the 2,014 heavy-bomber sorties dispatched against German targets for the missions of 2, 4, 8, 9, 10, and 14 October 1943 was 9.8 percent, more than double the 4 percent upper limit Eighth Air Force leaders believed dispatching 300 or more bombers at a time would guarantee. Worse, the losses per mission escalated dramatically over the course of these six missions as the Germans reacted to the deep-penetration raids.

But whatever the precise cause or causes, among them was surely the airmen's unfettered belief that, because bombing could directly attack the means and morale of the enemy state, strategic air attack constituted "a new form of warfare," whose "sphere of activity" was literally above and beyond that of armies and navies,[137] and therefore not bound by the same constraints. Certainly during the interwar years, this unconstrained vision of what bombers might ultimately achieve seems to have blinded British and American bomber enthusiasts to the considerable difficulties of successfully executing independent strategic air operations against a resourceful and determined enemy.

The second reasonably straightforward implication of the interwar experience concerns the unavoidable necessity of bureaucratic acceptance to successful peacetime innovation. Visionaries like Fuller or Admiral Fullam certainly had their places in goading the military institutions they sought to reform to reconsider accepted ways of doing business. But the complexity and high technological content of modern warfare indicates that one or two vocal visionaries will not suffice to bring about far-reaching, combined-systems innovations on which we have focused in this essay. Without the emergence of bureaucratic acceptance by senior *military* leaders, including adequate funding for new enterprises and viable career paths to attract bright officers, it is difficult, if not impossible, for new ways of fighting to take root within existing military institutions.

Bureaucracies, by their very nature, "are *not* supposed to innovate,"[138] and changes in weaponry do portend, as Elting Morison has noted, changes in military "societies." Consequently, it seems unlikely that any handful of visionaries, however dedicated and vocal, have much chance of forcing military institutions to adopt fundamentally new ways of fighting without the acquiescence or grudging cooperation implied by emerging bureaucratic recognition and acceptance. Even in Seeckt's case, it was undoubtedly crucial to the *persistence* of his reforms within the *Reichswehr* that he managed to ensure that his successor was of like mind on the need for a highly trained army and the importance of pursuing mobile warfare.[139] Equally important was the fact that Seeckt involved a substantial portion of the officer corps in the examination of the lessons of

---

[137] Sir Charles Webster and Noble Frankland, *The Strategic Air Offensive against Germany* (London, 1961), vol. 1, *Preparation* (London, 1961), pp. 6–8.

[138] Rosen, *Winning the Next War*, p. 2.

[139] Corum, *The Roots of Blitzkrieg*, p. 176. "Seeckt was dismissed as Reichswehr commander in October 1926 for initiating a political controversy: He had invited a Hohenzollern prince to attend the fall maneuvers. To ensure that his policies would be carried on . . . Seeckt arranged for General Heye, chief of the Truppenamt and a supporter of . . . Seeckt's policies, to be appointed commander after him" (ibid.).

World War I and the inclusion of those lessons into the doctrinal framework of the army.

As a corollary to the importance of bureaucratic acceptance among senior military leaders, the dynamics evident in the case studies suggest that the potential for civilian or outside leadership to *impose* a new vision of future war on a reluctant military service whose heart remains committed to existing ways of fighting is, at best, limited. A recent case in point can be seen in the steadfast refusal of the institutional air force to accept the critique of high-technology weaponry mounted by "military reformers" such as Franklin C. Spinney, William S. Lind, and John R. Boyd during the 1980s.[140] There is, however, another aspect to the inherent difficulties of reforming military institutions from the outside, whether by strident critics like Liddell Hart or by civilian leaders like service secretaries. If "outsiders" do entertain serious hopes of changing military institutions, their best chance for long-term success lies in "encouraging" bureaucratic acceptance by such means as changing officer career paths or creating organizations like the Bureau of Aeronautics.

Turning to a third major implication of the interwar history, institutional processes for exploring, testing, and refining conceptions of future war – in the specific sense of linking those inherently imprecise and ever-evolving visions to concrete decisions over time about new military systems, operational concepts, doctrines, and organizational arrangements – are literally a *sine qua non* of successful military innovation in peacetime. Especially in the case of *combined-systems innovations*, the critical issue is achieving a better "fit" between hardware, concepts, doctrine, and or-

---

[140] For the U.S. Air Force's side of this debate, see (then Colonel) Walter Kross, *Military Reform: The High-Tech Debate in Tactical Air Forces* (Washington, DC, 1985). For the reformers' side of what was often termed the "quality-quantity" debate, see Franklin C. Spinney, *Defense Facts of Life: The Plans/Reality Mismatch*, ed. James Clay Thompson (Boulder, CO, 1985). In retrospect, one of the ironies of this debate would appear to be that Richard Cheney, who was a congressional supporter of the military reformers' views on Capitol Hill in the mid-1980s, later became the secretary of defense on whose watch the 1991 Persian Gulf war occurred, a war which USAF leaders saw as broadly vindicating their investments in advanced systems like precision-guided weapons and low-observables technology. As then Chief of Staff General Merrill McPeak noted in the immediate aftermath of Desert Storm, "Stealth, in combination with precision-guided munitions, I think, has certainly the potential to revolutionize warfare" ("Transcript of McPeak Briefing on Air Force Air Power in the Iraqi War," *Inside the Air Force*, 22 March 1991, p. 16). Key air force figures in the planning and execution of the Desert Storm air campaign, including General Charles Horner (the Joint Forces Air Component Commander during the war), have made much the same point ("Can Bombing Win a War?," NOVA television show #2002, Journal Graphics transcript, 19 January 1993 [air date], pp. 1 and 6–7).

ganizations than do one's prospective adversaries. Furthermore, success-ful innovation demands some evidentiary-driven process such as the German Army and the U.S. naval aviation community created after World War I. Such a process, in turn, hinges crucially on the "intellectual atmosphere" of the military societies involved. A litmus test for any military institution confronted with the need for substantive peacetime innovation is a willingness to examine past military experience with something approaching the degree of objectivity, candor about shortcomings (or, even, outright failures), and openness to radical ideas that characterized the *Reichswehr* under Seeckt.

By way of underscoring this conclusion, several points are worth noting about the willingness of the American military to subject its performance in the Persian Gulf War to such scrutiny.

1. The U.S. military conducted no comprehensive battlefield survey of the territory occupied by Coalition forces on 1 March 1991.

2. While the U.S. Air Force in general, and former Secretary of the Air Force Donald Rice in particular, deserve much praise for the Gulf War Air Power Survey's thorough examination of the air campaign, the institutional air force has leaned toward ignoring some of the messier, less convenient aspects of what the survey uncovered, just as the founding fathers of the U.S. Air Force largely ignored the surprisingly parallel findings of the U.S. Strategic Bombing Survey concerning air power's strategic efficacy against Germany and Japan.[141] It should be discomforting to observe that a 1993 Defense Science Board panel portrayed the reluctance of both the air force and navy to make the needed investments in precision-guided weapons and associated target-engagement systems as one of the lessons most conspicuously "not learned" from the Persian Gulf War.[142]

3. There does not appear to be *any* precedent in the entire history of the American military for subjecting past combat experience to the kind of merciless institutional scrutiny manifest in the German examination of World War I under Seeckt that took place during the early 1920s or in 1939 after the Polish campaign.[143]

---

[141] Perry McCoy Smith, *The Air Force Plans for Peace: 1943–1945* (Baltimore and London, 1970), pp. 25–26.

[142] Office of the Under Secretary of Defense for Acquisition and Technology, "Report of the Defense Science Board Task Force on Tactical Air Warfare," November 1993, cover memorandum, p. 1.

[143] In late 1919 and early 1920 Seeckt ordered creation of no less than fifty-seven committees and subcommittees to examine World War I's battlefield lessons (Corum, *The*

4. Finally, as the case of Anglo-American theories of strategic bombardment underscores, the adverse consequences of military theories
unchecked by evidence, or based on a fundamental misunderstanding
of combat processes, can be extremely costly in both blood and treasure when put to the test of combat.

To push the third of these four points a bit further, the organizational
process of testing visions of future war at sea based on experimentation
and evidence that emerged in the U.S. naval aviation community during
the early 1920s cannot be attributed to the navy as a whole. One of the
more poignant demonstrations of this observation was the ineffectiveness
of American submarines during the early years of World War II that
stemmed from inadequacies in both torpedoes and submarine skippers.
What is striking about the problems with the Mark XIV torpedo and the
Mark VI exploder is how reluctant the bureaucracy responsible for those
devices was to admit even the possibility of design defects despite the ac-
cumulation of combat experience in the Pacific indicating that something
was extraordinarily wrong with either the torpedo, the fuse, or both. By
the end of March 1942, "almost every Pearl Harbor submariner who had
fired a torpedo in anger" believed one or both of these devices to be de-
fective.[144] It turned out there were three major design defects in this
weapon, but it was not until August 1943 – *twenty-one months* into the
Pacific war – that the navy, with little help from the engineers, finally isolated and solved all three design flaws. Worse, each defect had to be first
discovered and fixed *in the field*, in all cases "over the stubborn opposition of the Bureau of Ordnance."[145]

As for the "skipper problem," even though the American submarine
commanders who began the Pacific war were both Annapolis graduates
and "a handpicked, highly qualified group . . . by peacetime standards,"
almost 30 percent had to be relieved "for unfitness, or lack of results, dur-

Roots of Blitzkrieg, p. 37). For the most part, general staff officers led the committees
and, in the end, over 400 officers became involved in the work (ibid., p. 38). Because
the majority of these officers had firsthand experience with the tactical and doctrinal
developments of 1917 and 1918, the reports rested on solid, realistic assessments of
what had actually occurred, not on what generals might have believed to have happened. The result was the extraordinary Army Regulation 487 ("Leadership and Battle
with Combined Arms"), Part I of which appeared in 1921 and Part II in 1923 (ibid.,
p. 39). A similar effort to examine the experience of World War I in the air was initiated by the German Air Service (*Luftstreitkräfte*) on 13 November 1919 (ibid., p. 144).
[144] Clay Blair, Jr., *Silent Victory: The U.S. Submarine War against Japan* (New York, NY,
1975), p. 216.
[145] Ibid., p. 439.

ing 1942," and another 14 percent in 1943 and 1944.[146] What these problems suggest is that the naval aviation community's praiseworthy process for testing visions of future carrier warfare did not extend to all parts of the navy in the interwar years. We emphasize this point not for the sake of criticizing the interwar navy, but to underscore how fragile and isolated the *ad hoc* processes by which BuAer, the Newport war gamers, the General Board, the fleet, and war planners in the CNO's office tested a vision of carrier operations in the 1920s and 1930s were. One suspects that without explicit and forceful nurturing by senior flag officers, the kind of narrow-mindedness evident in the Bureau of Ordnance over torpedoes is, and will be, the rule rather than the exception.

The other aspect of organizational processes for testing visions of future war that bears underscoring is the paramount importance of empirical evidence. One could argue that, in a process or structural sense, the institutional arrangements for examining visions of the future that emerged at the Air Corps Tactical School after 1926 were every bit as admirable and sound as those evident in the *Truppenamt* during Seeckt's tenure, or within the American naval aviation community during Moffett's era at BuAer. Although strategic bombardment did eventually come to dominate thinking at the Tactical School in the late 1930s, it should be stressed that the bomber advocates won the debate fair and square.

> Theirs wasn't the only subject taught at Maxwell. And anyone at the Tactical School was free to make any argument he wanted. Even after [the fighter advocate Claire] Chennault left there were other fighter advocates, such as Earle Partridge and Hoyt Vandenberg, on the faculty. Each side put its case to the students, who were free to decide which, if either, they wanted to believe.[147]

Yet, based on the ruthless test of World War II combat,[148] the latter two processes seem to have been discernibly more realistic and successful

---

[146] Ronald H. Spector, *Eagle against the Sun: The American War with Japan* (New York, NY, 1985), pp. 481–482. While the reasons for the "skipper problem" were many, the bottom line was that many officers who had been perfectly satisfactory in peacetime were shown, by the test of combat, to lack "the combination of nerve, judgement, and calculated recklessness needed in a successful submarine commander" (ibid., p. 130; see also Blair, *Silent Victory*, pp. 199–201). As the 1933 Truppenführung (an updating of the 1921/1923 Army Regulation 487) rightly observed: "In war, character outweighs intellect. Many stand forth on the field of battle who in peace would remain unnoticed" (*Truppenführung*, U.S. Army Report No. 14,507, p. 1).

[147] Geoffrey Perret, *Winged Victory: The Army Air Forces in World War II* (New York, 1993), p. 27.

[148] As R.V. Jones has observed, "War is different from peace: in the latter fallacies can be

than the first. Why? The main difference we can see in the historical record is, once again, openness to real-world evidence. The Germans preserved their World War I combat experience and built on it with further experimentation that gave due attention to the difficulties, frictions, and nonlinearities of actual combat. The many strategic mistakes that the Germans subsequently made in their prosecution of World War II – starting with their failure to prepare Europe's economy for prolonged war or incorporate mass-production and managerial control techniques – largely offset the realism they displayed in developing the *Blitzkrieg*, but these mistakes should not blind us to what they got right and why. Similarly, American naval aviators were reasonably successful in substituting war games, fleet exercises, day-to-day flight operations from carriers, and related experimentation with the technical aspects of operating carriers and their aircraft for direct combat experience. But in the case of the Air Corps Tactical School, unchecked vision and pure theory increasingly took precedence over the unruly realities, nonlinearities, and the messiness of actual combat. The difference was a subtle one, yet of critical importance nonetheless when one considers some of the shortcomings that emerged during World War II bombing operations.

A related hypothesis – and we offer it as no more than that – is that military organizations which have trouble being scrupulous about empirical data in peacetime may have the same difficulty in time of war. The RAF's failure before and during the early years of World War II to deal with the problems of locating targets, much less accurately bombing them, would appear to be a graphic instance of this sort of intellectual "bad habit" carrying over from peacetime to wartime.

The last major implication that emerges from the interwar experience concerns the complex way in which contingency pervades peacetime military innovation. On the one hand, chance, luck, nonlinearity, and pure serendipity are part and parcel of the process.[149] An early lead, or having triumphed in the last conflict, by no means guarantees success in coping with the sort of fundamental changes in future wars that now appear to lie just over the temporal horizon, just out of clear view. On the other hand, the case studies also suggest that the role of chance by no means implies that peacetime innovation is tantamount to a series of random rolls of the dice. The appearance of a Moffett or Seeckt in the right place at the right time is hard to count upon, as is the decision of a Sims to re-

---

covered up more or less indefinitely and criticism suppressed, but with the swift action of war the truth comes fairly quickly to light." (*The Wizard War*, p. 139).

[149] Admiral William Owens succinctly captured this aspect of peacetime innovation when he noted, during the first of a series of OSD/NA innovation workshops held in the Pentagon in late 1993, that "innovation is a crap shoot."

turn to an intellectual post such as the Presidency of the Naval War College after commanding the U.S. fleet in Europe. But these kinds of "happenstance" events, once they occur, tend to have long-term consequences for peacetime innovation.

A metaphor for this aspect of innovation suggested by Andrew Marshall is that of walking into a gambling casino for an evening at the crap tables. Chance is clearly involved, but over any extended period of time, the house has a large advantage. Finding individuals with the capabilities of a Seeckt, willingly supporting them for the long haul, making emergent areas of military capability founded on new weapons or operational concepts attractive to the brightest career officers available, and encouraging evidentiary-driven institutional processes as a basis for making technical choices, changing doctrine, and evolving new organization arrangements for future combat are unquestionably among the things likely to confer a similarly large advantage to the American military in the decades ahead.

This conclusion further suggests that genuine innovation, like democratic government, is unlikely to be a tidy process – much less one that can be tightly or centrally controlled by senior defense managers. Indeed, attempts to eliminate the inherent messiness – including the tendency for adaptation to proceed in fits and starts[150] – may be one of the surest ways to kill innovation. At the same time, however, senior leaders who do manage to choose a fruitful vision of future war during periods of fundamental change in how wars are fought can certainly set the basic direction for long-term innovation. In fact, judging on the basis of the major turnarounds achieved by Ford in the 1980s and Chrysler in the early 1990s, major shifts in how military organizations envision future war are possible in a relatively few years, even though full maturation of the new way of fighting may still take a decade or two. And, if senior leaders also manage to inculcate the requisite intellectual atmosphere and institutional processes within the military societies involved, then they will greatly enhance the chances for long-term success.

---

[150] Recent research on how a sampling of U.S., European, and some successful Japanese companies have adapted to technological change suggests that such adaptation occurs in fits and starts: there is an initial, intense burst of adaptive activity followed by longer periods of relatively stable, routine use as participants return to more regular production and day-to-day tasks (Marcie J. Tyre and Wanda J. Orlikowski, "Exploiting Opportunities for Technological Improvement in Organizations," *Sloan Management Review*, Fall 1993, pp. 13–14). One major reason for this episodic pattern seems to be that there are limits to how much change people endure within a given period of time while neglecting the continuous fire fighting needed just to maintain normal operations (ibid., p. 23). We see no reason to suppose that military innovation during peacetime will exhibit a fundamentally different pattern over time.

INDEX

Printed in the United States
By Bookmasters